Hyphenated Techniques in Grape and Wine Chemistry

Hyphenated Techniques in Grape and Wine Chemistry

Edited by

Riccardo Flamini
CRA, Viticulture Research Center,
Conegliano Veneto, Italy

John Wiley & Sons, Ltd

Telephone (+44) 1243 779777

Email (for orders and customer service enquiries): cs-books@wiley.co.uk
Visit our Home Page on www.wiley.com

Other Wiley Editorial Offices

John Wiley & Sons Inc., 111 River Street, Hoboken, NJ 07030, USA

Jossey-Bass, 989 Market Street, San Francisco, CA 94103-1741, USA

Wiley-VCH Verlag GmbH, Boschstr. 12, D-69469 Weinheim, Germany

John Wiley & Sons Australia Ltd, 42 McDougall Street, Milton, Queensland 4064, Australia

John Wiley & Sons (Asia) Pte Ltd, 2 Clementi Loop #02-01, Jin Xing Distripark, Singapore 129809

John Wiley & Sons Ltd, 6045 Freemont Blvd, Mississauga, Ontario L5R 4J3, Canada

Wiley also publishes its books in a variety of electronic formats. Some content that appears in print may not be available in electronic books.

Library of Congress Cataloging in Publication Data

Flamini, Riccardo.
Hyphenated techniques in grape & wine chemistry / Riccardo Flamini.
p. cm.
Includes bibliographical references and index.
ISBN 978-0-470-06187-9 (cloth : alk. paper)
1. Wine and wine making—Chemistry. I. Title. II. Title: Hyphenated techniques in grape and wine chemistry.
TP548.F49 2008
663′.2—dc22 2007046642

British Library Cataloguing in Publication Data

A catalogue record for this book is available from the British Library

ISBN 978-0-470-06187-9

Typeset in 10.5/13pt Sabon by Integra Software Services Pvt. Ltd, Pondicherry, India
Printed and bound in Great Britain by TJ International Ltd, Padstow, Cornwall.

To Emanuela and Vittorio, my family

Contents

List of Contributors

Pierluigi Caboni, Università degli Studi di Cagliari, Dipartimento di Tossicologia, Via Ospedale, 72 – 09124 Cagliari, Italy

Paolo Cabras, Università degli Studi di Cagliari, Dipartimento di Tossicologia, Via Ospedale, 72 – 09124 Cagliari, Italy

Silvia Carlin, Istituto Agrario di San Michele all'Adige (IASMA), Centro Sperimentale, Via Mach 1, 38010 San Michele all'Adige, Italy

Andrea Curioni, Università degli Studi di Padova, Facoltà di Agraria, Viale dell'Università, 16 – 35020 Legnaro (PD), Italy

Eduardo Dellacassa, Facultad de Quimica, Departamento de Quimica Orgánica Catedra de Farmacognosia y Productos Naturales, Universidad de la República, Gral Flores 2124, Montevideo 11800, Uruguay

Mirko De Rosso, Università degli Studi di Padova, Facoltà di Agraria, Viale dell'Università, 16 – 35020 Legnaro (PD), Italy

Rocco Di Stefano, Dipartimento ITAF, Facoltà di Agraria, Università degli Studi di Palermo, Viale delle Scienze, 90128 Palermo, Italy

Bruno Fedrizzi, Istituto Agrario di San Michele all'Adige (IASMA), Centro Sperimentale, Via Mach 1, 38010 San Michele all'Adige, Italy; Dipartimento di Scienze Chimiche, Universita di Padova Via Marzolo, 1 – 35131 Padova, Italy; Unione Italiana Vini, Laboratorio Enochimico, Viale del Lavoro, 8 – 37135 Verona, Italy

Riccardo Flamini, Consiglio per la Ricerca e la Sperimentazione in Agricoltura (CRA), Centre di Ricerca per la Viticoltura, Viale XXVIII Aprile, 26 – 31015 Conegliano (TV), Italy

Roberto Larcher, Istituto Agrario di San Michele all'Adige (IASMA), Centro Sperimentale, Via Mach 1, 38010 San Michele all'Adige, Italy

Franco Magno, Dipartimento di Scienze Chimiche, Università di Padova, Via Marzolo 1, 35131 Padova, Italy

Giorgio Nicolini, Istituto Agrario di San Michele all'Adige (IASMA), Centro Sperimentale, Via Mach 1, 38010 San Michele all'Adige, Italy; Unione Italiana Vini, Laboratorio Enochimico, Viale del Lavoro, 8 – 37135 Verona, Italy

Giuseppe Versini, Istituto Agrario di San Michele all'Adige (IASMA), Centro Sperimentale, Via Mach 1, 38010 San Michele all'Adige, Italy

Simone Vincenzi, CIRVE, Università degli Studi di Padova, Facoltà di Agraria, Viale XXVIII Aprile, 14 – 31015 Conegliano (TV), Italy

Preface

In the last twenty years, the increased production and consumption of table grapes and wines has partly been encouraged by the amply demonstrated beneficial effects of these foods on human health. The effects of moderate wine consumption on some categories of illness, such as cardiovascular diseases, degeneration of the brain due to ageing, and certain carcinogenic diseases, have been studied. The *Organisation Internationale De la Vigne et Du Vin* (O.I.V.) reports that raisins are one source of food with which to fight hunger in the world.

The efforts of the largest grape- and wine-producing countries are focused on improving product quality, rather than on increasing production, in order to remain competitive with emerging countries by better systematic positioning of 'premium' and 'super premium' wines in market niches. In this sphere, viticulture aims at improving grape quality by means of appropriate growing techniques, selection of the best clones and varieties, and study of environmental influences on vineyards. The main efforts of enology are devoted toward optimizing industrial processes: enhancement of grape compound extraction in wine-making, the best ways of achieving alcohol and malolactic fermentation, and barrel- and bottle-ageing all aim at obtaining products with particular characteristics and well-defined identities.

The legislation of the European Community and of single countries is devoted to protecting consumer health and internal markets from the sometimes harmful effects which may be caused by low-quality products. As a consequence, in order to export wines and wine-derived products, quality certificates are required (analysis of contents of pesticides, heavy metals, toxins, etc), for which legal limits are defined. The current activity of researchers and control organisms is also devoted to detecting adulterated products and illegal additives, and to verifying the proper match between the true characteristics of products and their producers' declarations (variety, geographic origin, quality, vintage, etc).

For all these aims, knowledge of the chemical composition of grapes and wines is essential, and analytical chemistry is the proper tool with which to achieve it.

In my opinion, research in Italy has played and still plays an important role in the field of grape and wine chemistry. So when John Wiley & Sons asked me to edit this volume, I promptly contacted some of my Italian research colleagues and asked each of them to contribute one or more chapters on the main topics of their research. Surprisingly, in view of their workloads, they all gave me an enthusiastic 'Yes!'.

The result is in your hands: a work on the most recent analytical approaches and techniques applied to grape and wine chemistry, perceived not only as support to Research and Quality Control Laboratories, but also as an up-to-date source for students beginning work in the field of oenological analytical chemistry.

This volume is divided into eight chapters, by subject. Due to its multidisciplinary nature, analytical techniques are not described, apart from brief introductions to Chapters 7 and 8, to which readers are referred for more specific analytical chemistry books. Again, due to the enormous number of subjects discussed, only a brief summary of methods, including materials and instruments used, is given, and readers are referred to single publications. The main parameters of wine fermentation are treated in the first chapter, but discussion of volatile esters and higher alcohols were deemed to be more suitably located in Chapter 5 on aromatic compounds.

In order to avoid any form of advertising, wherever possible the trade names of products and producers have been omitted, and only a general product description is given.

I hope that this work may at least partially fill a gap in the literature on the subject, which has been particularly enriched in recent years by the development of new analytical techniques and technologies.

I would like to thank all the authors and contributors whose work made this volume possible – in particular, Mirko De Rosso and Antonio Dalla Vedova, for their support in organizing the text, and Rocco Di Stefano, Paolo Cabras, Andrea Curioni, Roberto Larcher and Giorgio Nicolini for their faith in me, which they demonstrated by so willingly writing their contributions.

A separate acknowledgment goes to Giuseppe Versini who, with his contribution, spirit of enterprise, and belief in the project right from the beginning, enormously assisted me in completing this work.

Lastly, I would like to thank John Wiley & Sons Ltd for giving us this opportunity to present Italian research in the field of oenology.

Riccardo Flamini
CRA – Centro Di Ricerca per la Viticoltura
Conegliano Veneto, Italy, September 2007

1

Principal Parameters of Grape Ripening and Wine Fermentation

Rocco Di Stefano and Riccardo Flamini

1.1. Introduction

1.1.1. Main Parameters of Grape Ripening

Determination of ripening parameters is important to evaluate the quality and maturity state of the grape. The principal parameters usually considered for this aim are pH, total acidity, tartaric, malic, and citric acid, fructose and glucose in the grape must, anthocyanins, tannins and other compounds in the grape whose contents change during ripening. The date to start the harvest is decided mainly on the basis of pH, total acidity, sugar and organic acid content of the must. An index utilized to estimate the technological maturity is the sugars/total acidity ratio of must, which becomes constant close to the full maturation of the grape. Trends of the organic acids and sugars content in the grape during ripening are shown in Figure 1.1.

The principal organic acids in grape are L(+)-tartaric, L(−)-malic, shikimic and citric acid. Gluconic, 2-keto-D-gluconic and mucic acids are present in rot grapes as metabolites of *Botrytis Cinerea*. The structures of these compounds are reported in Figure 1.2.

Hyphenated Techniques in Grape and Wine Chemistry Edited by Riccardo Flamini

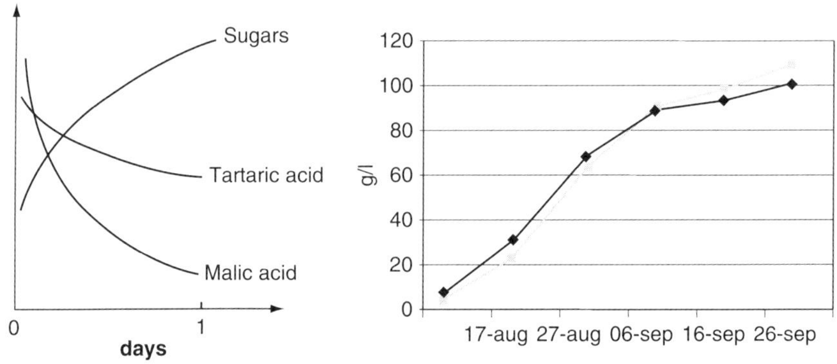

Figure 1.1 Trends of organic acids and sugars in grape during ripening (♦ glucose; ■ fructose)

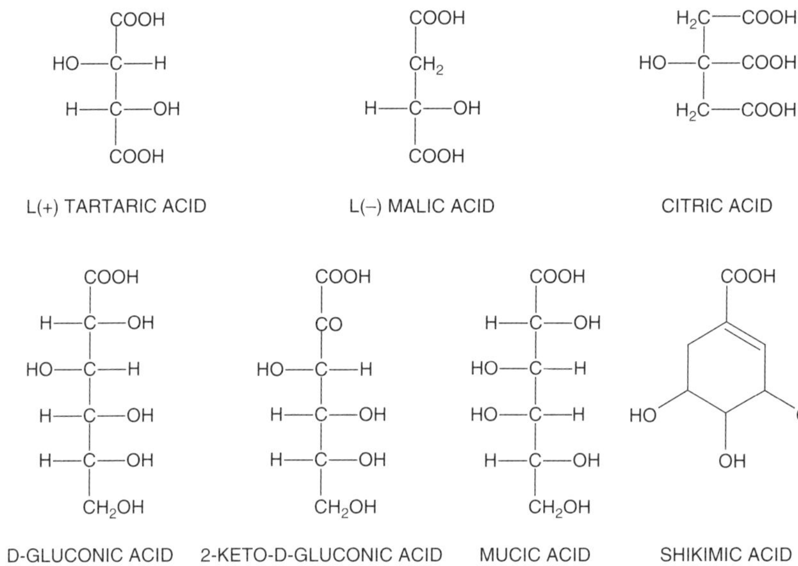

Figure 1.2 The main organic acids in grape

In the production of monovarietal wines, even if a single variety is used, grapes are usually collected from different vineyards which are often sited in zones characterized by different microclimates and are grown with different cultural practices. As a consequence, to correctly program the harvest it is necessary to know the ripening parameters of all the different grapes. Moreover, on the basis of the sugar levels of the must the alcoholic content of wine is predicted, while the anthocyanin profile and content of grape skins suggest the more suitable practices to be used in winemaking. Knowledge of the pH value, total acidity and organic acids of must provides important information about the acidic and sensorial characteristics of the final wine and about the need to correct acidic parameters, and allows planning of the technological practices such as tartaric stabilization and correction of the wine acidity, and controlling the course of malolactic fermentation. In particular, intensity and duration of acidic sensation play an important role as quality parameters of white wines, as well as in astringency perception of red wines. In grapes affected by *Botritys cinerea* (grey rot) attack a significant level of gluconic acid is usually present. Affection of grape by others moulds (e.g. *Aspergillus niger*) may induce formation of some organic acids such as glucuronic and citric acid.

The singular organic acids and sugars in the grape juice can be determined quickly and with high precision and accuracy by high-performance-liquid-chromatography (HPLC) analysis. HPLC analysis of skin extracts also provides the anthocyanin profile of red grapes. By titration and spectrophotometry methods total acidity, pH, total polyphenols and extractable phenolics indexes of the must are determined.

In wine, total organic acids content, total acidity (free organic acids) and ashes alkalinity (organic acids in salt form) are interrelated as follows:

$$\text{total organic acids (meq/L)} = \text{total acidity (meq/L)} + \text{ashes alkalinity (meq/L)}$$

It means the sum of organic acids in wine determined by HPLC have to be lower than the sum total acidity and wine ashes alkalinity determined by titration.

The grape sampling

By using HPLC coupled with modern sample preparation methods determination of tartaric, malic, and citric acids, of fructose and glucose and of the anthocyanin profile, can be performed with high precision and accuracy. Nevertheless, grape sampling in the vineyard is often affected by a high variability, due to the fact that the grapes collected may be at different stages of ripening. To minimize this undetermined error, particular attention has to be devoted to the choice of the method used to obtain a representative sample for analysis. The sampling method needs to be rapid and reproducible; on the other hand the clusters collection requires high quantities of grape to ensure a representative sample.

Picking of a suitable number of berries from some clusters directly in the vineyard, rather than part of clusters that require large samples and selection of berries in the laboratory, can in part overcome these problems and provide a representative sample. In the case of berry sampling, it is necessary that the plants selected for the collection are representative of the vineyard, and the size of the cluster and plants has to be the same in all sampling. Consequently, it is very important that the person collecting the samples has experience and knowledge of the vineyard in question – its cultural practices, treatments and general condition. Following these rules, a few hundred berries can be sufficient to have a representative sample for determination of the mean ripening state of the grapes.

1.1.2. Fermentation Parameters of Wine

Alcoholic fermentation

The microorganisms responsible for alcoholic fermentation are mainly *Saccharomyces cerevisiae* yeasts. Nevertheless, in the initial phase of the process the non-*Saccharomyces* yeasts can contribute, positively or negatively, to the composition and quality of wines. The principal event occurring in alcoholic fermentation is transformation of sugars into ethanol. Many volatile and non-volatile secondary metabolites, such as ethanol homologous alcohols, fatty acids and esters, aldehydes, ketones, hydroxyacids, ketoacids and dicarboxylic acids, are produced by the yeasts. Alcohols, ketoacids and hydroxyacids are by-products of proteic metabolism of yeasts. Even if yeasts can directly incorporate the must aminoacids, most of these compounds are synthesized by transfer of α–NH_2 amino group of the must aminoacids to specific ketoacids produced in the glyceropyruvic fermentation (Figure 1.3). Amination of ketoacids occurs by a transamination reaction in which the couples glutamic acid/pyridoxal phosphate and 2-ketoglutaric acid/pyridoxamine phosphate are –NH_2 transporters. As a result, glutamic acid transfers the α–NH_2 group from a must aminoacid to a ketoacid from glycolysis and a new aminoacid, that can enter in proteic synthesis, is produced. The excess ketoacids undergo loss of carboxyl group and the aldehydes generated are then reduced to alcohols such as 2-methyl-propan-1-ol, 2-methyl-butan-1-ol, 3-methyl-butan-1-ol, 3-methylthio-propan-1-ol, 2-phenylethanol, tyrosol, tryptophol. It is hypothesized that even γ-butyrolactone is formed from ketoglutaric acid through this pathway.

Propan-1-ol is directly originated by the loss of the 2-ketobutyric acid carboxyl group and subsequent reduction of propionic aldehyde formed.

It is interesting to note that in fermented products the same alcohols always form, even if different sources of nitrogen, but not singular aminoacids, are available. Ketoacids can also be directly reduced to hydroxyacids such as lactic, 2-hydroxy-3-methyl butyric, 2-hydroxy-3-methyl pentanoic, 2-hydroxy-4-methyl pentanoic, 2-hydroxy-3-phenyl propionic and 2-hydroxy glutaric acids. The corresponding ethyl esters can dramatically influence the wine aroma.

It was hypothesized that succinic acid, the principal fermentative dicarboxyl acid in wine, is formed in the tricarboxyl acids cycle (Figure 1.4). Even if this metabolism is not used by microorganisms in fermentation, the enzymes involved are able to operate liberation of succinic acid.

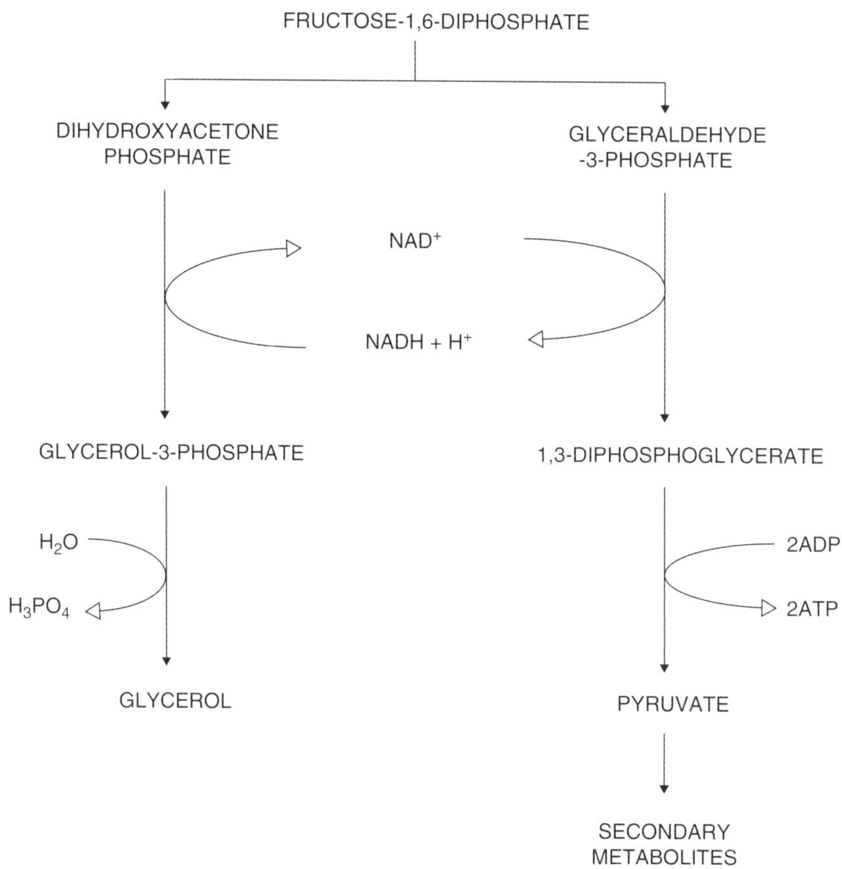

Figure 1.3 Glycero-pyruvic fermentation pathway

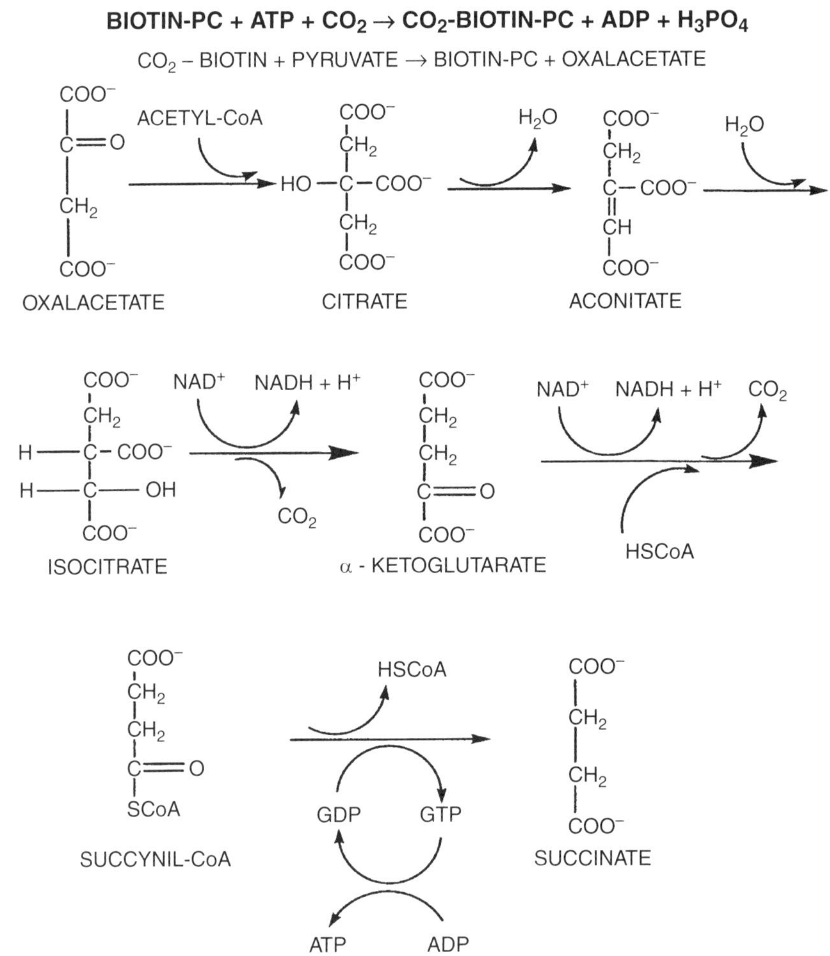

Figure 1.4 Synthesis of succinic acid. PC: pyruvate carboxylase

Production of oxalacetate by CO_2 addition to pyruvic acid, operated by pyruvate carboxylase (biotin coenzyme), allows the cycle to continue.

Short- and middle-chain fatty acids are formed in the lipids metabolism. Relevant levels of acetic acid are usually present in wine (several hundred mg/L) due to the fact that this compound can be formed by different metabolic pathways. On the contrary, the levels of short and middle-chain fatty acids are relatively low, with a maximum of a few mg/L in red wines and tens mg/L in white wines.

It was observed that yeast production of middle-chain fatty acid esters, and acetates of alcohols formed by reduction of the correspondent aldehydes derived from decarboxylation of ketoacids, is promoted by conditions such as strong clarification of the must, lower fermentation temperature, higher α-NH_2 nitrogen content, lower oxygen and lipids availability. This phenomenon can be due to the arrest of lipid synthesis at the level of middle-chain fatty acids: if synthesis would continue, the low availability of oxygen in the fermenting must, necessary for oxidation of C-16 and C-18 fatty acids would promote accumulation of long chain fatty acids with problem for the yeast cell membrane. As a consequence, part of the middle chain fatty acids are converted into esters in order to avoid their accumulation in the cell and the consequent citoplasmatic pH alteration.

After ethanol, glycerol is the main product of alcoholic fermentation. This compound forms at the start of fermentation due to low availability of pyruvate decarboxylase and alcohol dehydrogenase enzymes. As a consequence, re-oxidation of $NADH^+$ to NAD^+, necessary to convert glyceraldehyde-3-phosphate into 1,3-diphosphoglycerate to continue glycolysis, is too slow, and re-oxidation of NAD^+ occurs with the transformation of dihydroxyacetone phosphate into glycerol. Not all pyruvic acid molecules undergo decarboxylation into acetaldehyde, ketoacids, succinic acid and 2,3-butanediol form from these non-decarboxylated molecules. Pyruvic acid is also used in the production of acetyl-CoA, an important intermediate involved in the synthesis of most secondary metabolites of fermentation. The glycero-pyruvic fermentation pathway is reported in Figure 1.3. The scheme in Figure 1.4 shows synthesis of succinic acid.

Malolactic fermentation

Malolactic fermentation (MLF) is an important process, nowadays also conducted on an industrial scale, aimed at improving organoleptic characteristics and conferring microbiological stability to quality wines (Davis *et al.*, 1985). The main transformation of the wine occurring in this process operated by lactic bacteria, is decarboxylation of L(−)-malic acid with formation of L(+)-lactic acid (Figure 1.5).

Moreover, changes in the composition of carbonyl compounds in wine occur. Some of these compounds, such as diacetyl and acetoin, confer aromatic complexity to the wine; acetadehyde at high level confers a pungent note to the wine (Di Stefano and Ciolfi, 1982). Glyoxal, methylglyoxal and hydroxypropandial are produced by *Oenococcus oeni* (De Revel and Bertrand, 1993; Guillou *et al.*, 1997; Guillou, 1997),

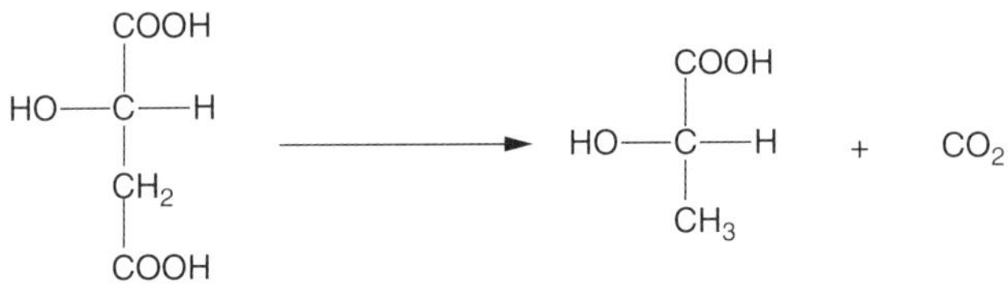

L(–)-MALIC ACID L(+)-LACTIC ACID

Figure 1.5 Transformation of L(–)-malic acid into L(+)-lactic acid occurring in malolactic fermentation (MLF)

the lactic bacterium responsible for MLF. Aliphatic aldehydes such as hexanal, (*E*)-2-hexenal, (*E*)-2-heptenal, octanal and (*E*)-2-octenal, produced by oxidation of unsaturated fatty acids by grape lipooxigenases, are responsible for herbaceous odours (De Revel and

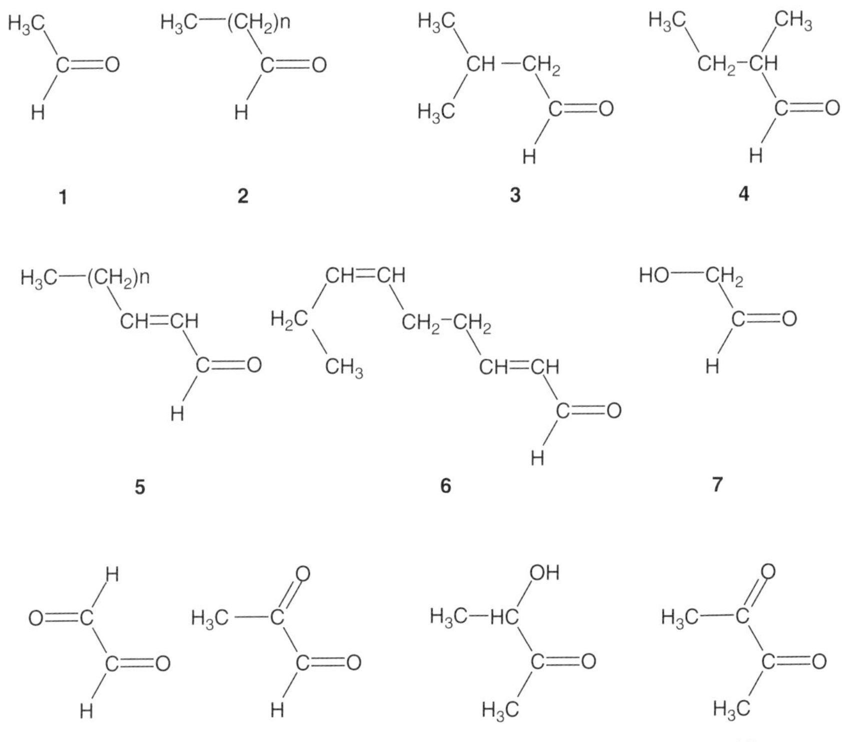

Figure 1.6 Principal carbonyl compounds involved in MLF. (1) acetaldehyde; (2) n = 2 butyraldehyde, n = 4 hexanal, n = 5 heptanal, n = 6 octanal, n = 7 nonanal, n = 8 decanal; (3) isovaleraldehyde; (4) 2-methylbutyraldehyde; (5) n = 2 (E)-2-hexanal, n = 5 (E)-2-nonenal; (6) (E,Z)-2,6-nonadienal; (7) glycolaldehyde; (8) glyoxal; (9) methylglyoxal; (10) acetoin; (11) diacetyl

Bertrand, 1994; Allen, 1995); decanal and (*E*)-2-nonenal, on the other hand, are associated with 'sawdust' or 'plank' odour (Chatonnet and Dubourdieu, 1996; 1998). The principal carbonyl compound formed in MLF is 2,3-butanedione (diacetyl), whose level can improve, or affect, the wine with its butter-like or 'fat' note (Davis *et al.*, 1985). Diacetyl and 3-hydroxy-2-butanone (acetoin, the reduced form of diacetyl) are produced by pyruvate metabolism of yeasts and lactic bacteria, and their levels may increase two or three fold with MLF depending on the lactic bacteria strain involved (Davis *et al.*, 1985; Martineau and Henick-Kling, 1995; Radler, 1962; Fornachon and Lloyd, 1965; Rankine *et al.*, 1969; Mascarenhas, 1984). For diacetyl in wine sensory thresholds ranging from 0.2 mg/L (in *Chardonnay*) to 0.9 mg/L (*Pinot noir*), and 2.8 mg/L (*Cabernet Sauvignon* wine), are reported (Martineau *et al.*, 1995).

Structures of the principal carbonyl compounds involved in MLF are reported in Figure 1.6 (Flamini *et al.*, 2002a).

1.2. Determination of Grape Ripening Parameters

1.2.1. Sample Preparation

Determination of pH, total acidity and organic acids is usually performed on the must sample prepared by hand pressing of the grape. Due to the fact that in white grape winemaking no contact between must and berry skins occurs, this method of must preparation is more suitable for white rather than for red grapes. Contact with skins induces a partial extraction of cations with a consequent change of pH and total acidity of must with respect to the sample prepared by hand pressing. As a consequence, for preparation of the must of red grapes it is better to homogenize the grapes and recover the must by centrifugation. To clarify the must, a treatment with pectolitic enzyme can be performed.

Clarification of the must in winemaking is made by treatments with silica gel, filtration, centrifugation, or by the use of enzymes. During filtration and centrifugation, oxidation of polyphenols may occur; loss of protective colloids occurring with enzymes can promote precipitation of potassium bitartrate affecting the tartaric acid, pH and total acidity data. For determination of organic acids either in skins or in the whole berry, the sample is prepared by performing extraction using a mineral acid solution.

For analysis of organic acids and sugars, about 400 grape berries are picked in vineyard as described in section 1.1.1 in order to ensure a well-representative sample. Two hundred berries are weighed and homogenized for a fixed time (1–2 min), the solution is centrifuged at 4000 *g* for 15 min and the must is used for LC analysis.

In general, to preserve musts from the sugars and organic acids degradation due to fermentation, addition of allyl isothiocyanate at a concentration of 50 mg/L in the sample is suitable to preserve the sample for three months, and allyl isothiocyanate or ethyl bromoacetate at 100 mg/L for six months' storage. The use of sodium azide does not prevent degradation of malic and tartaric acids, but at 75 mg/L inhibits fermentation for six months. Pasteurization or octanoic acid addition do not guarantee the preservation of a must sample (Flamini and Dalla Vedova, 2007).

1.2.2. Organic Acids Analysis

Sample preparation for analysis of organic acids in must is performed by solid-phase-extraction (SPE) in order to remove phenolic compounds. 1 mL of must is diluted 1:5 with a H_3PO_4 5×10^{-3} M solution and passed through a C_{18} 300-mg cartridge previously activated by passage of 2 mL methanol followed by 3 mL of H_3PO_4 5×10^{-3} M. The eluate is recovered in a 20-mL volumetric flask, the stationary phase is washed with 2 mL of H_3PO_4 5×10^{-3} M and eluate is collected in the same volumetric flask; finally, the volume is adjusted with the H_3PO_4 solution. HPLC is usually performed by a reverse-phase (RP) C_{18} column. An HPLC chromatogram from analysis of organic acids in a grape must sample is shown in Figure 1.7.

Analysis of organic acids can be also performed by using a polystyrene-divinylbenzene primary amine cation exchange resin (Aminex) column (Schneider *et al.*, 1987). The grape must sample is diluted 1:2 with the LC elution solvent (H_2SO_4 0.013 N for musts, 0.026 N for wines), filtered on membrane 0.22 μm and directly injected into the column. Also in this case, to remove phenolic compounds in particular in red grape musts, SPE can be performed as described above by using a C_{18} cartridge previously activated with passage of methanol and H_2SO_4 0.013 N (0.026 N for wines).

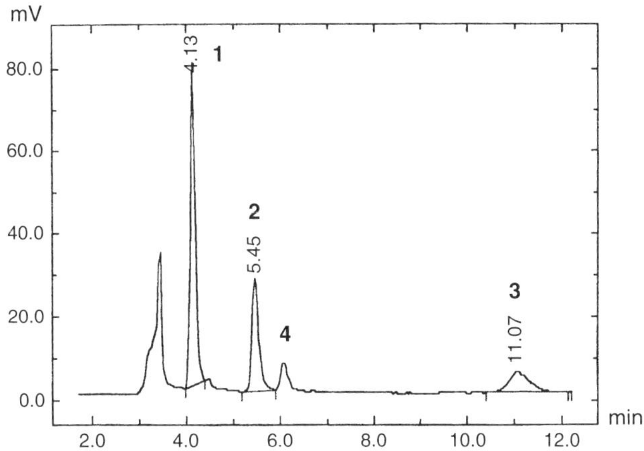

Figure 1.7 HPLC chromatogram of organic acids analysis of *Prosecco* grape must sample. 1. tartaric acid, 2. malic acid, 3. citric acid, 4. shikimic acid. Analytical conditions: Lichrospher 100 RP-18 (250 × 4 mm, 5 μm) column (Merck, Darmstadt, Germany) at room temperature, detection at wavelength 210 nm, sample volume injected 20 μL; solvent H_3PO_4 5×10^{-3} M with isocratic elution at flow rate 0.6 mL/min

1.2.3. Analysis of Sugars

Analysis of sugars in must is usually performed by using an Aminex column and refractometer as detector. The HPLC chromatogram from a grape must analysis performed in this way is shown in Figure 1.8.

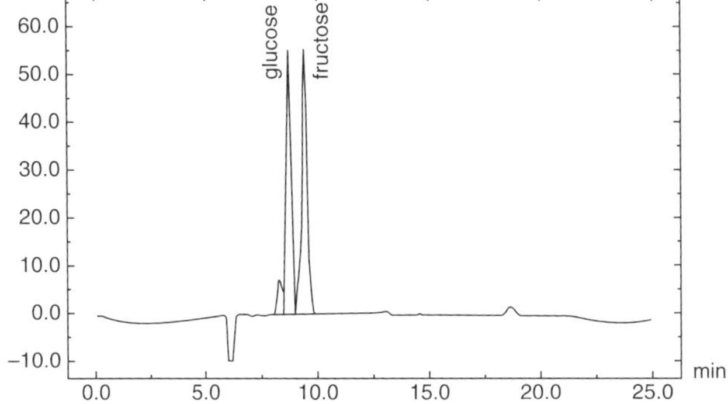

Figure 1.8 Chromatogram relative to HPLC analysis of sugars in a grape must sample. Analytical conditions: column Aminex HPX-87H (300 × 7.8 mm, 9 μm) (Bio-Rad Laboratories, Richmond, CA) at 60 °C; detector refractometer; sample volume injected 20 μL; solvent H_2SO_4 0.013 N with isocratic elution at flow rate 0.6 mL/min

The HPLC method for analysis of glucose, fructose, sucrose and glycerol in musts and wines proposed by the Organisation Internationale de la Vigne et du Vin (O.I.V.), uses an alkylamine (250×4 mm; 5 μm) column (Compendium of International Methods of Wine and Must Analysis OIV, 2006). A volume of 9 mL of must sample is diluted 1:5 with water and passed through a C_{18} cartridge previously activated. The first 3 mL of the eluate are lost, the last 6 mL are collected in a 10-mL volumetric flask. Calibration curves are calculated with a solution of glucose, fructose and glycerol at concentration 10 g/L. Table 1.1 summarizes the analytical conditions used. For glucose and fructose, limits of detection (LOD) of 0.18 and 0.12 g/L, respectively, and limits of quantification (LOQ) of 0.6 and 0.4 g/L respectively, are reported.

Table 1.1 Analytical conditions in the sugars and glycerol analysis by using alkylamine column (*Compendium of International Methods of Wine and Must Analysis OIV*, 2006).

column:	alkylamine (250×4 mm; 5 μm)
column temperature:	room
solvent:	acetonitrile/H_2O (80:20 v/v)
sample volume inj.:	20 μL
elution:	isocratic
flow:	1 mL min^{-1}
detector:	refractometer
	column retention time (min)
glycerol	5.0
fructose	8.5
glucose	9.5
sucrose	14.0

1.3. Analysis of Fermentation Parameters of Wine

1.3.1. Alcoholic Fermentation

Organic acids analysis

Preparation of the sample for analysis of organic acids in wine is analogous to the method reported for preparation of the must sample by using SPE C_{18} cartridge (paragraph 1.2.2) but, instead of collecting the eluate of a 1 mL diluted sample in a 20-mL volumetric flask, a 10-mL volumetric flask is used (the final sample is diluted 10-fold). To improve separation between acetic acid and ethanol peaks in the HPLC chromatogram using a C_{18} column, a second SPE step can be performed in order to isolate organic acids from the matrix. The solution from the first C_{18} SPE cartridge is passed through a second cartridge containing an anionic exchange quaternary-amine resin (e.g. 500 mg). Prior to the passage through the second cartridge, the pH of the solution is adjusted to 7–8 by addition of some drops of a concentrated NaOH solution, and the cartridge is activated by the passage of 2 mL methanol followed by 20 mL of NaF 0.5 M solution and 5 mL water (Schneider *et al.*, 1987; Hunter *et al.*, 1991; Flamini and Dalla Vedova, 1999). A 2.5 mL volume of sample solution is passed through the anionic exchange resin, then the cartridge is washed with 2 mL of water to remove inorganic compounds and analytes are recovered with 4 mL of HCl 1% (v/v) aqueous solution collecting the eluate in a 5-mL volumetric flask. Finally, the volume is adjusted by H_3PO_4 5×10^{-3} M and the solution is filtered on membrane 0.22 μm. A scheme of sample preparation is shown in Figure 1.9.

Quantitative recoveries of tartaric, malic, citric, citramalic and succinic acids are achieved, and are higher than 90% for acetic and

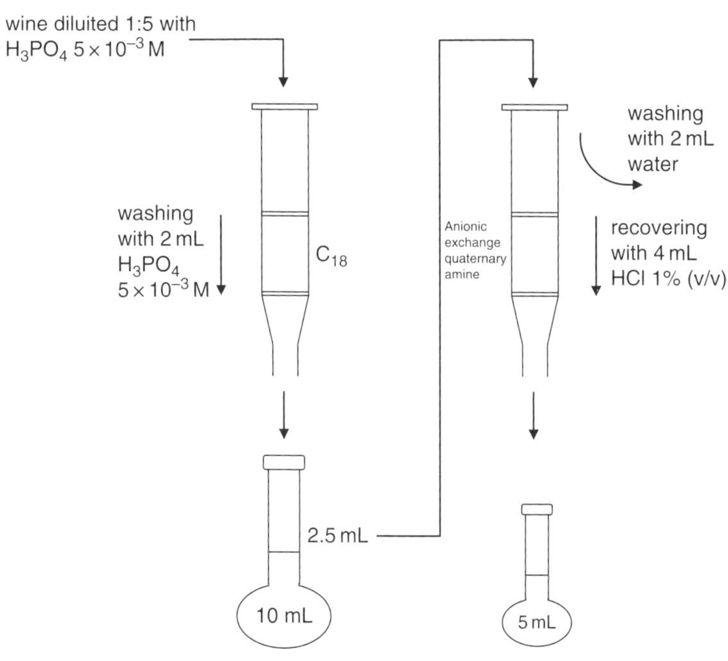

Figure 1.9 Isolation of organic acids from the wine by 2-step solid-phase-extraction

lactic acids; recoveries of pyruvic and shikimic acids are not quantitative. Figure 1.10 shows the HPLC chromatogram recorded in the analysis of a *Cabernet Sauvignon* wine after the 2-step SPE sample preparation.

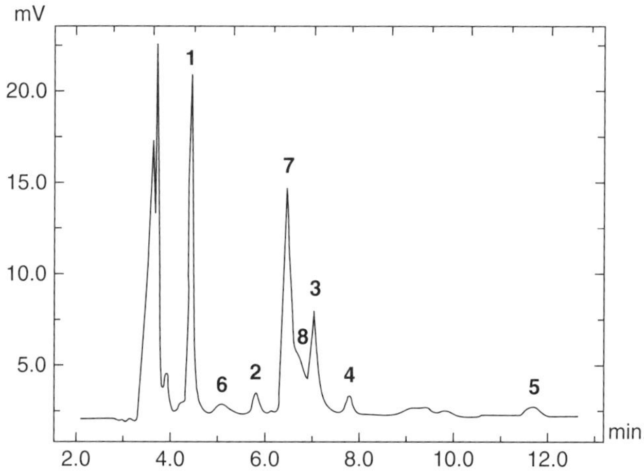

Figure 1.10 HPLC analysis of organic acids in *Cabernet Sauvignon* wine using after sample preparation by C_{18} SPE followed by isolation of organic acids a 500-mg amine-quaternary resin (Figure 1.9). 1. tartaric acid, 2. malic acid, 3. lactic acid, 4. acetic acid, 5. citric acid, 6. pyruvic acid, 7. shikimic acid. Analytical conditions: column C_{18} (250 × 4 mm, 5 μm) at room temperature, detection at wavelength 210 nm, sample volume injected 20 μL, solvent H_3PO_4 5×10^{-3} M with isocratic elution at flow rate 0.6 mL/min (Flamini and Dalla Vedova, 1999)

By performing HPLC analysis using the Aminex column, the chromatograms shown in Figure 1.11 (standard solution) and 1.12 (wine sample), are recorded.

By using both C_{18} and Aminex HPLC column, succinic, citramalic and fumaric acid can also be determined in the same run of the other organic acids. With C_{18} columns, these compounds exit in the chromatogram after citric acid in the sequence succinic-citramalic-fumaric; in the chromatogram using the Aminex column, succinic acid exits close to shikimic acid, the fumaric acid peak falls between those of lactic acid and acetic acid.

If a manual injection is performed, it may be useful to introduce an internal standard into the sample. For analysis using Aminex, formic acid can be used; this compound exits in the chromatogram between lactic acid and acetic acid. Using a C_{18} column, either maleic acid or acrylic acid

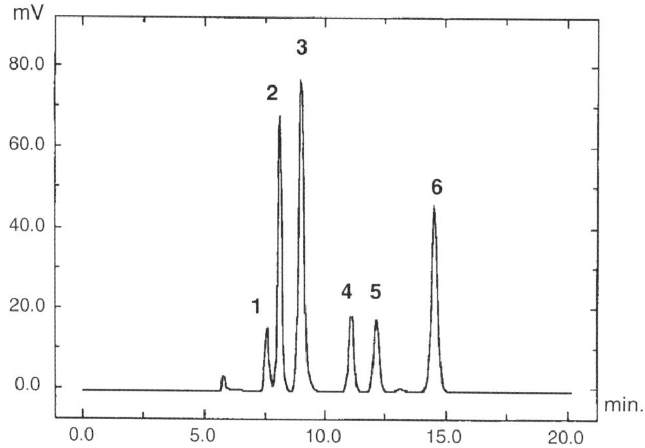

Figure 1.11 Analysis of an organic acids standard solution. 1. citric acid, 2. tartaric acid, 3. malic acid, 4. succinic acid, 5. lactic acid, 6. acetic acid. Analytical conditions: column Aminex HPX-87H (300 × 7.8 mm, 9 μm) at 65 °C (Bio-Rad Laboratories, Richmond, CA); detection at wavelength 210 nm; sample volume injected 10 μL; solvent H_2SO_4 0.026 N with isocratic elution at flow rate 0.8 mL/min

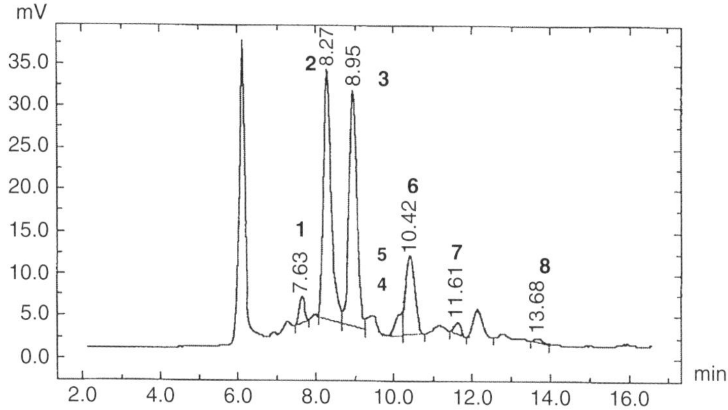

Figure 1.12 Chromatogram relative to analysis of organic acids in a *Pinot blanc* wine sample. 1. citric acid, 2. tartaric acid, 3. malic acid, 4. pyruvic acid, 5. succinic acid, 6. shikimic acid, 7. lactic acid, 8. acetic acid. Analytical conditions: column Aminex HPX-87H (300 × 7.8 mm, 9 μm) at 65 °C (Bio-Rad Laboratories, Richmond, CA); detection at wavelength 210 nm; sample volume injected 10 μL; solvent H_2SO_4 0.02 N/acetonitrile (91.5:8.5 v/v) with isocratic elution at flow rate 0.6 mL/min

can be used: the first elutes from the column after acetic acid (retention time ca. 7.5 min) and it is more nonsuitable for analysis of fermented musts and wines; acrylic acid exits from the column in about 16 min after succinic and fumaric acid and it is more suitable in wine analysis. At wavelength

210 nm maleic acid shows an absorbance 4-fold higher than acrylic acid (Flamini and Dalla Vedova, 1999).

Recently, an increased interest in shikimic acid (SHA) in grapes and wines as genotypic variety tracer, is observed (Fischleitner *et al.*, 2004; Symonds and Cantagrel, 1982; Holbach *et al.*, 2001). Methods for determination of SHA are to perform silylation and GC-FID analysis, or to perform direct injection in a reverse phase HPLC column and detection at 210 nm. The latter method was reported to be less time consuming and more robust, with a detection limit half of that obtained by GC, and a better repeatability of analyses at low concentrations (Versini *et al.*, 2003). Sample preparation for HPLC analysis is performed by diluting 5 mL of wine 1:4 (v/v) with H_3PO_4 0.07% solution and filtration on membrane 0.22 μm. Analysis is performed by two C_{18} (250 × 4 mm, 5 μm) columns in sequence operating at 40 °C and aqueous H_3PO_4 0.07% as eluent at flow rate 0.6 mL/min and detection at 210 nm. SHA exits from the columns after 10.8 min. By using a diode-array-detector (DAD), simultaneous detection at 223 and 243 nm is performed to verify the presence of malonic and ascorbic acid co-eluting with SHA. The latter compound can be eliminated from the sample prior to analysis by treating wine with H_2O_2. In these conditions, the detection limit in standard solution is 0.9 mg/L.

Gas chromatographic analysis of organic acid methyl esters

Another method for determination of the principal organic acids in musts and wines is by synthesis of their methyl (Di Stefano and Bruno, 1983) or silyl esters (De Smet *et al.*, 1981) and GC analysis of derivatives. For the sample preparation, a volume of 0.5 mL wine is transferred in a 50-mL distillation flask and some drops of ammonia 0.1 M solution are added. The solution is evaporated to dryness under vacuum, re-dissolved in 1 mL of methanol and 0.28 mL of HCl concentrate solution is added. One millilitre of a maleic acid 1 g/L methanolic solution is added as internal standard. The reaction is carried out at 55 °C for 4 h. After cooling at room temperature, 5 mL of $NaHCO_3$ 0.5 M aqueous solution is added, and transferred in a separatory funnel. Three consecutive liquid–liquid extractions using 5 mL of dichloromethane each, are performed. The organic phases are combined, the resulting solution is dried over Na_2SO_4 anhydrous and reduced to small volume prior to GC analysis. To improve the reaction yield, HCl can be substituted with a H_2SO_4 concentrate drop. Figure 1.13 shows the GC chromatograms from analysis of a standard solution, a *Chardonnay* wine and an *Asti* sparkling wine.

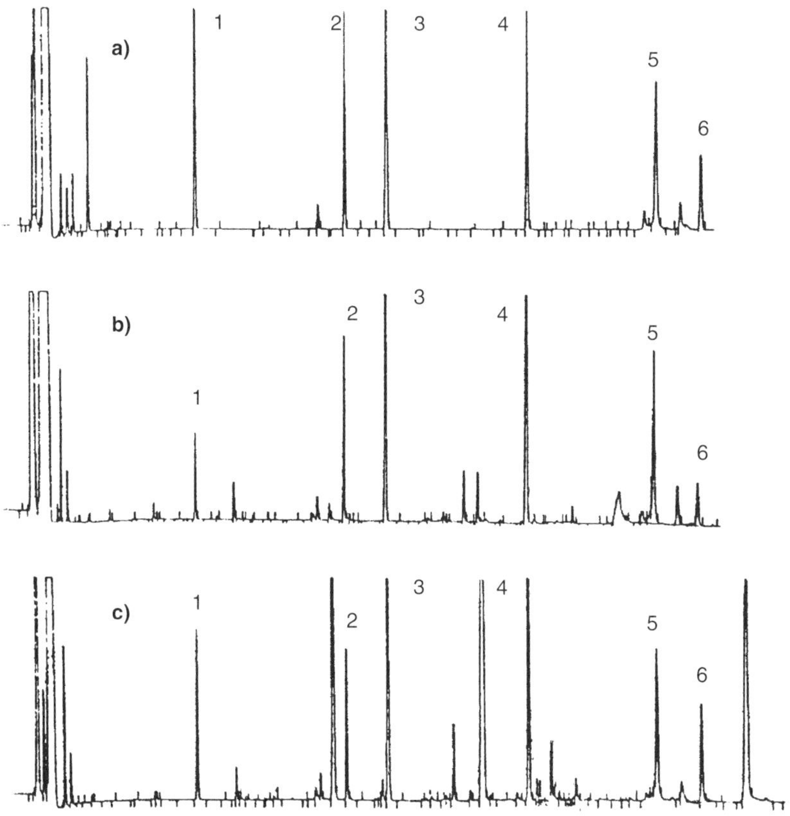

Figure 1.13 GC analysis of organic acid methyl esters in: a) standard solution, b) *Chardonnay* wine, c) *Asti* sparkling wine. 1. methyl lactate, 2. dimethyl succinate, 3. dimethyl maleate (i.s.), 4. dimethyl malate, 5. dimethyl tartrate, 6. trimethyl citrate. Chromatographic conditions: poly(ethylene) glycol fused silica capillary column (30 m × 0, 25 mm; 0.25 μm), injector and detector temperature 250 °C, flame ionization detector. Oven program: 2 min at 50 °C, from 50 to 200 °C at rate 4 °C/ min, 200 °C isotherm for 10 min (Di Stefano and Bruno, 1983)

HPLC analysis of sugars and alcohols

The method proposed by Shneider *et al.* (1987) for analysis of organic acids in wine can be applied to perform the simultaneous determination of organic acids, glucose and fructose, glycerol and ethanol by using the instrumental configuration showed in figure 1.14 containing two detectors – a spectrophotometer (UV) and a refractometer (RI) – on-line.

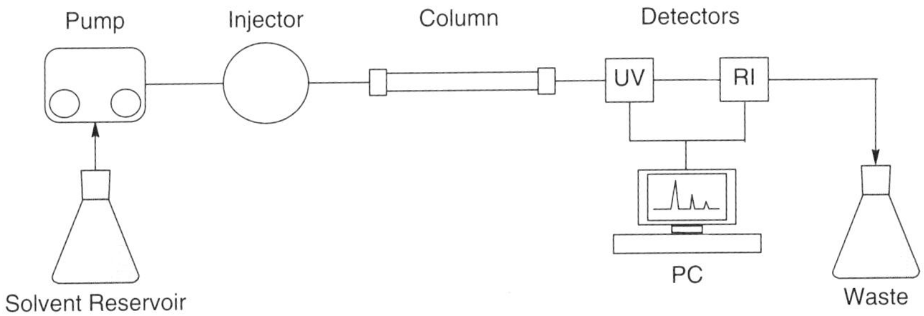

Figure 1.14 Scheme of HPLC system with detectors UV-Vis and RI on-line used for simultaneous determination of organic acids, sugars and alcohols in wine

Depending on the chromatographic conditions used, glucose or fructose co-elute from the column with malic acid and it is necessary to correct the malic acid contribution to the sugar peak area in the RI chromatogram. Due to the lower absorption at wavelength 210 nm of hexoses with respect to malic acid, the peak area measured in the UV chromatogram is, practically speaking, due to malic acid. To perform quantification of glucose, the area of the peak in RI chromatogram formed from overlapping of the two compounds is subtracted from the area value corresponding to half malic acid concentration calculated from the UV chromatogram (Testa, 1991). To avoid the sample dilution resulting from SPE sample preparation using a C_{18} cartridge, phenolic compounds can be removed by treating the wine with carbon 4–8 g/L. However as a consequence of this treatment, a significant loss only of succinic acid was reported (Shneider *et al.*, 1987).

An HPLC-RI chromatogram from analysis of wine sugars and glycerol by using an Aminex column is shown in Figure 1.15. In the same run, organic acids are determined by UV detection (chromatogram in Figure 1.12).

Traditional methods of analysis for determination of principal organic acids, glycerol and sugars in wine are based on enzymatic or colorimetric reactions; ethanol is determined by distillation of wine and density measurement of the distillate. In Table 1.2, data of organic acids, glycerol, glucose, fructose and ethanol determined by HPLC are reported as a percentage of results obtained by the traditional methods.

As may be seen from the data in Table 1.2, HPLC analysis well fit with those obtained by traditional methods in both musts and wines,

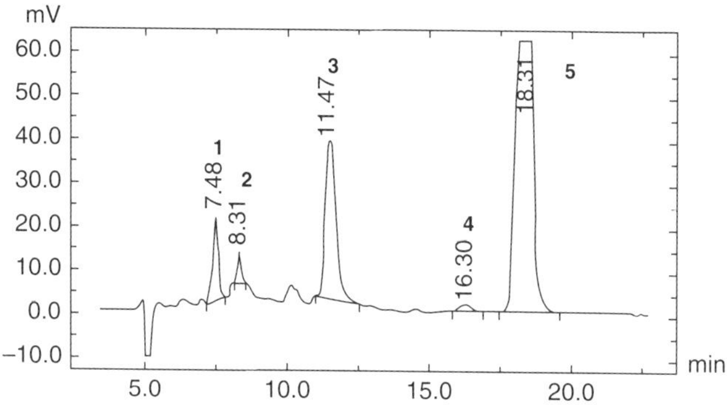

Figure 1.15 HPLC-RI chromatogram in the analysis of sugars, alcohols and glycerol in a *Pinot blanc* wine performed using Aminex HPX-87H (300 × 7.8 mm, 9 μm) (Bio-Rad Laboratories, Richmond, CA) column. Analytical conditions are the same reported in the caption of Figure 1.12. 1. glucose, 2. fructose, 3. glycerol, 4. methanol, 5. ethanol

Table 1.2 Data percentage of parameters determined by HPLC analysis (Aminex column) with respect to the traditional methods. (From Schneider *et al.*, 1987; Testa, 1991).

n° analysis	mean HPLC data (%)								
	citric acid	tartaric acid	malic acid	lactic acid	succinic acid	acetic acid	glycerol	glu+fru	ethanol
46 wines	93.8	97.0	99.2	217.0		100.5			
25 wines					111.5		101.6	101.1	99.2
6 musts	98.0	101.1	98.4						

except for lactic acid. For this compound, a good agreement between HPLC and enzymatic analysis was observed only in samples containing more than 1–2 g/L lactic acid.

Analysis of glycols in wines by gas chromatography

Ethylene glycol, 1,2-propanediol, 1,3-propanediol and 1,4-butanediol are compounds naturally present in wine. Contents of these compounds found in wines from different grape varieties and harvests are summarized in Table 1.3.

The origin of glycols is mainly linked to the fermenting yeast type, rather than to aminoacids metabolism or transamination reactions (Ciolfi *et al.*, 1990). *Saccharomyces bayanus* (ex *uvarum*) may produce 1,2-propandiol, the most abundant glycol in wine, until 200 mg/L.

Table 1.3 Contents of glycols determined in wines from four different grape varieties (*Barbera, Nebbiolo, Cortese, Chardonnay*) and five different harvests (year 1981–1986) (Di Stefano *et al.*, 1988).

Compound	content found ($mg \cdot L^{-1}$)	mean content ($mg \cdot L^{-1}$)	standard deviation
ethylene glycol	1.9–7.6	4.61	1.60
1,2-propanediol	6.9–45.3	27.37	12.22
1,3-propanediol	2.1–10.9	3.88	2.04
1,4-butanediol	0.4–2.1	1.25	0.51

Sample preparation is performed by passing a solution containing 5 mL of acetonitrile, 0.2 mL of wine and 0.2 mL of diethylene glycol 100 mg/L in acetonitrile as internal standard, through a $-NH_2$ Sep-Pak 300-mg cartridge previously activated with 2 mL of acetonitrile. The eluate is collected in a 50-mL distillation flask and the stationary phase is washed with 2 mL of acetonitrile collecting the eluate in the same distillation flask. The solution is reduced to a small volume (0.2–0.4 mL) under vacuum at 30 °C prior to GC analysis (Di Stefano *et al.*, 1992). This method is specific for analysis of glycols naturally present in wine and for diethylene glycol (in this case another internal standard must be used), but can also be applied for determination of 2,3-butanediol isomers and glycerol. Figure 1.16 shows the chromatograms relative to a standard solution and a wine analysis.

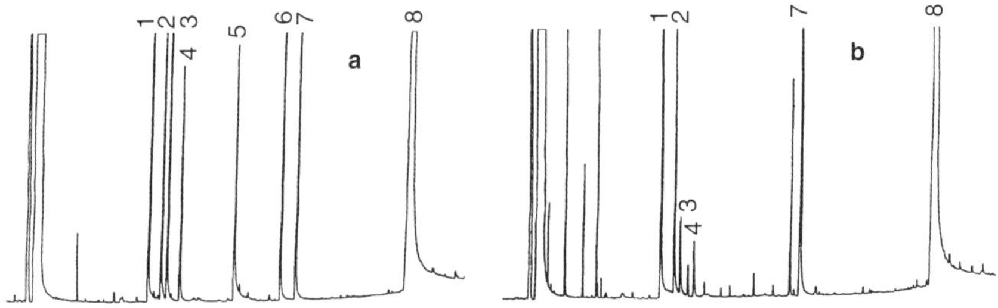

Figure 1.16 Glycols analysis of: a) standard solution, b) wine. 1. levo-2,3-butanediol, 2. meso-2,3-butanediol, 3. 1,2-propanediol, 4. ethylene glycol, 5. 1,3-propanediol, 6. 1,4-butanediol, 7. diethylene glycol (i.s.), 8. glycerol. Chromatographic conditions: poly(ethylene) glycol fused silica capillary column (30 m × 0.25 mm, 0, 25 μm), carrier gas He 1.5 mL/min flame ionization, detector. Oven program: 2 min at 50 °C, from 50 to 80 °C at rate 30 °C/min, from 80 to 200 °C at rate 3 °C/min, from 200 at 210 °C at rate 5 °C/min, 210 °C isotherm for 10 min (Di Stefano *et al.*, 1992)

1.3.2. Parameters Involved in MLF of Wine: Aliphatic Aldehydes, Acetaldehyde, Diacetyl and Acetoin

To perform gas chromatography-mass spectrometry (GC/MS) analysis of *O*-pentafluorobenzyl (PFB) oximes produced by reaction of carbonyl compounds with *O*-(2,3,4,5,6-pentafluorobenzyl)-hydroxylamine (PFBOA), provides a highly satisfactory method of detection of many carbonyl compounds in terms of selectivity and high sensitivity, achieved operating with mass spectrometer in Selected Ion Monitoring (SIM) mode registering the signal *m/z* 181, the characteristic base peak of the PFB-oxime derivatives fragmentation spectra (Cancilla and Que Hee, 1992; Lapolla *et al.*, 2003; 2006). Several studies on carbonyl compounds in wine, model wine solutions, spirits and beer were performed by synthesis of *O*-pentafluorobenzyl oximes in the diluted hydro-alcoholic matrix (Vanderlinde *et al.*, 1992; Vidal *et al.*, 1992; De Revel and Bertrand, 1993; 1994; Guillou, 1997; Guillou *et al.*, 1997; Flamini *et al.*, 2002a; 2002b; Flamini and Dalla Vedova, 2003, Ochiai *et al.*, 2003; Vesely *et al.*, 2003; Flamini, 2005; Flamini *et al.*, 2005a). By performing reaction with PFBOA at room temperature directly in the aqueous sample, the ketones derivatization may require longer reaction times with respect to aldehydes (Vidal *et al.*, 1993). On the other hand, by performing the organic solvent extraction losses of more polar compounds may occur. The simpler procedure, and good linearity and sensitivity observed for many carbonyl compounds, indicates direct derivatization of aqueous samples as the more convenient procedure. In general for each compound containing one carbonyl group the two geometrical isomers *syn* and *anti*, form (except for formaldehyde); for diacetyl three chromatographic peaks, corresponding to the (Z, Z), (E, E) and $(Z, E) + (E, Z)$ isomers, are observed. Figure 1.17 shows the reaction scheme of acetaldehyde PFB-oximes synthesis.

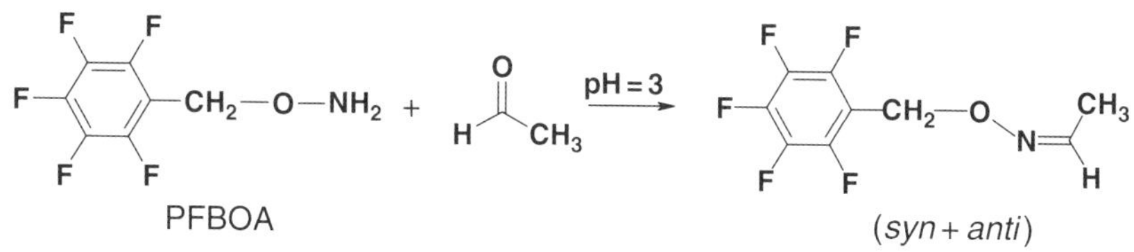

Figure 1.17 Reaction scheme of PFB-oximes synthesis of acetaldehyde

For analysis of carbonyl compounds in wine it can be useful to remove the large amount of pyruvic acid often present by sample passage through an ion exchange column. In fact PFBOA–pyruvate oxime derivatives leave the GC column as one broad peak covering a large part of the chromatogram. Furthermore, their formation subtracts a considerable amount of PFBOA from the reaction. The column is prepared with 10 g of a strong anionic exchange resin and is preconditioned by passage of 20 mL of HCl 1% (v/v) followed by 100 mL NaF 0.5 M, finally rinsed with water. The pH of 50 mL of wine is adjusted to 7.5 by addition of 1 mL NaOH 4 M and the solution is passed through the column. The stationary phase is washed with 10 mL ethanol and the two fractions from the column are combined. After the addition of *o*-chlorobenzaldehyde as internal standard (e.g. 200 μL of a 0.1 mM ethanolic solution), the pH of the resulting solution is adjusted to 3.0 by HCl. After the addition of 20 mg PFBOA, reaction is carried out under stirring for 1 h at room temperature. PFB-oximes are recovered by three consecutive extractions with 3 mL of ethyl ether/hexane (1:1 v/v) under stirring for 5 min, organic phases are combined and the volume is reduced to 0.4 mL before GC/MS analysis (Flamini *et al.*, 2002a).

GC/MS analysis of PFBOA-derivatives is performed using a poly(ethylene glycol) (30 m × 0.25 mm i.d.; df 0.25 μm) capillary column. PFB-oximes are characterized as highly volatile and by the loss of the m/z 181 fragment ions, corresponding to the pentafluorobenzyl ion, as the base peak of electron-impact (EI 70 eV) mass spectrum (Figure 1.18). To distinguish the aliphatic saturated aldehydes from the unsaturated ones, signals at m/z 239 and 250 are recorded. The first one is typical of non α-substituted compounds and probably corresponds to the *N*-vinyl pentafluorobenzyl oxime cation formed by 1-vinyl neutral aliphatic chain loss; the second is typical of α, β unsaturated compounds and corresponds to the probable loss of the aliphatic radical chain with formation of the isoxazoline ring. Figure 1.18 shows chromatograms relative to GC/MS-EI analysis of *Cabernet Sauvignon* wine recorded at m/z 181, 239 and 250 with schemes proposed for formation of the ions. Compounds numbered in chromatograms Figure 1.18 are reported in Table 1.4 with GC retention times (Flamini *et al.*, 2002a).

The average contents of the carbonyl compounds identified in a *Chardonnay* and a *Cabernet Sauvignon* wine before and after MLF performed by inoculation of two different *Oenococcus oeni* commercial strains, are reported in Table 1.5.

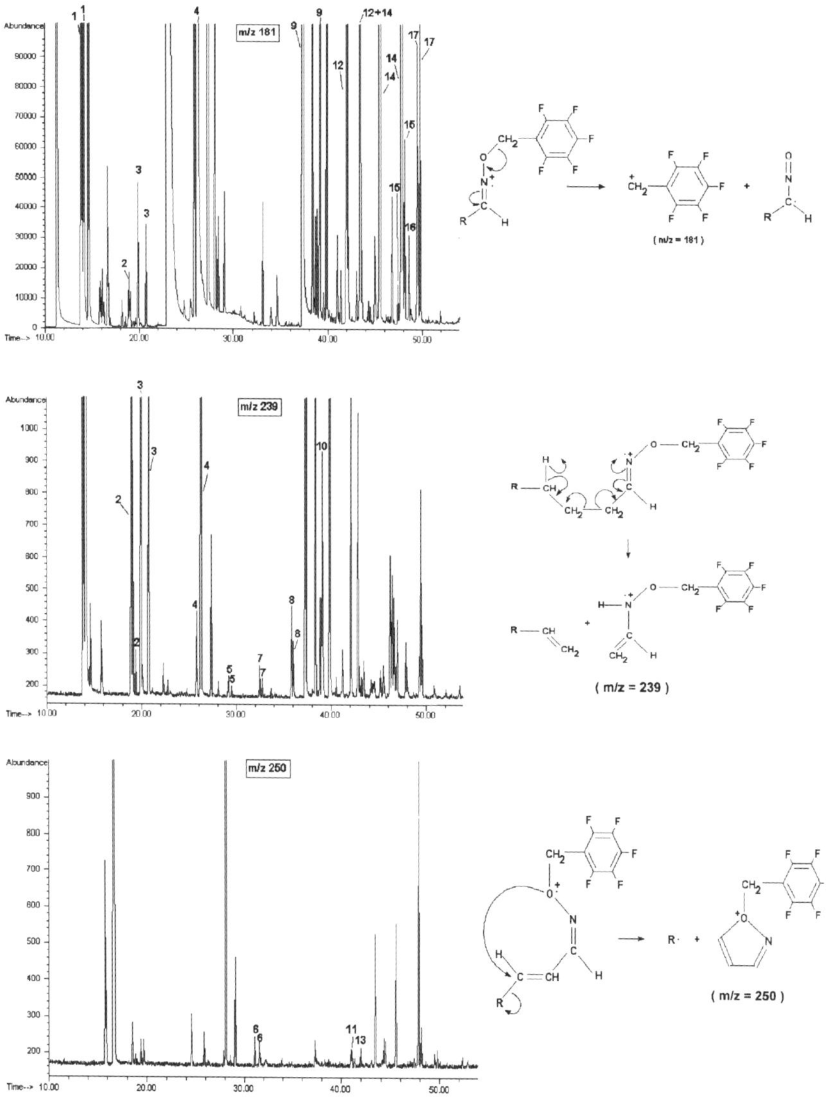

Figure 1.18 Chromatograms of *Cabernet Sauvignon* wine after MLF obtained by recording signals at m/z 181, 239 and 250 in SIM mode. Spectra for m/z 181 show signals of PFBOA-derivatives; m/z 239 allows signals of saturated aldehyde derivatives to be distinguished from the others; m/z 250 shows signals of α, β unsaturated aldehyde derivatives. Compounds corresponding to numbers on chromatograms are listed in Table 1.4. (Reprinted from *Vitis*, **41**, (2002), Flamini *et al.*, Changes in carbonyl compounds in *Chardonnay* and *Cabernet Sauvignon* wines as a consequence of malolactic fermentation, p. 110. Reproduced by permission of VITIS, journal of grapevine research)

Table 1.4 Carbonyl compounds identified as PFBOA-derivatives in the chromatogram Figure 1.18 relative to GC/MS analysis of a *Cabernet Sauvignon* wine. I.S., internal standard. (Reprinted from *Vitis*, **41**, 2002, Flamini *et al.*, Changes in carbonyl compounds in Chardonnay and Cabernet Sauvignon wines as a consequence of malolactic fermentation, p. 109. Reproduced by permission of VITIS, journal of grapevine research).

Peak number	Compound	Identification mode	RT (min)
1	acetaldehyde	MS/EI;RT	13.70,13.93
2	butyraldehyde	m/z 181,239;RT	18.68,19.25
3	isovaleraldehyde+ 2-methylbutyraldehyde	m/z 181,239;RT	19.79,20.63
4	hexanal	m/z 181,239;RT	25.67,26.20
5	heptanal	m/z 181,239;RT	29.12,29.45
6	(*E*)-2-hexenal	m/z 181,250;RT	30.99,31.47
7	octanal	m/z 181,239;RT	32.43,32.69
8	nonanal	m/z 181,239;RT	35.77,35.95
9	3-hydroxy-2-butanone	MS/EI;RT	37.24,39.00
10	decanal	m/z 181,239;RT	38.94,39.09
11	(*E*)-2-nonenal	m/z 181,250;RT	41.04
12	glycolaldehyde	MS/EI;RT	41.86,43.20
13	(*E, Z*)-2,6-nonadienal	m/z 181,250;RT	42.06
14	2,3-butanedione	MS/EI;RT	43.29,45.04,47.75
15	*o*-chlorobenzaldehyde (I.S.)	MS/EI;RT	47.24,47.98
16	methylglyoxal	lit.	48.53
17	glyoxal	MS/EI;RT	49.41,49.70

A selective method to determine acetaldehyde, diacetyl and acetoin in wines is to perform headspace solid-phase microextraction (HS-SPME) and GC/MS analysis of their PFBOA-derivatives (Flamini *et al.*, 2005b). Derivatization with PFBOA is carried on in a 4-mL vial: 100 μL of wine are added of 200 μL of 3.4 mg/L *o*-chlorobenzaldehyde solution in ethanol/water 1:1 (v/v) as internal standard and 1 mL of 2 g/L PFBOA aqueous solution. Finally, the volume is adjusted to 2 mL with water. The best conditions to achieve the highest sensitivity are to perform the reaction for 1 h at room temperature, followed by SPME for 15 min of the sample added of 50 mg NaCl. Because these compounds are significantly present in the wine (ppm level), a good compromise between the method sensitivity and analysis time is to carry on the reaction at 50 °C for 20 min, followed by SPME for 5 min at the same temperature. SPME is performed with a 65-μm poly(ethylene glycol)/divinyl benzene (PEG/DVB) fibre; analytes are thermally desorbed from the fibre in the GC injection port at 240 °C for 1 min. In general, the PFB-residue confers higher volatility to the molecule and good affinity for the

Table 1.5 Average contents of carbonyl compounds identified in *Chardonnay* and *Cabernet Sauvignon* wines before and after MLF operated by two different *Oenococcus oeni* bacterial strains. n.f. = not found. *Amounts expressed as internal standard *o*-chlorobenzaldehyde (I.S.). **Quantified on basis of 1 of 2 *syn/anti* oxime peaks. ***Amounts calculated as 1-heptanol (I.S.) in the GC/MS analysis performed without synthesis of the PFBOA-derivatives.

Compound	*Cabernet Sauvignon*		*Chardonnay*	
	before MLF	after MLF	before MLF	after MLF
(μg L^{-1})*				
butyraldehyde	12.36	19.99	8.48	5.08
isovaleraldehyde+ 2-methylbutyraldehyde	25.69	45.51	45.85	24.46
hexanal	94.63	91.67	10.79	89.94
heptanal	3.36	0.56	0.35	1.99
(*E*)-2-hexenal	3.43	0.58	0.82	1.09
octanal	0.82	0.77	0.39	0.80
nonanal	1.30	1.85	1.05	2.92
decanal	0.34	1.56	1.84	4.13
(*E*)-2-nonenal	trace	0.18	0.18	0.29
glycolaldehyde**	48.43	391.60	38.51	110.87
(*E*, *E*)-2,6-nonadienal	13.78	0.27	n.f.	n.f.
methylglyoxal**	10.96	14.37	34.73	71.55
glyoxal	41.33	106.01	49.11	95.92
(mg L^{-1})*				
acetaldehyde	52.01	5.02	46.54	67.04
acetoin	1.62	4.56	5.03	10.91
diacetyl	0.25	2.82	0.05	0.94
2,3-butanediol***	8.97	20.73	17.25	21.75

PEG/DVB fibre. In these conditions, the method shows good linearity, and sufficient accuracy and repeatability for all three compounds.

GC/MS analysis of PFBOA-derivatives can be performed by either EI or chemical ionization. Chemical ionization is performed in positive ion mode (PICI) using methane as a reagent gas. In this analysis, abundant protonated $[M+H]^+$ ion of acetaldehyde-derivatives at m/z 240, diacetyl-mono derivatives at m/z 282 and of internal standard *o*-chlorobenzaldehyde derivatives at m/z 336, form, and an abundant formation of $[M+H-18]^+$ ion of acetoin-derivatives at m/z 282, is observed. Mass spectra of acetaldehyde, diacetyl monooxime, acetoin and *o*-chlorobenzaldehyde PFB-derivatives recorded in the PICI analysis of a standard solution are reported in Figure 1.19.

To perform the analysis using an ion trap system allows the operation of normal scan recording, and collision induced dissociation (CID) experiments by selecting $[M+H]^+$ ion of diacetyl-mono derivatives at *m/z* 282 and $[M+H-18]^+$ ion of acetoin-derivatives at *m/z* 266 to confirm their identification, simultaneously. CID experiments are performed using helium as a collisional gas. Principal ions produced in CID experiments of diacetyl, acetoin and *o*-chlorobenzaldehyde derivatives are reported in Table 1.6.

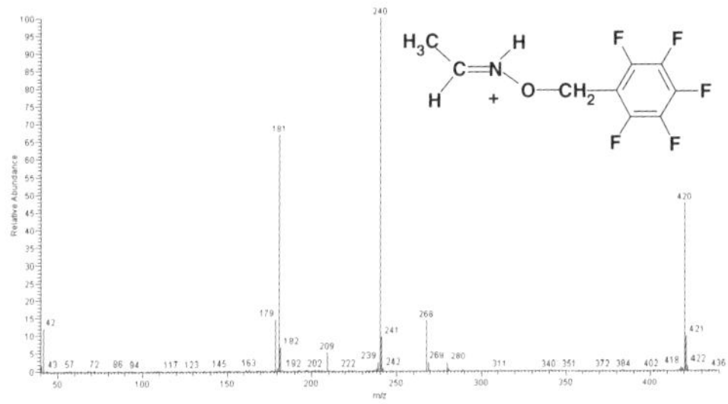

(a)

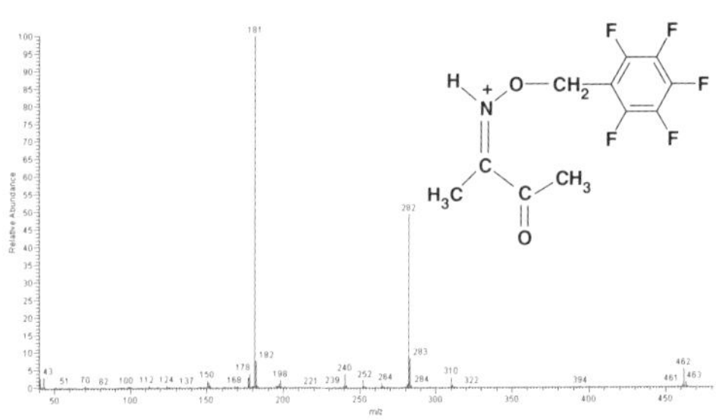

(b)

Figure 1.19 Mass spectra of acetaldehyde PFB-oxime (a), diacetyl mono PFB-oxime (b), acetoin PFB-oxime derivative (c), and *o*-chlorobenzaldehyde PFB-oxime (d) recorded in the GC/MS analysis of standard solution performed in positive ion chemical ionization mode using methane as reagent gas (reagent gas flow 1 mL/min; ion source temperature 200 °C). Flamini *et al.*, (2005) Monitoring of the principal carbonyl compounds involved in malolactic fermentation of wine by synthesis of O-(2,3,4,5,6-pentafluorobenzyl) hydroxylamine derivatives and solid-phase-micro-extraction positive-ion-chemical-ionization mass spectrometry analysis, *Journal of Mass Spectrometry*, 40, p. 1561. Copyright John Wiley & Sons, Ltd. Reproduced with permission

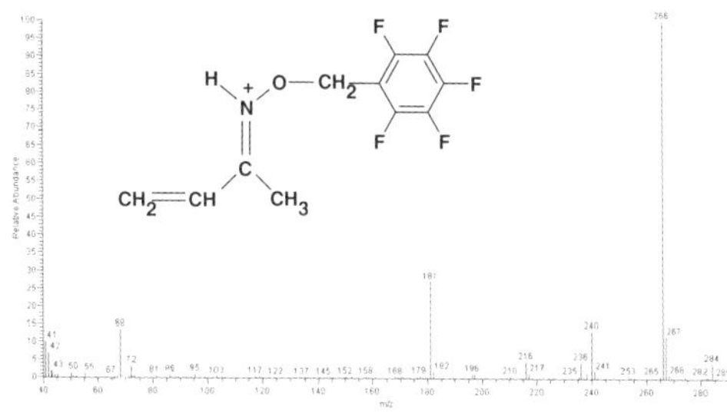

(c)

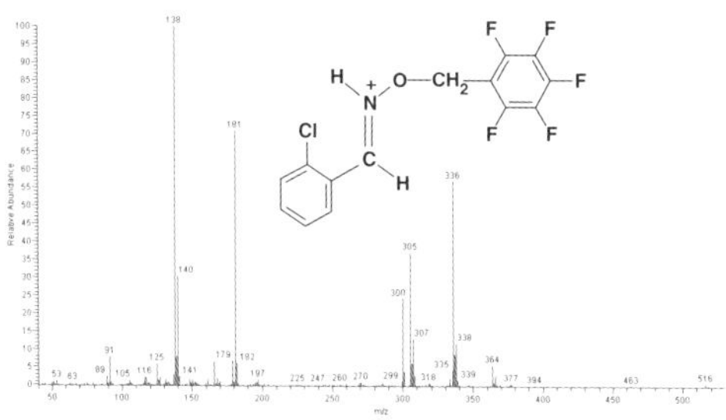

(d)

Figure 1.19 (Continued)

Quantitative analysis is performed operating in SCAN mode on the signal of $[M+H]^+$ ions at m/z 240 for acetaldehyde, at m/z 282 for diacetyl and at m/z 336 for internal standard, and on the signal at m/z 266 of $[M+H-18]^+$ ion for acetoin. The method shows a lower sensitivity in the detection of acetoin with respect to the other two compounds, probably due to the lower volatility of acetoin-derivatives and the higher polarity; the higher steric hindrance of this compound might influence the derivatization reaction yield.

Figure 1.20 shows the GC/MS chromatograms recorded in the analysis of a *Merlot* wine at the beginning of MLF (a) and after 5 days of fermentation (b) when the diacetyl and acetoin increases are evident.

Table 1.6 Principal ions produced by CID experiments of diacetyl mono-PFB-oxime $[M+H]^+$ (m/z 282), acetoin PFB-oxime $[M+H-18]^+$ (m/z 266), and *o*-chlorobenzaldehyde PFB-oxime $[M+H]^+$ (m/z 336) precursor ions. Collision gas He (dumping gas flow 0.3 mL/min). Flamini *et al.*, (2005) Monitoring of the principal carbonyl compounds involved in malolactic fermentation of wine by synthesis of *O*-(2,3,4,5,6-pentafluorobenzyl) hydroxylamine derivatives and solid-phase-micro-extraction positive-ion-chemical-ionization mass spectrometry analysis, *Journal of Mass Spectrometry*, **40**, p. 1561. Copyright John Wiley & Sons, Ltd. Reproduced with permission.

	precursor ion (P)		fragment ion					
	$[M+H]^+$	$([M+H]-H_2O)^+$	$(P-H_2O)^+$	$(P-CH_2O)^+$	$(P-CH_3O)^+$	$(P-C_2H_2O)^+$	$(P-CH_3OF)^+$	$(P-C_7H_3OF_5)^+$
mono-PFB-diacetyl	282		264	252		240		
PFB-acetoin		266	248		235		216	
PFB-*o*-chlorobenzaldehyde	336		318		305			138

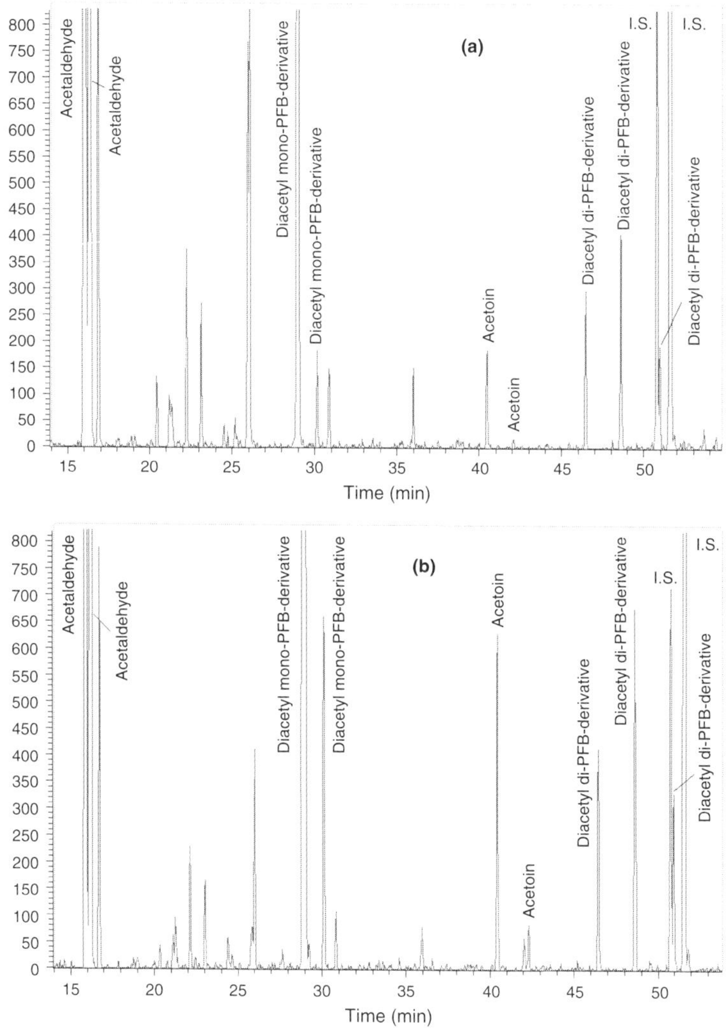

Figure 1.20 Reconstructed ion chromatograms (RIC) of $[M+H]^+$ species at m/z 240 (acetaldehyde-PFB-derivatives), m/z 266 ($[M+H-18]^+$ ion of acetoin-PFB-derivatives), m/z 282 (diacetyl mono-PFB-derivatives), m/z 336 (IS-PFB-derivatives) and m/z 477 (diacetyl di-PFB-derivatives) of a *Merlot* wine at the first day of MLF (**a**), and after 5 days of MLF (**b**). Analytical conditions: poly(ethylene glycol) (30 m × 0.25 mm i.d.; df 0.25 μm) fused silica capillary column; programmed oven temperature: 5 min at 60 °C, 3 °C/min to 210 °C, 5 min at 210 °C; transfer line temperature: 280 °C; carrier gas He; flow mode constant pressure 16 psi. (Reprinted from *Journal of Mass Spectrometry* **40**, Flamini *et al.*, Monitoring of the principal carbonyl compounds involved in malolactic fermentation of wine by synthesis of O-(2,3,4,5,6-pentafluorobenzyl) hydroxylamine derivatives and solid-phase-micro-extraction positive-ion-chemical-ionization mass spectrometry analysis, p. 1563, Copyright 2005, with permission from John Wiley & Sons Ltd)

References

Allen, M. (1995) What level of methoxypyrazines is desired in red wines? The flavour perspective of the classic red wines of Bordeaux, *Australian Grapegrower & Winemaker*, **381**, 7–9.

Cancilla, D.A. and Que Hee, S.S. (1992) O-(2,3,4,5,6-Pentafluorophenyl) methylhydroxylamine hydrochloride: a versatile reagent for the determination of carbonyl-containing compounds, *J. Chromatogr. A*, **627(1–2)**, 1–16.

Chatonnet, P. and Dubourdieu, D. (1996) Odeur de 'planche' dans les bois de chêne: les responsables identifiés, *Revue des Oenologues*, **82**, 17–19.

Chatonnet, P. and Dubourdieu, D. (1998) Identification of substances responsible for the sawdust aroma in oak wood, *J. Sci. Food Agr.*, **76(2)**, 179–188.

Ciolfi, G., Mariscalco, G. Borsa, D. and Di Stefano, R. (1990) Origine dei glicoli presenti nei vini, *Biologia Oggi*, **4(2–3)**, 41–48.

Compendium of international methods of wine and must analysis – OIV, Dosage of sugars in wine by HPLC (Resolution 23/2003), edition 2006, Vol. 1, MA-E-AS311-03-SUCRES, p. 14.

Davis, C.R., Wibowo, D., Eschenbruch, R., Lee, T.H. and Fleet, G.H. (1985) Practical implications of malolactic fermentation: A review, *Am. J. Enol. Vitic.*, **36(4)**, 290–301.

De Smet, P., Liddle, P.A.P., Cresto, B. and Bossard, A. (1981) The analysis of non volatile constituents of Wine by glass capillary gas chromatography, *J. Inst. Brew.*, **87**, 349–351.

De Revel, G. and Bertrand, A. (1994) Dicarbonyl compounds and their reduction products in wine. Identification of wine aldehydes, *in Trends in Flavour Research*, H. Maarse and D.G. van der Heij (Eds), Elsevier Science B.V.

De Revel, G. and Bertrand, A. (1993) A method for the detection of carbonyl compounds in wine: glyoxal and methylglyoxal, *J. Sci. Food Agr.*, **61(2)**, 267–272.

Di Stefano, R., Borsa, D. and Garcia Moruno, E. (1988) Glicoli naturalmente presenti nei vini, *Vini d'Italia*, **30(5)**, 39–44.

Di Stefano, R. and Bruno, C. (1983) Applicazione della gascromatografia-spettrometria di massa all'analisi degli esteri metilici degli acidi fissi del vino, *Vignevini*, **10(9)**, 43–45.

Di Stefano, R., Garcia Moruno, E. and Borsa, D. (1992) Proposta di un metodo di preparazione del campione per la determinazione dei glicoli dei vini, *Vini d'Italia*, **34(4)**, 61–64.

Di Stefano, R. and Ciolfi, G. (1982) Produzione di acetaldeide da parte di stipiti di lieviti di specie diverse, *Riv. Vitic. Enol.*, **35**, 474–480.

Fischleitner, E., Korntheuer, K., Wendelin, S. and Eder, R. (2004) Shikimic acid content as parameter for the verification of varietal authenticity with wines, *Mitteilungen Klosterneuburg, Rebe und Wein, Obstbau und Früchteverwertung*, **54(7–8)**, 234–238.

Flamini, R. (2005) Some advances in the knowledge of grape, wine and distillates chemistry as achieved by mass spectrometry. *J. Mass Spectrom.*, **40(6)**, 705–713.

Flamini, R. and Dalla Vedova, A. (1999) Solid phase extraction of major organic acids from grape musts and wines and HPLC analysis, *Riv. Vitic. Enol.*, **52(4)**, 3–13.

Flamini, R. and Dalla Vedova, A. (2003) Glyoxal/glycolaldehyde: a redox system involved in malolactic fermentation of wine, *J. Agric. Food Chem.*, **51(8)**, 2300–2303.

Flamini, R., Dalla Vedova, A. (2007) Preservation of Cabernet Sauvignon Grape Must Samples Destined For Chemical Analysis: Addition of Sodium Azide, Allyl Isothiocyanate, Octanoic Acid, Ethyl Bromoacetate, and effect of Pasteurization, *J. Food Process. Preserv.*, **31**, 345–355.

Flamini, R., Dalla Vedova, A. and Panighel, A. (2005a) Study of carbonyl compounds in some Italian marc distillate (grappa) samples by synthesis of O-(2,3,4,5,6-pentafluorobenzyl)-hydroxylamine derivatives. *Riv. Vitic. Enol.*, **58(1)**, 51–63.

Flamini, R., Dalla Vedova, A., Panighel, A., Perchiazzi, N. and Ongarato, S. (2005b) Monitoring of the principal carbonyl compounds involved in malolactic fermentation of wine by synthesis of O-(2,3,4,5,6-pentafluorobenzyl) hydroxylamine derivatives and solid-phase-micro-extraction positive-ion-chemical-ionization mass spectrometry analysis. *J. Mass Spectrom.*, **40(12)**, 1558–1564.

Flamini, R., De Luca, G. and Di Stefano, R. (2002a) Changes in carbonyl compounds in Chardonnay and Cabernet Sauvignon wines as a consequence of malolactic fermentation, *Vitis*, **41(2)**, 107–112.

Flamini, R., Tonus, T. and Dalla Vedova, A. (2002b) A GC-MS method for determining acetaldehyde in wines, *Riv. Vitic. Enol.*, **55(2–3)**, 15–21.

Fornachon, J.M.C. and Lloyd, B. (1965) Bacterial production of diacetyl and acetoin in wine, *J. Sci. Food Agr.*, **16(12)**, 710–716.

Guillou, I. (1997) Study of low molecular weight substances combining sulfur dioxide in white wines stemming from botrytised harvest. Revelation of the existence and importance of the role of hydroxypropanedial, *Bulletin de l'O.I.V.*, **70(791–792)**, 140–141.

Guillou, I., Bertrand, A., de Revel, G. and Barbe, J.C. (1997) Occurrence of hydroxypropanedial in certain musts and wines, *J. Agric. Food Chem.*, **45(9)**, 3382–3386.

Holbach, B., Marx, R. and Zimmer, M. (2001) RP-18+cationic exchange column, *Lebensmittelchemie*, **55**, 32–34.

Hunter, J.J., Visser, J.H. and De Villiers, O.T. (1991) Preparation of grapes and extraction of sugars and organic acids for determination by high performance liquid chromatography, *Am. J. Enol. Vitic.*, **42(3)**, 237–244.

Lapolla, A., Flamini, R., Tonus, T., Fedele, D., Senesi, A., Reitano, R., Marotta, E., Pace, G., Seraglia, R. and Traldi, P. (2003) An effective derivatization method for quantitative determination of glyoxal and methylglyoxal in plasma samples by gas chromatography/mass spectrometry, *Rapid Commun. Mass Sp.*, **17(8)**, 876–878.

Lapolla, A., Flamini, R., Aricò, C.N., Rugiu, C., Reitano, R., Ragazzi, E., Seraglia, R., Dalla Vedova, A., Lupo, A. and Traldi, P. (2006) The fate of glyoxal and methylglyoxal in peritoneal dialysis, *J. Mass Spectrom.*, **41(3)**, 405–408.

Martineau, B., and Henick-Kling, T. (1995) Formation and degradation of diacetyl in wine during alcoholic fermentation with Saccharomyces cerevisiae Strain EC 1118 and malolactic fermentation with Leuconostoc oenos Strain MCW, *Am. J. Enol. Vitic.*, **46**, 442–448.

Martineau, B., Acree, T.E. and Henick-Kling, T. (1995) Effect of wine type on threshold for diacetyl, *Food Res. Int.*, **28(2)**, 139–143.

Mascarenhas, M.A. (1984) The occurrence of malolactic fermentation and diacetyl content of dry table wines from northeastern Portugal, *Am. J. Enol. Vitic.* **35**, 49–51.

Ochiai, N., Sasamoto, K., Daishima, S., Heiden, A.C. and Hoffmann, A. (2003) Determination of stale-flavour carbonyl compounds in beer by stir bar sorptive extraction with in-situ derivatization and thermal desorption-gas chromatography-mass spectrometry, *J. Chromatogr. A.*, **986(1)**, 101–110.

Radler, F. (1962) Die bildung von acetoin und diacetyl durch die bakterien des biologischen säureabbaus. *Vitis*, **3(3)**, 136–143.

Rankine, B.C., Fonachon, J. and Bridson, D.A. (1969) Diacetyl in Australian dry red wines and its significance in wine quality, *Vitis*, 8, 129–134.

Schneider, A., Gerbi, V. and Redoglia, M. (1987) A rapid HPLC method for separation and determination of major acids in grape musts and wines, *Am. J. Enol. Vitic.*, **38**(**2**), 151–155.

Symonds, P. and Cantagrel, I. (1982) Anionic exchange column after cations elimination from wine, *Ann. Fals. Exp. Chim.* **75**, 63–70.

Testa, S. (1991) Metodica per il dosaggio contemporaneo mediante HPLC di acidi organici, glucosio, frutosio, glicerina ed etanolo, *L'Enotecnico*, (**10**), 81–87.

Vanderlinde, R., Bertrand, A. and Segur, M.C. (1992) Dosage des aldehydes dans les eaux-de-vie, in *Proceeding of 1er Symposium Scientifique International du Cognac, «Elaboration et connaissance des spiritueux»*, Cognac 11–15 May, Lavoiser TEC&DOC, Paris.

Versini, G., Mattivi, F., Moser, S., Pisoni, A. and Volonterio, G. (2003) Shikimic acid quantification in experimental monovarietal white wines produced in Italy, *in Oenologie 2003 – 7th Intern. Symposium*, TEC&DOC, Paris.

Vesely, P., Lusk, L., Basarova, G., Seabrooks, J. and Ryder, D. (2003) Analysis of aldehydes in beer using solid-phase microextraction with on-fiber derivatization and gas chromatography/mass spectrometry, *J. Agric. Food Chem.*, **51**(**24**), 6941–6944.

Vidal, J.P., Estreguil, S. and Cantagrel, R. (1993) Quantitative analysis of Cognac carbonyl compounds at the PPB level by GC-MS of their O-(pentafluorobenzyl amine) derivatives, *Chromatographia*, **36**(**1**), 183–186.

Vidal, J.P., Mazerolles, G., Estreguil, S. and Cantagrel, R. (1992) Analyse quantitative de la fraction carbonylée volatile des eaux-de-vie de Cognac, *in Proceeding of 1er Symposium Scientifique International du Cognac, «Elaboration et connaissance des spiritueux»*, Cognac 11–15 May, Lavoiser TEC&DOC, Paris.

2

High Performance Liquid Chromatography Analysis of Grape and Wine Polyphenols

Rocco Di Stefano and Riccardo Flamini

2.1. Introduction

2.1.1. Phenolics in Grapes and Wines

The phenolics in the grape berry are monomeric and polymeric molecules and are located in the juice (hydroxycinnamoyl tartaric acid esters), the solid part of the pulp (proanthocyanidins, hydroxybenzoic acids with structures reported in Figure 2.1), seeds (flavan-3-ols, proanthocyanidins, gallic acid) and the skin (anthocyanins, flavan-3-ols, proanthocyanidins, flavonols, dihydroflavonols, hydroxycinnamoyl tartaric acid esters, hydroxybenzoic acids, hydroxystilbens). Their levels in the grape are mainly linked to the variety, but can also be influenced by environmental variables, cultural techniques and the ripening state of the grape.

For different *Vitis vinifera* varieties, differences of phenols in the juice are mainly quantitative, and caffeyl tartaric acid (denominated caftaric acid, CTA) always prevails on *p*-coumaryl tartaric (coutaric acid, *p*-CuTA) and ferulyl tartaric (fertaric acid, FTA) acids. In the skins, the principal cinnamic acid is CTA, or alternatively *p*-CuTA, while FTA

Hyphenated Techniques in Grape and Wine Chemistry Edited by Riccardo Flamini

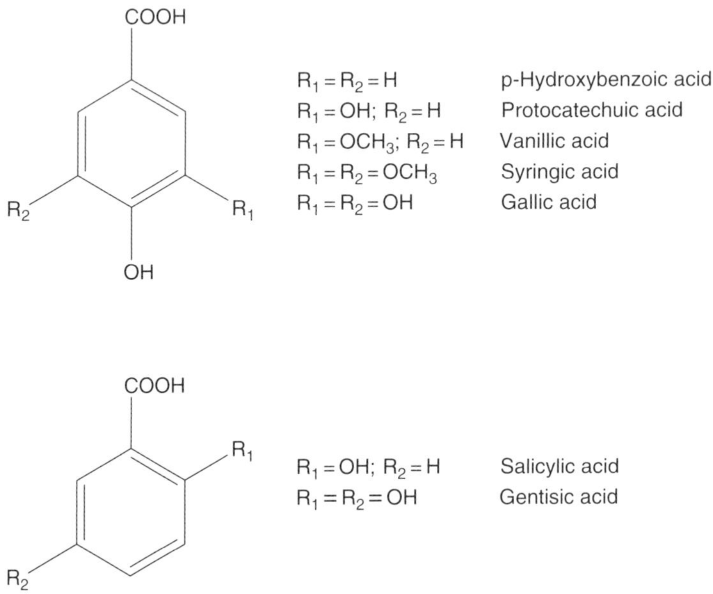

Figure 2.1 Hydroxybenzoic acids in grape

is always present in low amounts. Structures of these compounds are reported in Figure 2.2.

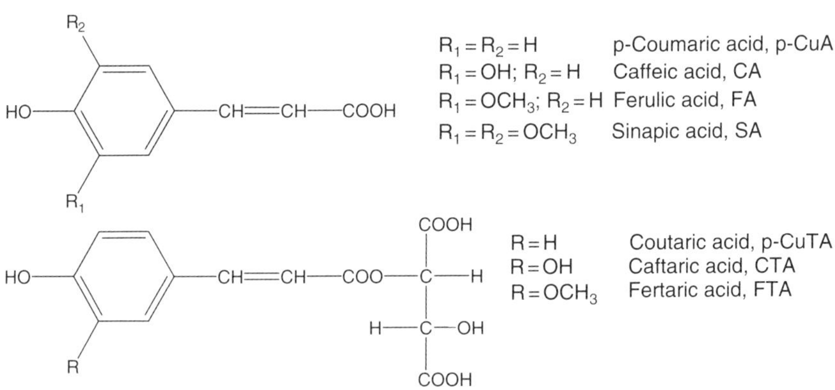

Figure 2.2 Hydroxycinnamoyl tartaric acid (CTA, *p*-CuTA, FTA) esters in grape and the correspondent hydroxycinnamic acids (HCA) found in wine

The principal polyphenols in the solid part of the berry pulp are proanthocyanidins, even if their content is always much lower than it is in seeds and skin. Principal flavan-3-ols in the seeds are (+)-catechin and

(−)-epicatechin (structures in Figure 2.3), the main dimeric procyanidins are the B-type procyanidin B1 and B2 (Figure 2.4). Figure 2.5 shows the structure of A-type dimer procyanidins.

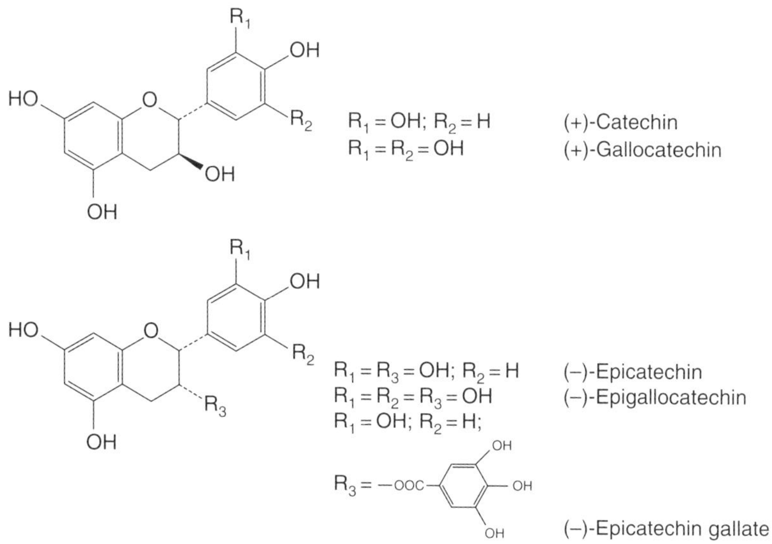

Figure 2.3 Structure of flavan-3-ols in grape

There are discordant reports regarding the distribution of the molecular weights (MW) of proanthocyanidins in the seeds. Nevertheless, it has been reported that their mean degree of polymerization (mDP) is lower than 10 and that they are mainly constituted of (+)-catechin and (−)-epicatechin with a minor extent of (−)-epicatechin gallate. (+)-Catechin, (−)-epicatechin and (−)-epicatechin gallate are the terminal monomers (Kennedy and Jones, 2001; Downey *et al.*, 2003; Pastor del Rio and Kennedy, 2006); (−)-epigallocatechin is absent in the seed proanthocyanidins.

In the skin, flavanol polymers prevail on the oligomers and catechins monomer are poorly present (Mateus *et al.*, 2001). Among the extension units of the skin proanthocyanidins, (−)-epicatechin and (−)-epigallocatechin prevail on (−)-epicatechin gallate and (+)-catechin (Souquet *et al.*, 1996; Pastor del Rio and Kennedy, 2006); while small quantities of (+)-gallocatechin and (−)-epigallocatechin gallate were found (Downey *et al.*, 2003), (+)-catechin is the most terminal unit (Downey *et al.*, 2003; Pastor del Rio and Kennedy, 2006). The ratio

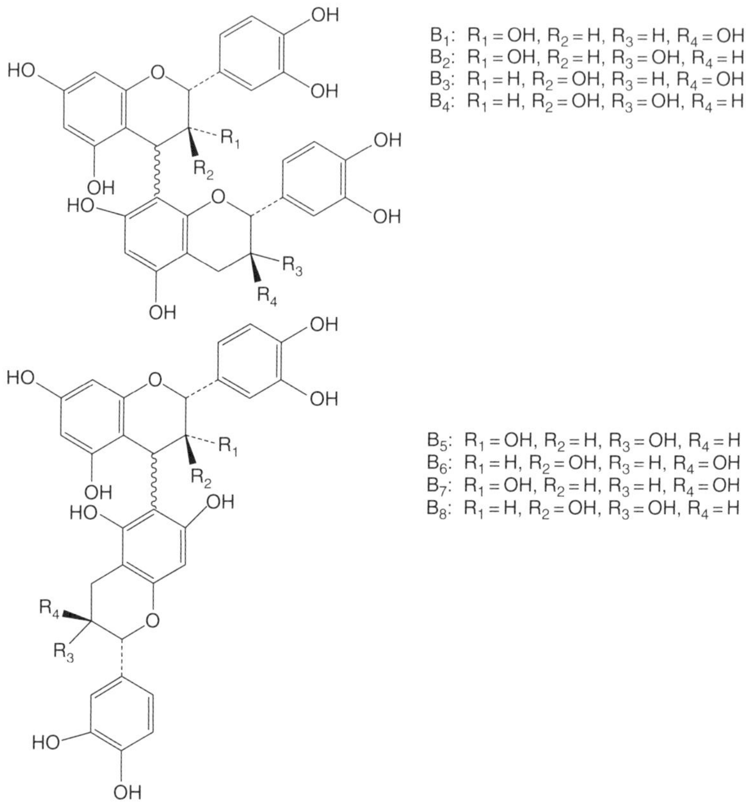

Figure 2.4 Structures of B-type dimer procyanidins

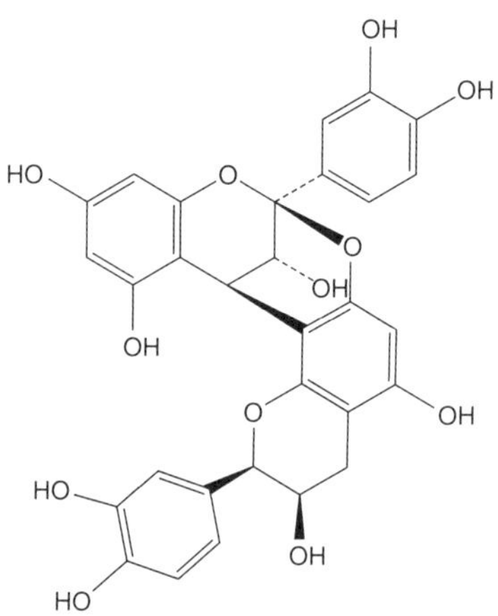

Figure 2.5 Structure of A-type dimer procyanidin

of (−)-epigallocatechin to (−)-epicatechin in the skin proanthocyanidins seems to depend on the variety.

The grape skin flavonols are kaempferol, quercetin and myricetin and are always present in the glycoside form, mainly as glucoside, glucuronide and rutinoside; isorhamnetin was also found (Cheynier and Rigaud, 1986). The content of quercetin and myricetin depends mainly on the variety of the grape (Squadrito *et al.*, 2007), but kaempferol is always the lowest flavonol. White grape varieties contain only traces of myricetin (Borsa *et al.*, 2005). Structures of the aglycone flavonols are reported in Figure 2.6.

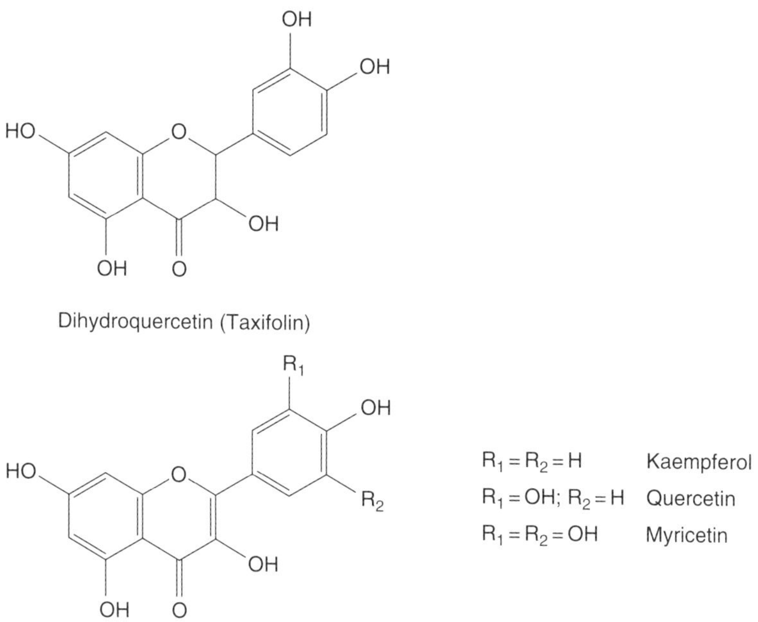

Figure 2.6 Aglycone form of the grape flavonols

Anthocyanins are present in the skin of red grapes and confer the color to them. In some varieties, the anthocyanins are present even in the pulp of the berry. The five anthocyanins that may be present in the *Vitis vinifera* grape varieties are cyanidin, peonidin, delphinidin, petunidin and malvidin, always as 3-glucosides. In the hydroxyl group in the position 6 of glucose an acetic, *p*-coumaric or caffeic acid residue can be also present. Degree and type of acylation of anthocyanins are characteristics of the variety; as a consequence, the anthocyanin profile is used for chemotaxonomic studies in the differentiation of red grape varieties (Castia *et al.*, 1992). Structures of the grape anthocyanidins (the aglycone form of anthocyanins), are reported in Figure 2.7.

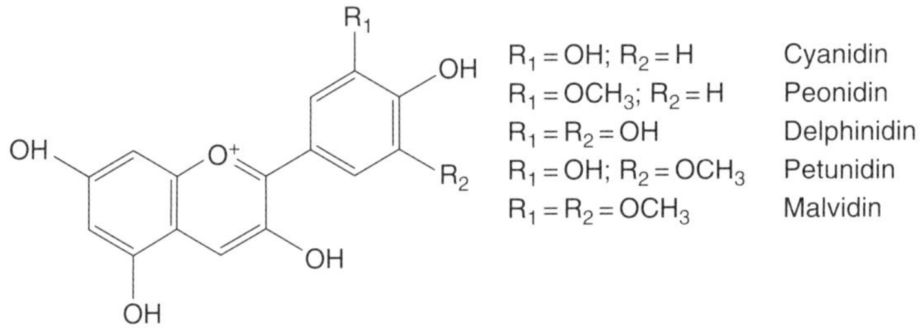

Figure 2.7 Anthocyanidins of *Vitis vinifera* red grape varieties

Hydroxystilbens (*trans*-resveratrol, piceatannol, pterostilbene, viniferines, structures in Figure 2.8) are present in the grape mainly as glycosides. These compounds are classified as fitoalexins and their synthesis and accumulation in the different parts of the plant, besides being linked to the variety, is induced by biological and chemical elicitors (Bavaresco and Fregoni, 2001).

In winemaking, all the different classes of polyphenols are, to different extents, transferred from the grape to the wine and form the sensorial characteristics of the product such as colour and astringency.

Monomer anthocyanins give the red grapes and wines colour: at the end of alcoholic fermentation they make up about 50% of the total pigments, the others are polymeric forms and pyranoanthocyanins. In aged red wines, pigments are, on the other hand, almost exclusively polymeric anthocyanins (Cheynier *et al.*, 2006). Tannins are responsible for bitterness and mouth astringency of certain wines. Bitterness seems to be mainly linked to the low molecular weight flavanols, such as monomer catechins and oligomer procyanidins; astringency is linked to all classes of procyanidins and to their galloylation degree (polymers constituted with epicatechin gallate units). Finally, glycoside flavonols in wine can hydrolyze and cause formation of some precipitates (Di Stefano, unpublished data).

Trans- and *cis*-resveratrol and hydroxystilbens are important qualitative parameters of wine because of the several beneficial effects on human health reported in a number of biological and clinical studies,

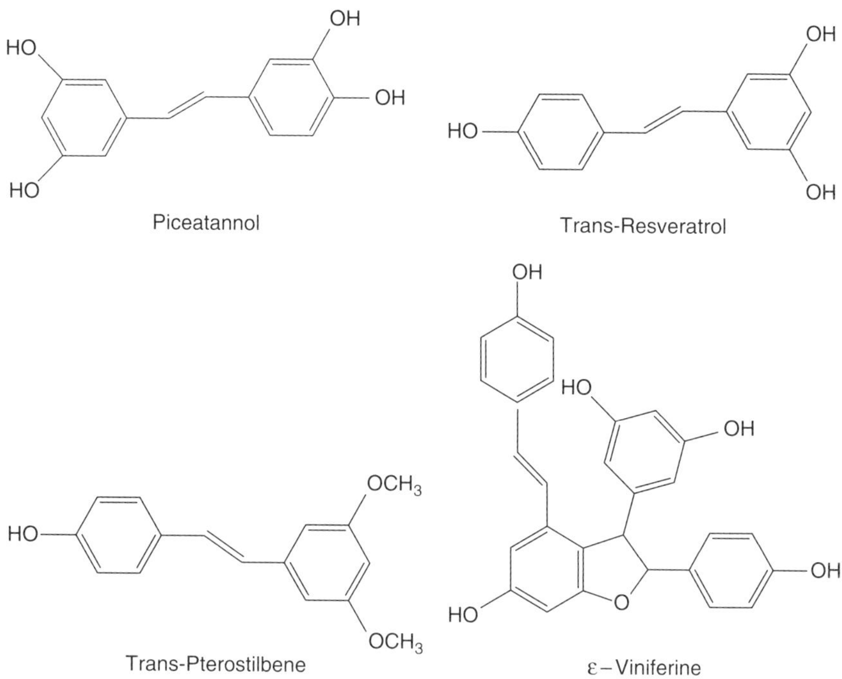

Figure 2.8 Structures of hydroxystilbens in grape

such as anticancer activity, cardioprotection, antioxidant activity, inhibition of platelet aggregation and anti-inflammatory activity (Jang *et al.*, 1997; Hung *et al.*, 2000; Frankel *et al.*, 1993; Fremont *et al.*, 1999; Bertelli *et al.*, 1995; Pace-Asciak *et al.*, 1995). The total resveratrol content in wine can be between 10–20 mg/L and more than 30 mg/L (Flamini and Dalla Vedova, 2004; Di Stefano et coll., unpublished data).

2.1.2. Biosynthetic Pathways of Phenolics in Grapes

The common progenitor of flavonols, anthocyanis, catechins and proanthocyanidins is tetrahydroxy calcone which, in turn, is produced by the reaction of three molecules of malonyl-CoA with one molecule of *p*-coumaroyil-CoA, catalyzed by the enzyme calcone synthetase (CHS) (Boss *et al.*, 1996; Jeong *et al.*, 2006). In addition to this, pentahydroxy calcone, generated from malonyl-CoA and caffeoyl-CoA, could be the progenitor of flavonoids, but this pathway has not yet been confirmed in the grape (Adams, 2006). Cyclization of tetrahydroxy calcone to naringenin (a flavanone) is mediated by the enzyme calcone isomerase (CHI). Naringenin can be monohydroxylated (enzyme F3'H) or dihydroxylated (enzyme F3'5'H) at the β-ring, forming eriodictyol and pentahydroxyflavanone, respectively. By trans hydroxylation in the C3 position in the heterocycle of the latter two compounds, catalyzed from enzyme flavonoid-3-hydroxylase (F3H), dihydroflavonols (flavanonols) dihydroquercetin and dihydromyricetin form. Alternatively, naringenin can be hydroxylated into dihydrokaempferol (F3H) then transformed into the mono- (enzyme F3'H) or dihydroxylate (enzyme F3'5'H) compound, forming dihydroquercetin or dihydromyricetin, respectively.

From dihydrokaempferol, dihydroquercetin and dihydromyricetin – by action of enzyme flavonol synthetase (FLS) – kaempferol, quercetin and myricetin respectively are formed. As flavonols are present in the glycoside form, analogous to anthocyanins, it is supposed that the beginning of their synthesis is concomitant with the flavonol-3-O-glycosyltransferase enzyme formation, at the veraison beginning.

By reduction of dihydroquercetin and dihydromyricetin, operated by the enzyme dihydroflavonol-4-reductase (DFR) (in vine the reduction of dihydrokaempferol does not occur), 2,3-*trans* leucocyanidin and leucodelphinidin form. Successive synthesis of (+)-catechin and (+)-gallocatechin (2,3-*trans* isomers) is mediated by the enzyme leucoanthocyanidin reductase (LAR). On the other hand, the synthesis of (−)-epicatechin and (−)-epigallocatechin (2,3-*cis* isomers) proceeds via oxidation of 2,3-trans leucoanthocyanins into cyanidin and delphinidin (accompanied with chirality loss) mediated by the enzyme leucoanthocyanidin dioxygenase (LDOX), which is followed by the reduction catalyzed by the enzyme anthocyanidin reductase (ANR) with reacquisition of chrality (Fujita *et al.*, 2005).

Synthesis of anthocyanins is catalyzed by the enzyme UDP-glucose: flavonoid 3-O-glucosyltransferase (UFGT). Their acylation at the C6 hydroxyl group of glucose is operated by the anthocyanin acyltransferase enzyme (AAT) (Gong *et al.*, 1997). From cyanidin and delphinidin, synthesis of methylated anthocyanins is mediated by the enzyme methyltransferase (MT). Both acyltransferase and methyltransferase enzymes would operate in the skin cells on the anthocyanin substrate (not on the aglicone anthocyanidin) and their activity is concomitant with UFGT production. Galloylation of epicatechin occurs in the seeds, where only the synthetic pathway starting from dihydroquercetin is active (Jeong *et al.*, 2006).

If, as supposed, epicatechin synthesis starts from cyanidin, then the production of proanthocyanidins in the skins and seeds does not occur by condensation of leucoanthocyanins but by condensation of catechins with oligomer proanthocyanidins and catechins catalyzed by specific enzymes.

Synthesis of caffeic and ferulic acids also needs hydroxylase and methyltransferase enzymes; transformation into hydroxycinnamic tartaric acid esters (HCTA) is operated by an acyltransferase enzyme. The scheme shown in Figure 2.9 summarizes the biosynthetic pathways described.

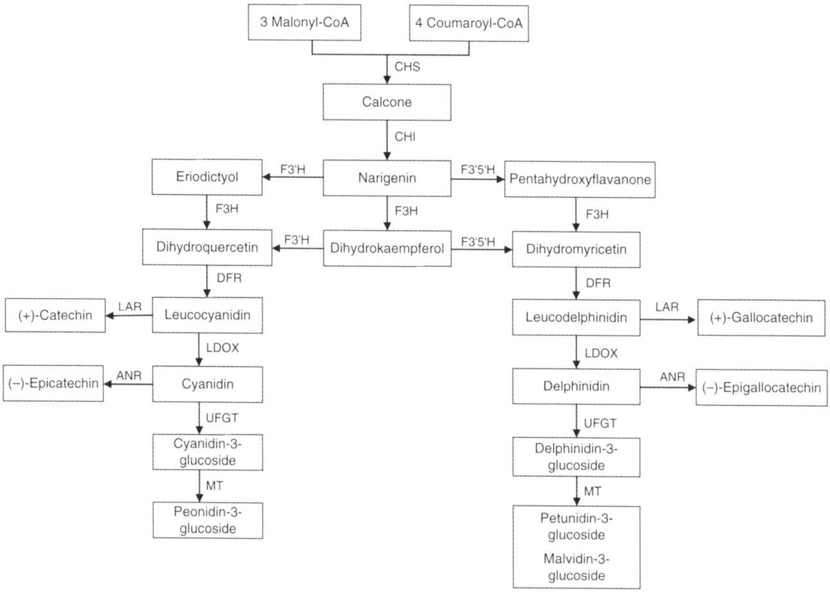

Figure 2.9 Biosynthetic pathways of phenolic compounds in grape

As with flavonoids, the biosynthetic pathway of stilbenes starts from malonyl-CoA and 4-coumaroyl-CoA. Nevertheless, there are some differences in genes and enzymes implicated in the biosynthesis of the different stilbenes (Bavaresco *et al.*, 2003). It has been reported, in *Vitis* the gene stilbenesynthetase (StSy) is not constitutively expressed, while is inducible by some pathogens; among them *Botrytis cinerea* plays a major role as well as chemical agents. As the constitutive gene, StSy is present in stems, seeds, roots and cluster stems and is involved in the mechanism of wood resistance to decay. In the maceration performed during fermentation of red wines, stilbenes spread from the berries to the must. Therefore, their amounts in wine are influenced by the oenological practices used (Bavaresco *et al.*, 2003).

Evolution of polyphenols in the grape ripening

In warm climates, factors such as high temperatures, low water availability for long periods, and exposure of the clusters to the sun induce inhibition of the phenolics synthesis, in particular anthocyanins. A similar behaviour is observed when the grape production of plants is high (Di Stefano, personal observations). In this last case, there is no interruption of the phenolic synthesis but a lack of the primary metabolites necessary to synthesize the secondary ones. The difference between the day and night temperature also plays a significant role in these syntheses (Mori *et al.*, 2005a; 2005b; Yamane *et al.*, 2006; Spayed *et al.*, 2002; Jeong *et al.*, 2004).

The synthesis of proanthocyanidins, and probably of HCTA, begins when the berry growing starts and goes on until the beginning of varaison. The production of (+)-catechin, (−)-epicatechin, (−)-epicatechin gallate and (−)-epigallocatechin slows and sometimes stops. In this period, the synthesis of anthocyanins begins and continues until the end of the physiological maturation of the grape (Kennedy *et al.*, 2000; Downey *et al.*, 2003; Pastor del rio and Kennedy, 2006). At the end of veraison, the anthocyanin profile of the grape is practically definitive and undergoes few changes during the ripening. Environmental variables can have a greater influence on the anthocyanin profile of varieties where cyanidin and peonidin prevail (Di Stefano, unpublished data).

From veraison, a decrease of HCTA in pulp and skin is observed (Di Stefano, unpublished data) probably due to how they are used in the acylated anythocynins synthesis.

Production of flavonols begins before the veraison; in the first period of veraison quercetin is the main flavonol present (the dihydroquercetin biosynthesis prevails) and production of myricetin is slower but, in varieties where biosynthesis of dihydromyricetin prevails, myricetin can be the main flavonol (Borsa, personal communication; Squadrito *et al.*, 2006).

Determination of polyphenols and tannins in the grape during the ripening is very important to determine the right moment to harvest. Glories (1998) proposed that the phenolic maturity of grape occurs when anthocyanins and tannins show the maximum extractability from grapes in winemaking. This state of the grape can be either concomitant or not with the maximum level of sugars in the berry. To estimate the phenolic maturity, a method of determining extractable and total polyphenols with the use of extractive buffers, was proposed (Glories, 1998). Trends of the principal phenols in grape during ripening are shown in Figure 2.10 and reported in Table 2.1.

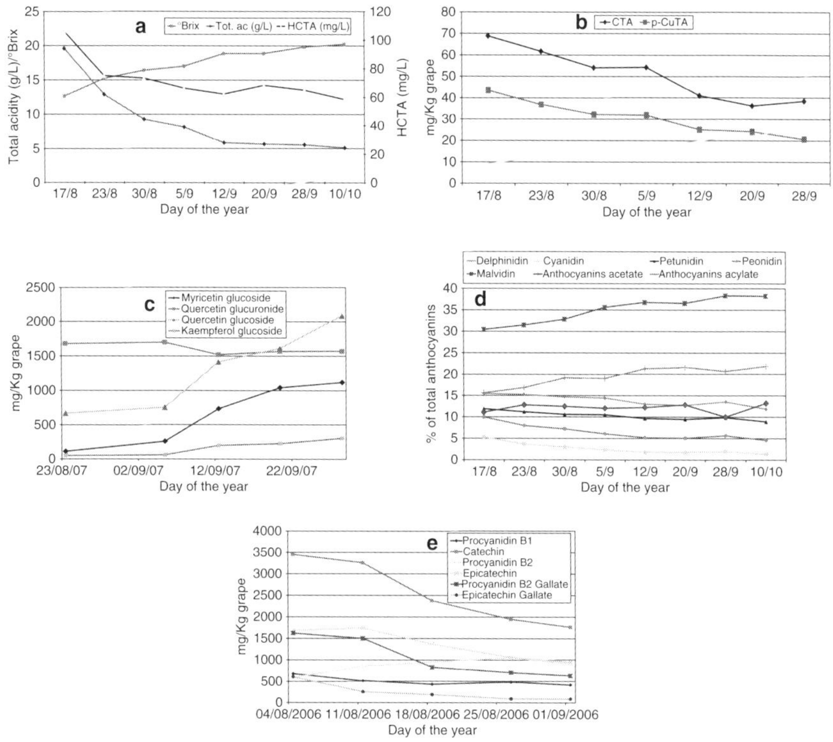

Figure 2.10 Trends of the principal phenols in grape during ripening: a) sugars (°Brix), total acidity (g/L of tartaric acid) and hydroxycinnamoyl tartaric acids (HCTA, mg/L) in a *Carménère* grape must; b) caftaric (CTA) and coutaric (p-CuTA) acids (mg/Kg grape) in skins extract of *Carménère* grapes; c) flavonols in skins extract of *Carménère* grape (mg/Kg grape); d) percentage of the anthocyanins in a *Carménère* grape skins extract; e) catechins and dimer procyanidins (mg/Kg grape) in *Nero d'Avola* grapes

Table 2.1 Changes of catechins and procyanidins (chromatographic peak area) in *Nero d'Avola* grape seeds during ripening.

Day of the year	Procyanidin B1	Catechin	Procyanidin B2	Epicatechin	Procyanidin B2 gallate	Epicatechin gallate
04/08	679.1	3449.9	595.7	1684.5	1620.8	608.7
11/08	512.6	3258.3	848.6	1743.4	1501.9	253.3
18/08	429.0	2380.7	949.7	1370.3	830.6	188.3
26/08	481.8	1949.8	1059.2	1063.7	699.9	90.4
01/09	415.5	1767.6	900.6	940.6	628.4	87.8

2.1.3. Polyphenolic Changes in the Winemaking

The principal phenols and polyphenols in musts from white grapes are HCTA from the juice, and catechins and proanthocyanidins, mainly from the skins. These compounds can undergo oxidation; consequently, the grape must often has to be protected by limiting the contact with oxygen and inhibiting PPO by SO_2 addition. As a consequence of oxidation, the phenolic profile of musts is characterized from lower levels of HCTA and flavanols, and high 2-S-glutationyl caffeyl tartaric acid (GRP). In fact, the presence of GRP in wine suggests that enzymatic reactions have occurred. Figure 2.11 shows the HCTA profiles of a must added of SO_2 to inhibit PPO, and the same must without addition.

Red wines are produced by fermentation of the must in the presence of the grape skins and seeds, sometimes also of stalks (maceration). During fermentation and storage of wine, phenols and anthocyanins can undergo several reactions, such as: enzymatic oxidation of CTA into orto-quinone; electrophilic substitution to the orto-quinone ring by sulphur-containing compounds such as reduced glutatione; anthocyanins degradation; cycloaddition of carbonyl compounds to anthocyanins and formation of vitisins; hydrolysis of proanthocyanidins and formation of carbocation in C4 position which attacks the C6 and C8 positions of proanthocyanidins and anthocyanins; attack of C4 carbocation of anthocyanin to C6 or C8 of proanthocyanidins (structure in the Figure 3.14a, see Chapter 3); attack of acetaldehyde to C6 and C8 of proanthocyanidins (or anthocyanins) and successive attack of anthocyanins (or proanthocyanidins), and formation of structures with flavanol linked to anthocyanin by ethyl bridge (Figure 3.14b); degradation of flavanol-ethyl-flavanol structures with formation of vinyl proanthocyanidins followed by reaction with anthocyanins; oxidation of polyphenols and formation of polymers with higher molecular weights; and formation of xanthylium salts. Structures of several anthocyanin derivatives formed through these mechanisms are reported in Figure 3.13 (see Chapter 3).

Finally, oxidation of hydroxycinnamoyl tartaric acid (HCTA) into hydroxycinnamic acids (HCA), and the successive decarboxylation operated by *Saccharomyces* yeasts, or reduction and decarboxylation by *Brettanomyces* yeasts, induce off-flavours due to vinyl phenols and ethyl phenols formation, respectively.

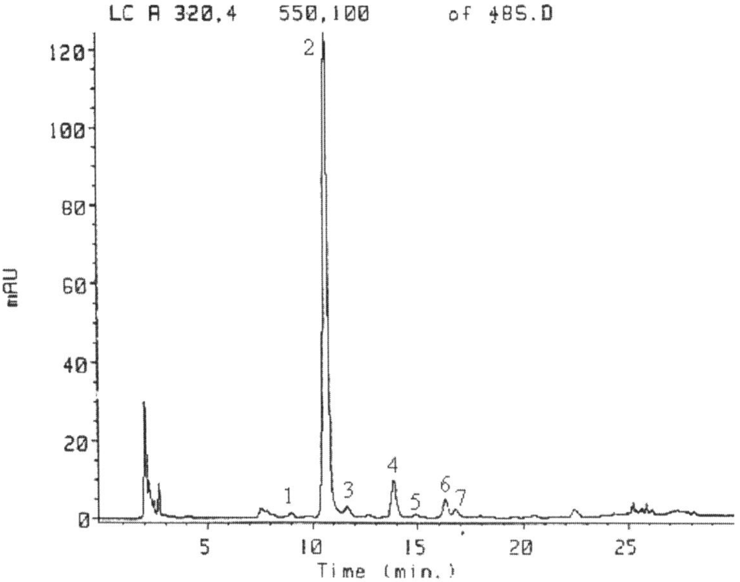

(a)

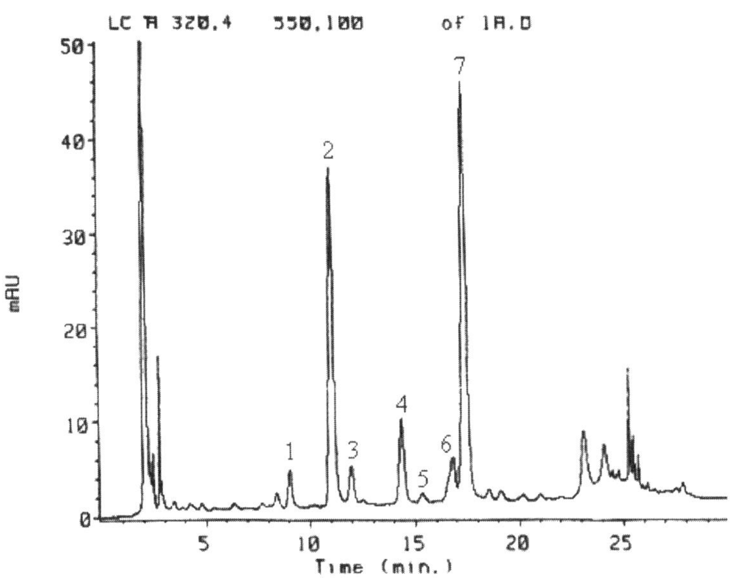

(b)

Figure 2.11 HCTA profile of a *Chardonnay* must sample (**a**) SO_2 added and (**b**) SO_2 not added, in order to inhibit the polyphenol oxidase (PPO) enzymes. (1) *cis*-CTA, (2) *trans*-CTA, (3) *cis*-p-CuTA, (4) *trans*-p-CuTA, (5) *cis*-FTA, (6) *trans*-FTA, (7) 2-S-glutathionyl caffeyl tartaric acid (GRP)

2.2. Extraction of Polyphenols from the Grape

Extraction of polyphenols from grape can be performed by using alcohols or hydroalcoholic solutions either containing, or not containing, mineral acids, alternatively using acetone/water solutions. The use of solvents containing a mineral acid (usually methanol containing HCl 1% by volume or an ethanol/water/HCl 70:30:1 v/v/v solution) allows extraction of all phenolic classes, but artefacts due to hydrolysis reactions of proanthocyanidins and acylated anthocyanins (in particular acetates) – even if the sample is stored at −20 °C – may occur. Moreover, condensation of the alcohol molecules of solvent with HCTA and formation of esters, may occur. On the other hand, the use of nonalcoholic acid containing solvents does not guarantee satisfying extraction of HCTA (mainly present as salts), nor of other phenolic compounds. Other possible artefacts occurring in the extraction are polyphenols polymerization and activity of the polyphenol oxydase (PPO) enzymes. Oxidations operated by these enzymes can be prevented by keeping extracts under liquid nitrogen, using extractive solvents containing high alcohol or acetone, or by addition of SO_2. Freezing of the grape at the harvest followed by heating at room temperature before performing extraction allows PPO to be in touch with and to oxidize the substrates. In any case, when the different parts of the berry are separated from each other it is not possible to completely prevent these reactions that, even if only very slightly, always occur.

A satisfying extraction procedure for all polyphenol classes where risks of artefacts are reduced to the minimum, is by the use of a model wine solution composed of tartrate buffer pH 3.2 containing 12% ethanol (v/v) and sodium metabisulphite 2 g/L (corresponding to about 1 g/L of SO_2 free in solution). This solution is prepared by adding 5 g tartaric acid, 22.2 mL of a NaOH 1 M solution and 2 g of $Na_2S_2O_5$ to 500 mL of distillate water whilst stirring, then adding 125 mL ethanol and adjusting the final volume to 1 L with water. This extractive solvent avoids hydrolytic and oxidative artefacts on polyphenols, and no transformations as a consequence of freezing of the extract occur. Unfortunately, this tartrate buffer solution is less effective than the acidic hydroalcoholic solvents in the extraction from seeds and skins of high molecular weight proanthocyanidins, and their recoveries can be improved increasing the ethanol percentage. Often, extracts have to be concentrated by solid phase extraction (SPE) beforehand to perform analysis by high-performance-liquid-chromatography (HPLC) or spectrophotometry. An advantage of the tartrate buffer solution is that it does not require removal of the organic solvent under vacuum before sample preparation, avoiding any consequent risk of artefacts.

Extraction of seeds and skins by tartrate buffer

As soon as the grape has been picked from the plant, the seeds and skins of 50 berries are separated from the others parts of the berry, separately immersed in two solutions of 50 mL tartrate buffer solutions containing 1 g/L SO_2 and immediately frozen. Before analysis, samples are heated at room temperature and so kept for 4 h in order to allow the extraction to occur. Extracts are then homogenized and centrifuged, the supernatant is transferred in a 100-mL volumetric flask, pellets are re-suspended twice in 15–20 mL of tartrate buffer and, after centrifugation, the liquid phases are collected in the same volumetric flask and the final volume is adjusted to 100 mL by tartrate buffer.

Extraction of seeds and skins by acetone/water solutions

After separation from the flashes, skins from 300 berries are rinsed with distilled water, immersed in 100 mL of aqueous solution containing 100 mg/L of SO_2 and stored at 4 °C (adapted from Kennedy *et al.*, 2002). To perform the extraction, the sample – made up of skins and aqueous solution – is transferred to a 500 mL Erlenmayer flask, and 200 mL of acetone is added. The flask is covered and placed on a shaker for 24 h at 20 °C at 100 rpm. After extraction, acetone is removed under vacuum at 35–40 °C, the solution is centrifuged, the supernatant is recovered, diluted at 100 mL with distilled water and stored at −20 °C until analysis.

Seeds of 20 berries are extracted with 50 mL of a acetone/water 2:1 (v/v) solution at 20 °C for 2 h (Kennedy *et al.*, 2000). Extracts are filtered, acetone is removed under vacuum at 35 °C, and the residual aqueous solution is diluted to 25 mL with distilled water. Ten millilitres of the resulting solution is transferred in a sample tube, frozen, and then freeze-dried. The solid residue is weighed and then dissolved in methanol (10 mg solid in 1 mL methanol).

Another method of extracting polyphenols from seeds and skins is by immersion of the fine powder, obtained by grinding the grape skins and seeds in liquid nitrogen, in an acetone/water 7:3 (v/v) solution containing 0.1% ascorbic acid added to prevent oxidation (Downey *et al.*, 2003).

Jordão *et al.* (2001) applied the method previously proposed by Bourzeix *et al.* (1986) to the extraction of polyphenols from grape seeds (6 g) and skins (20 g). Each sample is immersed in 100 mL methanol containing 5 mg of ascorbic acid and is stored at −24 °C for 24 h under nitrogen atmosphere. The solid residue is then extracted with 100 mL of a 80% methanol aqueous solution followed by 100 mL of a 50% methanol aqueous solution for 4 h each. After extraction, the sample is added of distilled water and is placed at −24 °C for 15 h. Finally, the residue is extracted with 75% acetone aqueous solution for 1 h.

Other methods of seeds and skins extraction

Others methods for seeds and skins extraction were proposed.

(1) By using methanol (Mori *et al.*, 2005): 250 mg of skins are stored at −80 °C, then ground with mortar and pestle in liquid nitrogen and soaked in 30 mL of methanol at 4 °C. The extract is concentrated to almost dryness and dissolved in 1 mL of methanol.
(2) In the presence of an organic acid (Jeong *et al.*, 2004): skins of 40 berries are peeled, immediately frozen in liquid nitrogen and kept at −80 °C. One gram of the sample is ground with mortar and pestle and extracted twice with 5 mL of a formic acid/methanol 1:9 (v/v) solution for 1 h under shaking at room temperature in the dark. The extracts are combined and centrifuged at 18.000 g for 10 min. In an alternative, 50% (v/v) acetic acid in water was proposed as solvent in the extraction of anthocyanins from skins (Yamane *et al.*, 2006).
(3) By ethanol/water/chloroform (De Freitas *et al.*, 2000): skins and seeds are freeze-dried. Skins (1 g dry weight) and grape seeds (0.5 g dry weight) are extracted with 20 mL of ethanol/water/chloroform 1:1:2 (v/v/v) for 2 min. The 50%-ethanol upper layer containing the polyphenolic compounds is separated from the chloroform lower layer containing chlorophylls, lipids and other undesirable compounds. Ethanol is removed under vacuum and the aqueous solution is filtered through a glass membrane, diluted with water up to 100 mL and stored at −20 °C until analysis.
(4) By ethanol/HCl (Spayed *et al.*, 2002): the pulp of 10 berries is removed from the skin and, using a 8 mm diameter cork borer, two disks are removed from each berry for a total of 20 disks per sample,

representing about 10 cm^2 of skin surface. Disks are placed in a 10-mL test tube and added 2 mL of 100% ethanol containing 0.2% HCl solution and homogenized. The homogenizer is rinsed with an additional 2 mL of acidified ethanol that is added to the 10-mL tube containing the homogenized grape skins. The sample is placed in the refrigerator overnight, then is centrifuged at 1200 g for 15 min. The supernatant is transferred in a 10 mL volumetric flask. The sample pellet is extracted twice more by mixing the pellet with 2 mL of acidified ethanol, holding for 2 h under refrigeration, centrifuging, and transferring the supernatant to the 10-mL volumetric flask. 3 mL of water is added to the volumetric flask. The extract is taken to 10 mL with acidified ethanol. By this method, the Authors reported that 97% of anthocyanins and flavonols are extracted.

(5) By ethanol/water (Mateus *et al.*, 2001): 1 g of freeze-dried grape seeds and skins are extracted 5×20 mL with ethanol/water 1:1 (v/v) using a blender. All extracts are adjusted to a final volume of 100 mL and pH 4.0. Prior to analysis, extracts are purified and fractionated on a Toyopearl Gel HW-40(s) column (250×5 mm) using methanol as eluent.

Comparison among methods of extraction

Data of polyphenols extracted from skins of *Pinot noir* and *Barbera* grapes using ethanol/H_2O/HCl 70:30:1 (v/v/v), tartrate buffer-12% ethanol containing $Na_2S_2O_5$ 2 g/L and ethanol/H_2O/acetic acid 70:30:1 (v/v/v) solutions, are reported in Tables 2.2 and 2.3, respectively.

All the solvents show a satisfactory extraction of anthocyanins; better extraction of flavans is achieved with solutions containing more ethanol. HCTA are mainly present in the skin as salts; as a consequence they are not extracted by solvents containing weak acids, even if in the presence of high ethanol. Only using solvents containing a strong acid and high alcohol can a good recovery of these compounds be achieved. In extraction of both grape types, FTA recovery resulted little better with tartrate buffer than using ethanol/H_2O/HCl. Tartrate buffer is less efficient in the extraction of flavanols and HCTA but the ratios among the singular compounds are not changed. In any case, by these three solutions a quantitative recovery for all the different classes of phenolic compounds is not possible without the use of large volumes of the solvent (Revilla *et al.*, 1998). By aqueous acetone solvents (acetone/H_2O 70:30 v/v) a good extraction of all polyphenols can be achieved, but this method requires the solvent be removed prior to performing the sample

Table 2.2 Polyphenols extracted from *Pinot noir* grape skins using three extractive solvents: CTA, caftaric acid; *p*-CuTA, coutaric acid; FTA, fertaric acid. n.d., not detected.

	extractive solvent		
	ethanol/H_2O/HCl 70:30:1 (v/v/v)	tartrate buffer (pH 3.2) 12% ethanol – $Na_2S_2O_5$ 2 gL^{-1}	ethanol/H_2O/ acetic acid 70:30:1(v/v/v)
Total anthocyanins (mg kg^{-1} grape)	1009	1287	1366
Total flavonoids (mg kg^{-1} grape)	2581	2010	2487
CTA (mg kg^{-1} grape)	105.50	80.40	n.d.
p-CuTA (mg kg^{-1} grape)	61.1	32.2	n.d.
FTA (mg kg^{-1} grape)	1.40	2.40	n.d.
% Delphinidin-3-glu	8.5	8.3	4.2
% Cyanidin-3-glu	2.9	2.9	1.9
% Petunidin-3-glu	10.1	9.4	8.5
% Peonidin-3-glu	19.2	17.6	19.2
% Malvidin-3-glu	59.4	61.9	66.2

Table 2.3 Polyphenols extracted from *Barbera* grape skins using two extractive solvents: CTA, caftaric acid; *p*-CuTA, coutaric acid; FTA, fertaric acid.

	extractive solvent	
	ethanol/H_2O/HCl 70:30:1 (v/v/v)	tartrate buffer (pH 3.2) 12% ethanol – $Na_2S_2O_5$ 2 gL^{-1}
Total anthocyanins (mg kg^{-1} grape)	759	843
Total flavonoids (mg kg^{-1} grape)	1594	1294
Total phenolics (mg kg^{-1} grape)	1916	1122
Proanthocyanidins (mg kg^{-1} grape)	1310	993
CTA (mg kg^{-1} grape)	117.3	104.3
p-CuTA (mg kg^{-1} grape)	99.5	91.9
FTA (mg kg^{-1} grape)	1.5	2.9
% Delphinidin-3-glu	17.7	19.2
% Cyanidin-3-glu	11.8	13.3
% Petunidin-3-glu	16.0	15.8
% Peonidin-3-glu	6.3	6.9
% Malvidin-3-glu	30.9	28.6
% anthocyanins acetate	8.8	9.6
% anthocyanins *p*-coumarate	8.5	6.6

preparation with a consequent increase in the risk of artefacts in the sample.

2.3. Sample Preparation for HPLC Analysis

2.3.1. Fractionation of the Different Classes of Polyphenols

Methods of analysis by direct injection in the column of sample were proposed (Lea *et al.*, 1979; Nagel *et al.*, 1979; Ong and Nagel, 1978; Wulf and Nagel, 1976; Roggero *et al.*, 1989; Lamuela-Raventós and Waterhouse, 1994), but usually, prior to analysis, the different classes of compounds are fractionated on absorbent polymers such as polyamide, Sephadex LH20 or C_{18}. The stationary phase C_{18} is also used for concentration and purification of the sample by solid phase extraction (SPE) prior to analysis.

A fractionation method for separation of neutral phenols from the acidic ones by using two 300-mg C_{18} cartridges was proposed. The operative procedure is summarized in the scheme in Figure 2.12 (Lee and Jaworski, 1987).

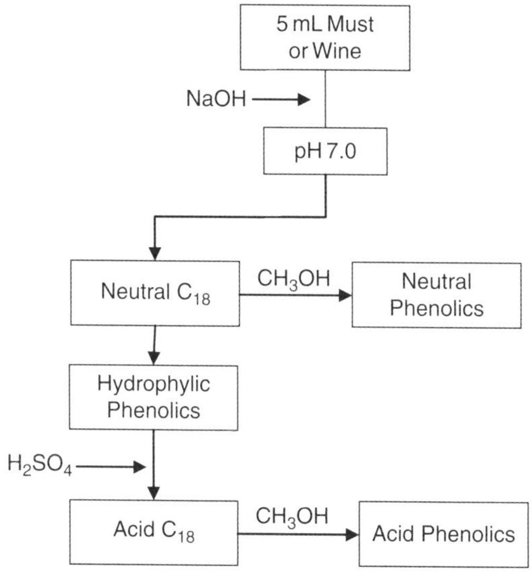

Figure 2.12 Fractionation of neutral and acidic phenols by using two 300-mg C_{18} cartridges (Lee and Jaworski, 1987)

For neutral phenolics the cartridges are preconditioned by sequential passing of 2 mL of methanol and 2 mL of distilled water; for acidic phenolics distilled water is substituted with 2 mL of H_2SO_4 5×10^{-3} M. Five millilitres of the grape juice are adjusted to pH 7.0 with NaOH 1 M and passed through the preconditioned neutral cartridge where the neutral phenolic compounds are adsorbed. The effluent solution is then adjusted to pH 2.5 with H_2SO_4 0.5 M and passed through the second cartridge to adsorb acidic phenolics. The adsorbed phenolics are eluted from both cartridges with methanol and the first 2 mL of the eluate are collected and used for HPLC analysis.

This method was applied to the grape extracts (Oszimianski and Lee, 1990): four millilitres of the extract (25 g of flesh and skins extracted with 50 mL of solution) are passed through a C_{18} cartridge preconditioned by successive passage of 5 mL ethyl acetate, 5 mL methanol and 5 mL H_2SO_4 5×10^{-3} M. After the sample loading, the cartridge is washed with 3 mL H_2SO_4 5×10^{-3} M and then dried. Phenolic compounds, but not anthocyanins, are recovered with 20 mL of ethyl acetate. The solvent is removed under vacuum, the residue redissolved in 20% acetonitrile and this solution is used for HPLC analysis. Anthocyanins adsorbed in the C_{18} cartridge are eluted with a methanol containing 0.1% HCl solution, the anthocyanin content can be determined by spectrophotometric (Niketsic and Hrazdina, 1972) or HPLC analysis.

A method for fractionation of cinnamic acids in must and wine was proposed (Di Stefano and Cravero, 1992). The method was developed using a XAD2 resin nowadays no longer commercially available, alternatively a C_{18} stationary phase can be used by performing the operative procedure summarized in the scheme in Figure 2.13.

A volume of 5 mL of must or wine is added to 15 mL of a pH 6.0 buffer and passed through a 1-g C_{18} cartridge, previously activated by the passage of 3 mL methanol followed by 5 mL distilled water, collecting the eluate in a flask. After the sample passage, the cartridge is washed with 5 mL of distilled water collecting the eluate in the same flask. The solution is adjusted to pH 2.0 by H_2SO_4 0.5 M and passed through a second 1-g C_{18} cartridge activated by the passage of 3 mL methanol followed by 3 mL of H_2SO_4 5×10^{-3} M. The cartridge is washed with 3 mL H_2SO_4 5×10^{-3} M; then HCA and HCTA are recovered, together with 10 mL diethyl ether; GRP and other compounds are recovered by the subsequent passage of 5 mL methanol through the cartridge. Monomer catechins adsorbed on the first cartridge are eluted by 10 mL diethyl ether recently distilled on $FeSO_4$. From the same

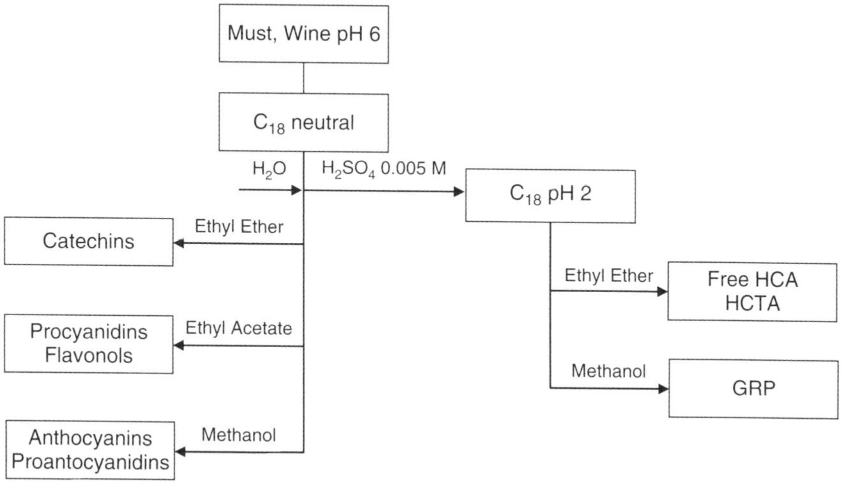

Figure 2.13 Fractionation of cinnamic acids in must and wine using two C_{18} cartridges (Di Stefano and Cravero, 1992). HCA: hydroxycinnamic acids; HCTA: hydroxycinnamic tartaric acid esters; GRP: 2-S-glutathionyl caffeyl tartaric acid

cartridge, dimer and oligomer procyanidins and flavonols are recovered by 10 mL ethyl acetate, monomer anthocyanins and polymer proanthocyanidins and anthocyanins by 5 mL of methanol (Figure 2.13). After evaporation of the solvent under vacuum, the sample residue is redissolved with 1 mL of methanol/H_2SO_4 5×10^{-3} M 4:6 (v/v) solution and used for HPLC analysis.

Flavans in seeds and skin extracts can be fractionated on a Toyopearl TSK HW-50(F) (200 × 25 mm) column (Vidal *et al.*, 2002). Monomers and oligomers are eluted with 5 bed volumes of ethanol/water 55:45 (v/v) containing 0.05% trifluoroacetic acid (TFA) solution. The proanthocyanidins are desorbed from the column using 3 bed volumes of acetone/water 60:40 (v/v). Prior to the HPLC analysis, for fractionation of monomeric and polymeric proanthocyanidins, the sample is loaded into a preconditioned C_{18} cartridge. Elution from the cartridge starts by passing through 5 mL of water. After drying of the stationary phase with a nitrogen stream, flavanols monomer are eluted with 5 mL of diethyl ether, then proanthocyanidin oligomers and polymers are recovered with 5 mL of methanol.

Separation of flavans monomer and oligomers from polymers and anthocyanins can be achieved by using a polyamide (1.7 g) column (Bourzeix *et al.*, 1986). A volume of 2–10 mL of the sample (wine, seeds or skins extract) is passed through the column, and the more polar compounds are removed by washing of the stationary phase with water. Catechins and procyanidins are eluted from the column with 35 mL of water/acetone 25:75 (v:v) solution.

Methods for fractionation of flavans by chromatography on Sephadex LH20 were also proposed (Salagoty-Auguste and Bertrand, 1984; Oszmianski and Sapis, 1989).

2.3.2. Sample Preparation for Determination of the Mean Polymerization Degree (mDP) of Proanthocyanidins in Skins and Seeds Extracts and in the Wine

Analysis of intact proanthocyanidins provides information on their number average molecular weight and weight average molecular weight while providing distribution information. Analysis after acid-catalyzed cleavage instead provides information on their subunit composition as well as the interflavonoid bond location. From proanthocyaninds, under acidic conditions electrophilic flavan-3-ol intermediates form, the two most common nucleophilic reagents used to generate analyzable adducts are phloroglucinol and benzyl mercaptan (Kennedy and Jones, 2001).

The polyphenols in the crude extract of grape seeds and skins or in the wine (e.g. 30 mL) are adsorbed on a C_{18} cartridge (1 g). The stationary phase is washed with water and the adsorbed compounds are eluted with the minimal volume of methanol. After addition of an equal volume of water containing 0.1% TFA, proanthocyanidins are separated from catechins by performing chromatography on a Toyopearl TSK HW 40-F column. The column (270 × 28 mm) is equilibrated with methanol/water 1:1 solution containing 0.1% (v/v) TFA. After applying the solution containing polyphenols to the Toyopearl column, the monomer flavonols are removed by rinsing with five column volumes of the mobile phase. The proanthocyanidins are then eluted with 3 column volumes of acetone/water 2:1 containing 0.1% v/v TFA solution. The eluent is concentrated under reduced pressure at 35 °C and the residue is redissolved in a volume of 0.1 M HCl methanolic solution containing 50 g/L ploroglucinol and 10 g/L ascorbic acid in order to obtain a concentration of at least 5 g/L in proanthocyanidins. Reaction is carried out at 50 °C for 20 min, then 5 volumes of 40 mM aqueous sodium acetate solution are added to stop the reaction and HPLC analysis of the mixture is performed (Kennedy and Jones, 2001).

Alternatively, determination of the mean degree of polymerization (mDP) of proanthocyanindins and of the monomers contained in these molecules can be done by performing their hydrolysis using benzyl mercaptan as reagent (Kennedy *et al.*, 2002). The proanthocyanidins of grape seeds and skins extract (or wine) are separated from flavan-3-ols as described above. 800 μL of the proanthocyanidin extract is placed into a 1.8 mL vial and added to 800 μL of methanolic HCl 0.4 M solution containing 12% benzyl mercaptan. The reaction is carried out by immersing the closed vial in a water bath at 60 °C for 10 min; then the solution is added of 200 μL of 5.2 M ammonium acetate solution in methanol and the HPLC analysis is performed.

2.3.3. Sample Preparation for Analysis of Anthocyanins Monomer in Skins Extract

When concentration of anthocyanins in the extract is sufficiently high, after acidification by addition of 0.5 mL of H_3PO_4 1 M solution to 4.5 mL of extract and filtration 0.2 μm, the sample can be directly injected into the HPLC column. If the anthocyanin content of the extract is low, it can be useful to perform concentration of the sample prior to analysis. A 300-mg C_{18} cartridge is activated by the passage of 2 mL methanol followed by 3 mL H_2SO_4 5×10^{-3} M. A volume of 2.5 mL extract is added to 7.5 mL H_2SO_4 5×10^{-3} M and the solution is slowly passed through the cartridge. After washing with 2 mL of H_2SO_4 5×10^{-3} M, the cartridge is dried under a nitrogen stream, then anthocyanins are eluted with methanol. After removing the solvent under vacuum, the residue is redissolved in 1–2.5 mL of methanol/H_3PO_4 10^{-3} M 40:60 (v/v) and the resulting solution is used for HPLC analysis.

For extracts containing a very low level of anthocyanins, a higher volume of extract can be diluted 1:4 and passed through the C_{18} cartridge, and a larger C_{18} cartridge (e.g. 1 g) can be used.

2.3.4. Sample Preparation for Analysis of Anthocyanins and Derivatives in Wines

Anthocyanins monomer

To perform analysis of young wines, where anthocyanins are mainly present as monomers, a satisfying HPLC separation can be achieved by direct injection into the column of the sample previously acidified. In older wines, anthocyanins are mainly present polymerized with other polyphenols; consequently, when performing detection at wavelength 520 nm an increase of the HPLC chromatogram baseline is observed. Therefore, for determination of anthocyanins monomers it is necessary to perform separation from the polymeric forms as well as the sample concentration. Ten millilitres of wine sample are diluted 1:1 (v/v) with H_2SO_4 5×10^{-3} M and 1 g or more of polyvinyl pyrrolidone (PVPP) is added to completely adsorb polyphenols. The solution is stirred for 1 min, then is centrifuged and the supernatant eliminated. The pellet is resuspended twice in 10 mL of ethanol/water/HCl 70:3:1 (v/v/v) solution; after centrifugation the supernatants are combined, transferred to a 100-mL distillation flask and ethanol is removed under vacuum. The resulting aqueous solution is passed through a C_{18} cartridge (0.3–1 g) previously activated by passage of 2 mL methanol and 3 mL H_2SO_4 5×10^{-3} M. After the sample passage, the cartridge is washed with 2 mL H_2SO_4 5×10^{-3} M, then anthocyanins are recovered by 3 mL of methanol collecting the eluate in a 50-mL distillation flask. After evaporation of methanol to dryness under vacuum, the residue is redissolved in 2 mL of methanol/H_3PO_4 10^{-3} M 40:60 (v/v) and the solution is filtered with a membrane 0.45 μm prior to HPLC analysis.

Anthocyanin derivatives

The wine sample has to be concentrated prior to analysis. Five millilitres of wine are diluted by addition of 20 mL H_2SO_4 5×10^{-3} M and passed through a 1-g C_{18} cartridge previously activated by passage of 4 mL methanol followed by 5 mL di H_2SO_4 5×10^{-3} M. After washing with 4 mL of H_2SO_4 5×10^{-3} M solution, polyphenols are eluted with methanol. The solvent is evaporated to dryness at 30 °C, the residue redissolved with 2 mL of methanol/H_3PO_4 10^{-3} M 40:60 (v/v) and filtered with a membrane 0.45 μm.

To perform analysis of the wine added of H_3PO_4 1M (4.5 mL wine + 0.5 mL H_3PO_4 1M) and centrifuged (12.000 g for 4 min), provides information on the total pigments present in the sample. The difference between the two analyses provides the profile of the non-polymeric pigments (linked to polysaccharides or peptides) of the wine which are not eluted from the C_{18} cartridge with methanol.

2.3.5. Sample Preparation for Analysis of HCTA and Flavonols in Grape Extracts and Juice

HCTA and flavonols can be analyzed after acidification of the extract as described in paragraph 2.3.3 for anthocyanins: 4.5 mL of the extract is added to 0.5 mL of a H_3PO_4 1 M solution and filtered with a membrane 0.45 μm. The two classes of compounds are determined in the same chromatographic run by performing detection of HCTA at wavelength 320 nm and of flavonols at wavelength 360 nm. If it is necessary, a concentration of analytes by a C_{18} cartridge can be performed prior to analysis. Alternatively, because the other compounds in the skins extract do not interfere in the analysis, a simple method of sample preparation can be used: 5 mL of the extract previously diluted with H_2SO_4 5×10^{-3} M (in order to have an alcoholic content lower than 4%) is passed through a 1-g C_{18} cartridge activated by the passage of 2 mL of methanol and 3 mL of H_2SO_4 5×10^{-3} M. The stationary phase is washed with 2 mL of H_2SO_4 5×10^{-3} M then analytes are recovered by elution with methanol. The solvent is evaporated under vacuum, the residue is redissolved in 1 mL of a methanol/H_2SO_4 5×10^{-3} M 40:60 (v/v) solution and filtered 0.45 μm before HPLC analysis.

2.4. HPLC Analyses

Monomer phenols in grape and wine are usually analyzed by HPLC using a reverse phase C_{18} column (usually 250 × 4 mm, 5 μm) operating close to or at room temperature.

2.4.1. Analysis of Anthocyanins and Derivatives

The anthocyanin monomers profile of grape and wine

The chromatogram in Figure 2.14 shows the anthocyanin profile of a *Cabernet sauvignon* grape skins extract. Analysis was performed by a C_{18} (250 × 4 mm, 5 μm) column at 45 °C, using a binary solvent mixture composed of A) formic acid/water 10:90 (v/v) and B) formic acid/methanol/water 10:50:40 (v/v/v) and detection at wavelength 520 nm. Elution of compounds from the column is performed by the gradient program reported in Table 2.4. The sample volume injected is 20 μL.

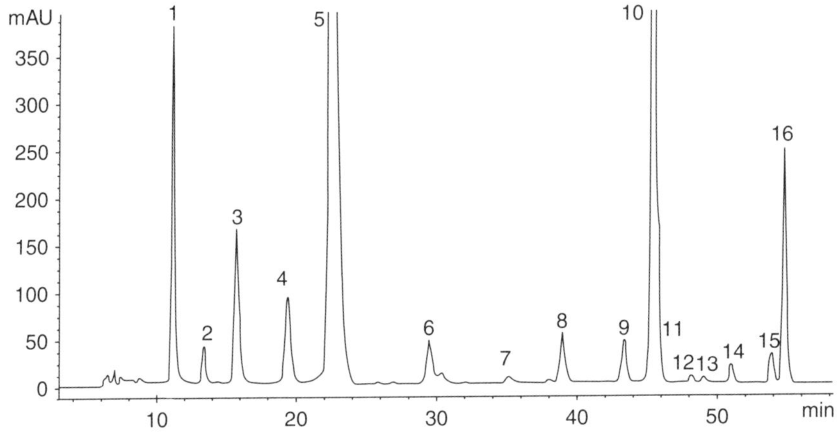

Figure 2.14 HPLC anthocyanins profile of *Cabernet sauvignon* grape skins extract. 1. delphinidin-3-glucoside (glu), 2. cyanidin-3-glu, 3. petunidin-3-glu, 4. penidin-3-glu, 5. malvidin-3-glu, 6. delphinidin-3-glu-acetate (ac.), 7. cyanidin-3-glu-ac., 8. petunidin-3-glu-ac., 9. peonidin-3-glu-ac., 10. malvidin-3-glu-ac., 11. delphinidin-3-glu-*p*-coumarate (*p*-coum.), 12. malvidin-3-glu-caffeate, 13. cyanidin-3-*p*-coum., 14. petunidin-3-*p*-coum., 15. peonidin-3-*p*-coum., 16. malvidin-3-*p*-coum

Table 2.4 HPLC gradient program used for analysis of anthocyanins in grape skins extract by C_{18} (250×4 mm, 5 μm) column (chromatogram in Figure 2.14) (flow rate 0.5 mL/min).

Time (min)	formic acid/water 10:90 %	formic acid/methanol/water 10:50:40 (v/v/v) %
0	55	45
15	55	45
35	30	70
45	10	90
50	0	100
54	0	100
56	55	45
60	55	45

Figure 2.14 shows the peaks of the 16 anthocyanins typical of the *Vitis vinifera* varieties: 3-glucosides, 3-glucoside acetates, 3-glucoside *para*-coumarates and malidin 3-glucoside caffeate. The elution sequence of compounds from a reverse phase column is due to their polarity and is always the same, even if a different column is used.

Concentration of the single anthocyanin is expressed in mg/Kg grape of malvidin-3-glucoside, the compound used as external standard to calculate the calibration curve, or as area percentage on the area sum of the peaks of all anthocyanins identified. Standard of malvidin-3-glucoside is commercially available, or can be isolated from a grape skins extract by TLC on silica gel, or by performing a semipreparative HPLC (Favretto and Flamini, 2000).

Other chromatographic methods for analysis of anthocyanins in skins extracts were proposed: by C_{18} (150×4.6 mm, 5 μm) column using a binary solvent mixture composed of A) acetic acid/water 10:90 (v/v) and B) acetonitrile with a linear gradient program of elution starting from 0% of B, then from 0% to 20% B in 30 min, finally isocratic B for 10 min at a flow rate of 1 mL/min (Mori *et al.*, 2005); by a C_{18} (250×3.6 mm, 5 μm) column using a binary solvent composed of A) formic acid/water 10:90 (v/v) and B) acetonitrile with gradient program of elution starting from 95% A and 5% B; then from 5% to 8% of B in 5 min; from 8% to 9% of B in 10 min; from 9% to 10% of B in 5 min; from 10% to 20% of B in 10 min; from 20% to 25% B in 5 min; finally from 25% to 30% B in 15 min at a flow rate of 1 mL/min (Mori *et al.*, 2005).

Anthocyanins derivatives in wines

In general, by HPLC and UV-Vis detection it is possible to analyze some anthocyanin derivatives in wine, such as vitisins, several pyranoanthocyanins and flavanol-ethyl-anthocyanin derivatives. On the other hand, due to the absence of standards commercially available, identification and quantification of these compounds is difficult.

In the HPLC analysis of old wines using a C_{18} column, the polymeric pigments are eluted as very broad peaks and monomer anthocyanins appear in the chromatogram as small peaks on the broad baseline. Alternatively, the use of a polystyrene divinylbenzene HPLC column was proposed (Peng *et al.*, 2002). The column was a reverse phase polystyrene divinylbenzene (100 Å, 250×4.6 mm, 5 μm) operating at 30 °C; elution was performed by a binary mobile phase mixture composed of A) aqueous H_3PO_4 1.5% (w/w) and B) solvent A/acetonitrile 20:80 (v/v) with the gradient program reported in Table 2.5. Usually detection is performed at wavelength 520 nm and compounds are quantified as malvidin-3-glucoside. Analysis can be performed by direct injection of the sample or, to increase the sensibility, after concentration on a C_{18} cartridge as described in paragraph 2.3.4.1.

Table 2.5 HPLC gradient program used for analysis of polymeric pigments in wine using a polystyrene divinylbenzene (100 Å, 250 × 4.6 mm, 5 μm) column (flow rate 1.0 mL/min) (Peng *et al.*, 2002).

Time (min)	solvent A aqueous H_3PO_4 1.5% (w/w) %	solvent B solvent A/acetonitrile 20:80 (v/v) %
0	92	8
55	73	27
59	73	27
64	30	70
69	30	70
76	92	8

2.4.2. Analysis of HCTA and Flavonols

HCTA and flavonols in the skins extract

HCTA and flavonols can be simultaneously detected in a single run using a C_{18} (250 × 4 mm, 5 μm) column operating at 40 °C performing elution of analytes with a binary solvent mixture composed of A) H_3PO_4 10^{-3} M and B) methanol, and the elution gradient program reported in Table 2.6. Figures 2.15 and 2.16 show the chromatograms of HCTA and flavonols, respectively, recorded in the analysis of a *Cabernet Sauvignon* grape skins extract. Detection wavelengths are at 320 nm for HCTA, and 360 nm for flavonols (sample volume injected 25 μL). Identification of compounds is performed on the elution order from the column and by recording UV-Vis spectra on the basis of maximum wavelengths reported in Table 3.4.

Table 2.6 HPLC gradient program used for analysis of HCTA and flavonols in skins extract by a C_{18} (250 × 4 mm, 5 μm) column (chromatograms in Figures 2.17 and 2.18) (flow rate 0.5 mL/min).

Time (min)	$H_3PO_4 10^{-3}$ M%	methanol %
0	95	5
5	90	10
20	70	30
30	40	60
40	0	100
45	0	100
53	95	5
58	95	5

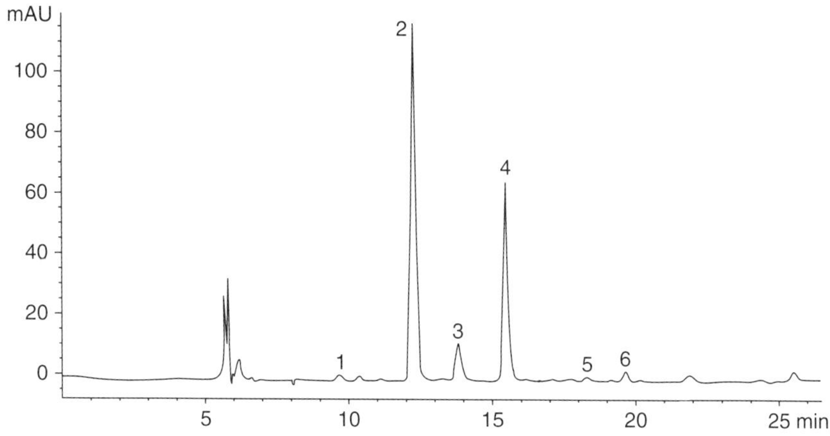

Figure 2.15 HPLC profile of HCTA of a *Cabernet Sauvignon* grape skins extract recorded at wavelength 320 nm. (1) *cis*-CTA, (2) *trans*-CTA, (3) *cis*-p-CuTA, (4) *trans*-p-CuTA, (5) *cis*-FTA, (6) *trans*-FTA

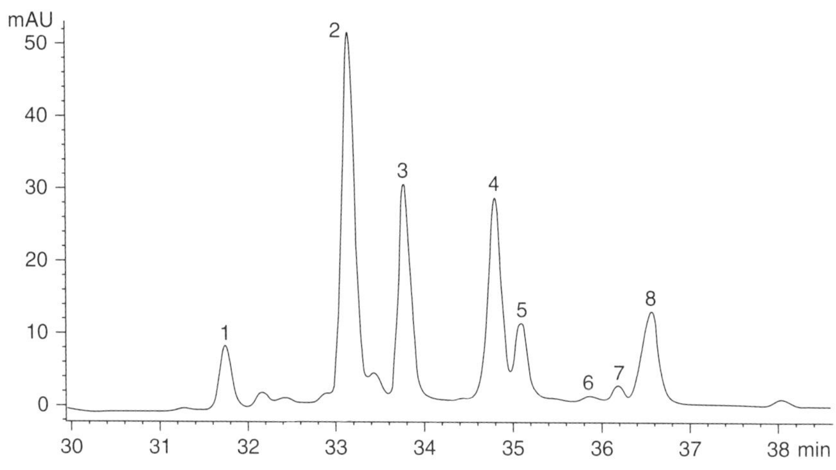

Figure 2.16 HPLC profile of flavonols of a *Cabernet Sauvignon* grape skins extract recorded at wavelength 360 nm. (1) myricetin glucuronide, (2) myricetin glucoside, (3) quercetin glucuronide, (4) quercetin glucoside, (5) quercetin rutinoside, (6) kaempferol glucuronide, (7) kaempferol glucoside, (8) unidentified flavonol

Because standards of HCTA are not commercially available, for quantitative analysis chlorogenic acid or free hydroxycinnamic acids are used as external standards. Standards of flavonols commercially available are quercetin and myricetin glucoside and other flavonols glycoside (e.g. rutin). Alternatively, the single HCTA and flavonols can be isolated from a grape skins extract by performing semipreparative HPLC.

Analysis of flavonols in skins extract after hydrolysis of the glycoside forms

Fifteen millilitres of skin extract are added of 30–40 mL of 50% (v/v) methanol aqueous solution containing HCl (0.5 M) and morin as the internal standard; the hydrolysis reaction is carried on under reflux for 1 h at 85 °C (Vuorinen *et al.*, 2000). After cooling, the sample is filtered through a 0.45 μm membrane and analyzed by HPLC.

Concentration of the sample for analysis can be performed on a C_{18} cartridge (1 g) activated by passage of 2 mL methanol, followed by 3 mL of H_2SO_4 5×10^{-3} M, performing the recovery of analytes from the cartridge with methanol. Analysis is performed by a C_{18} (125 × 3 mm, 5 μm) column using a binary solvent mixture composed of A) acetonitrile and B) aqueous 1% formic acid and the gradient program reported in Table 2.7. Quantification is performed on the ratio of the UV responses of the internal standard and of the analytes recorded at wavelength 360 nm. Limits of quantification and of detection are 5 mg/L and 1 mg/L, respectively (Vuorinen *et al.*, 2000).

Table 2.7 HPLC gradient program used with a C_{18} (125 × 3 mm, 5 μm) column for analysis of free flavonols in the hydrolyzed skin extract.

Time (min)	acetonitrile %	aqueous 1% formic acid %	flow rate (mL min^{-1})
0	5	95	0.4
10	40	60	0.4
20	70	30	0.5
22	90	10	0.5
25	5	95	0.4

HCTA and flavonols in the juice

Pulps are separated from skins and seeds and are transferred to a flask containing some milligrams of sodium metabisulphite ($Na_2S_2O_5$). After homogenization the juice is separated from the solid parts by centrifugation and the supernatant is diluted with H_2SO_4 0.5 M (4 mL of juice added of 1 mL of H_2SO_4 solution). The resulting sample can be frozen; prior to analysis it has to be heated at 30 °C to dissolve eventual precipitates. Analysis is performed by a C_{18} (250 × 4 mm, 5 μm) column at 30 °C using a binary solvent mixture composed of A)

H_3PO_4 10^{-3} M and B) methanol and the elution gradient program reported in Table 2.8. Detection is performed at wavelength 320 nm (sample volume injected 25 μL).

Table 2.8 HPLC gradient program used for analysis of HCTA in grape juice by a C_{18} (250 × 4 mm, 5 μm) column (flow rate 1.0 mL/min).

Time (min)	H_3PO_4 10^{-3} M %	methanol %
0	95	5
5	90	10
20	70	30
30	40	60
32	0	100
34	0	100
39	95	5
44	95	5

Alternatively, isocratic HPLC analysis can be performed using the same column described above and an eluent composed of aqueous 0.5% formic acid/acetonitrile 90:10 (v:v). To determine HCA, concentration of the sample on C_{18} cartridges can be performed as described in paragraph 2.3.1 (Lee and Jawroski, 1987; Di Stefano and Cravero, 1992).

HCTA and flavonols in wines

Analysis is performed on the acidified sample prepared by addition of 0.5 mL H_3PO_4 1 M to 4.5 mL of wine and filtration on membrane 0.45 μm. The HPLC conditions used are the same reported for skins extract.

Figures 2.17 and 2.18 show the chromatograms relative to analysis of HCTA and flavonols in *Raboso* and *Brunello di Montalcino* wines, respectively.

2.4.3. Analysis of Catechins and Procyanidins

Catechins and procyanidins dimer in skins extract

Due to the relatively low levels of these compounds in the grape skins, direct analysis of the extract usually does not provide sufficient sensibility. As a consequence, prior to analysis it is necessary to concentrate the sample by passage through a C_{18} cartridge. Five millilitres of

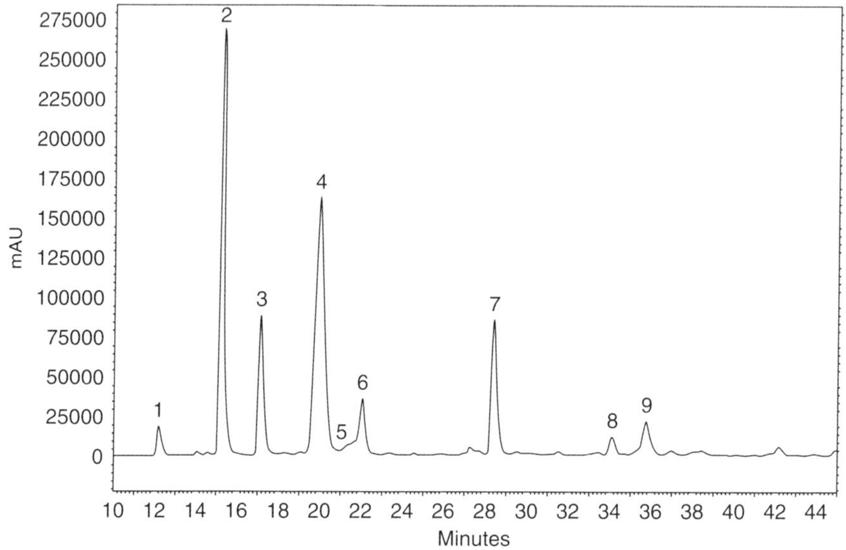

Figure 2.17 HPLC profile of HCTA of a in a *Raboso* wine recorded at wavelength 320 nm. (1) *cis*-CTA, (2) *trans*-CTA, (3) *cis*-p-CuTA, (4) *trans*-p-CuTA, (5) *cis*-FTA, (6) *trans*-FTA, (7) *trans*- caffeic acid, (8) *trans-p*-coumaric acid, (9) *trans*-ferulic acid

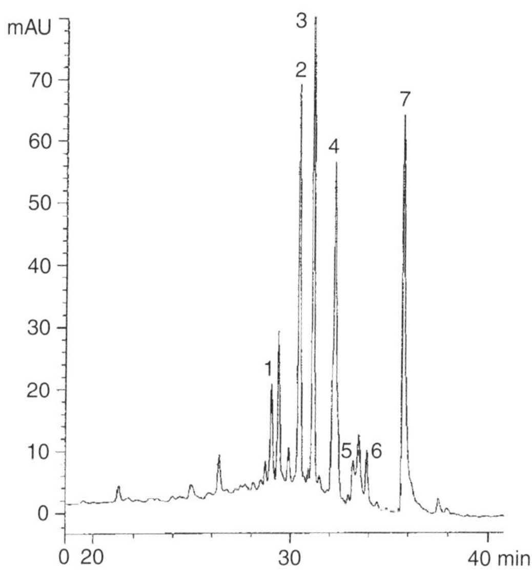

Figure 2.18 HPLC profile of flavonols of a *Brunello di Montalcino* wine recorded at wavelength 360 nm. (1) myricetin glucuronide, (2) myricetin glucoside, (3) quercetin glucuronide, (4) quercetin glucoside + quercetin rutinoside, (5) kaempferol glucuronide, (6) kaempferol glucoside, (7) quercetin aglycone

extract are diluted by 15 mL of H_2SO_4 5×10^{-3} M and the resulting solution is passed through the C_{18} (1 g) cartridge activated with 5 mL of methanol and 6 mL of H_2SO_4 5×10^{-3} M. After the sample passage, the stationary phase is washed with 3 mL of H_2SO_4 5×10^{-3} M and catechins are recovered with 15 mL of diethyl ether, then procyanidins dimer and oligomer are recovered by passage of 20 mL of ethyl acetate. After evaporation of the solvents to dryness, the residues are redissolved in 2 mL of methanol/H_3PO_4 10^{-3} M 40:60 (v/v) solution and used for analysis.

Analysis is performed by a C_{18} (250 × 4 mm, 5 μm) column operating at 40 °C using a binary solvent mixture composed of A) H_3PO_4 10^{-3} M and B) acetonitrile and the elution gradient program reported in Table 2.9. Detection is performed at wavelength 280 nm. For the quantitative analysis, standards of (+)-catechin and C4-C8 B-type procyanidins (see Figure 2.4) are commercially available. Figures 2.19 and 2.20 show the chromatograms relative to analysis of catechins monomer (fraction diethyl ether) and procyanidins dimer (fraction ethyl acetate), respectively.

Table 2.9 HPLC gradient program used for analysis of catechins and procyanidins dimer in the skins extract by C_{18} (250 × 4 mm, 5 μm) column (flow rate 0.5 mL/min).

Time (min)	H_3PO_4 10^{-3} M%	acetonitrile %
0	91	9
10	86	14
20	82	18
30	60	40
45	40	60
55	91	9
60	91	9

Catechins and procyanidins dimer in seeds extract

HPLC analysis can be performed by direct injection of the tartrate buffer extract. The chromatographic peaks of catechin and epicatechin are well resolved, but an overlapping of procyanidins and epicatechin gallate with other compounds, may occur. To improve the separation among them, a fractionation of the sample on a C_{18} cartridge can be performed: 5 mL of extract are added of 15 mL H_2SO_4 5×10^{-3} M and passed

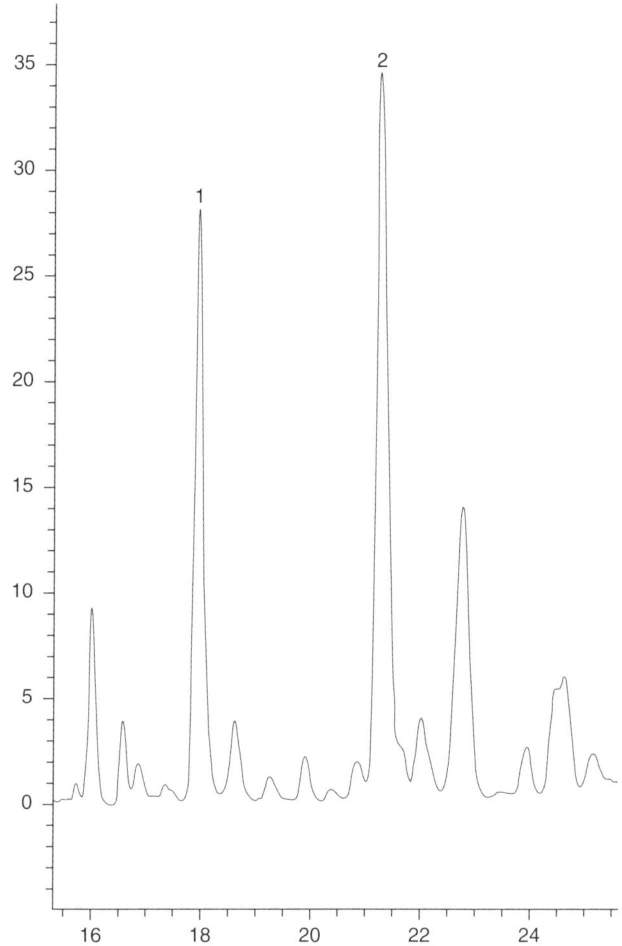

Figure 2.19 Chromatogram relative to analysis of monomer catechins in the skins extract (fraction diethyl ether from C_{18} cartridge) (sample volume injected 10 μL). 1. (+)-catechin, 2. (−)-epicatechin

through a 1-g C_{18} cartridge previously activated with 2 mL of methanol and 3 mL H_2SO_4 5×10^{-3} M. The stationary phase is washed with 3 mL of H_2SO_4 5×10^{-3} M, then catechin, epicatechin and epicatechin gallate are eluted from the cartridge with 10 mL of ethyl ether, and procyanidins dimer with 20 mL ethyl acetate, collecting the eluates in two 100-mL distillation flasks. After the solvent is removed, the diethyl ether residue is redissolved in 2.5 mL of methanol/H_3PO_4 10^{-3} M 40:60 (v/v) and the ethyl acetate residue in 2 mL H_3PO_4 10^{-3} M 40:60 (v/v), then the two

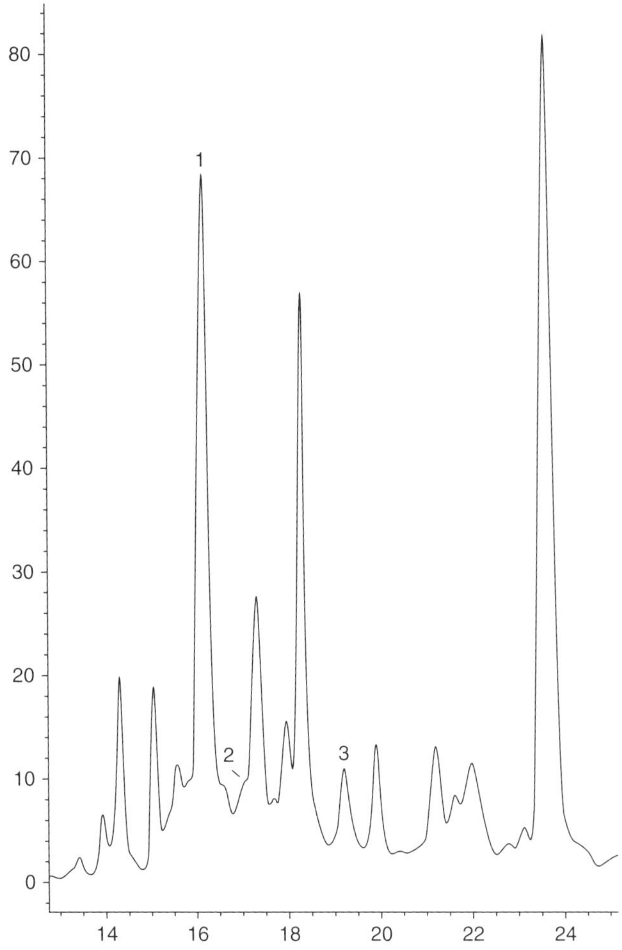

Figure 2.20 Chromatogram relative to analysis of dimer procyanidins in the skins extract (fraction ethyl acetate from C_{18} cartridge) (sample volume injected 10 μL). 1. procyanidin B1, 2. procyanidin B3, 3. procyanidin B2

solutions are filtered through a 0.45 μm membrane and HPLC analysis is performed using the conditions described in Table 2.9.

Figures 2.21, 2.22 and 2.23 show the chromatograms relative to analysis of catechins and procyanidins prior to fractionation in the cartridge, of catechins monomer (fraction diethyl ether) and of procyanidins dimer (fraction ethyl acetate), respectively. As evidenced by Figures 2.20 and 2.23, the main procyanidin in seeds is procyanidin B2, however in skins procyanidin B1 prevails.

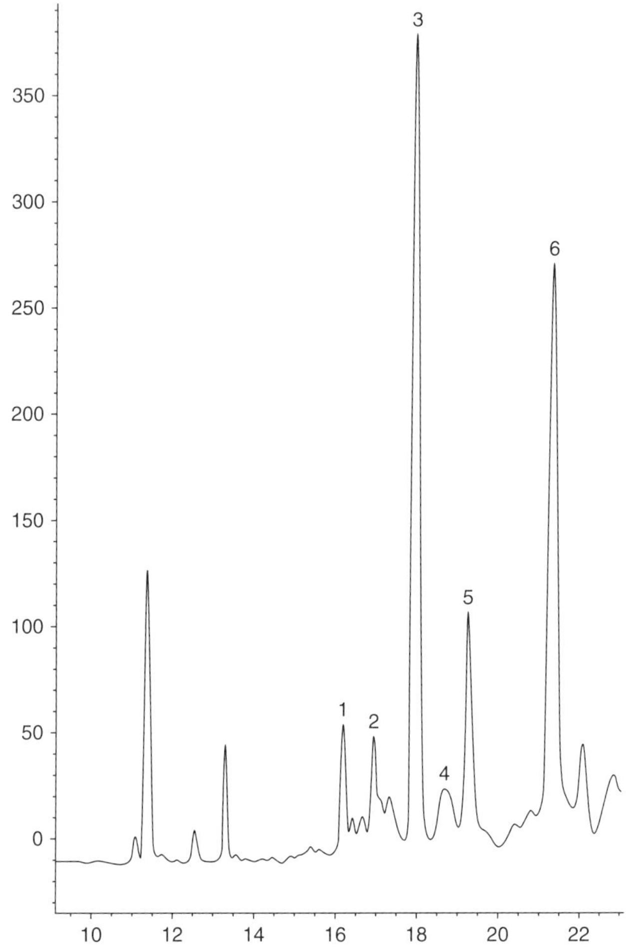

Figure 2.21 Chromatogram relative to analysis of catechins and procyanidins in the seeds extract prior fractionation on the C_{18} cartridge (sample volume injected 10 μL). 1. procyanidin B1, 2. procyanidin B3, 3. (+)-catechin, 4. procyanidin B4, 5. procyanidin B2, 6. (−)-epicatechin

Determination of the mean polymerization degree (mDP) of proanthocyanidins in skins and seeds extracts and in the wine

The phloroglucinol adducts prepared as described in paragraph 2.3.2, are analyzed by HPLC using a C_{18} (250 × 4.6 mm, 5 μm) column. The method utilizes a binary solvent mixture composed of A) 1% aqueous acetic acid and B) methanol. Eluting peaks are detected at wavelength 280 nm. The gradient program of elution is reported in Table 2.10 (Kennedy and Jones, 2001).

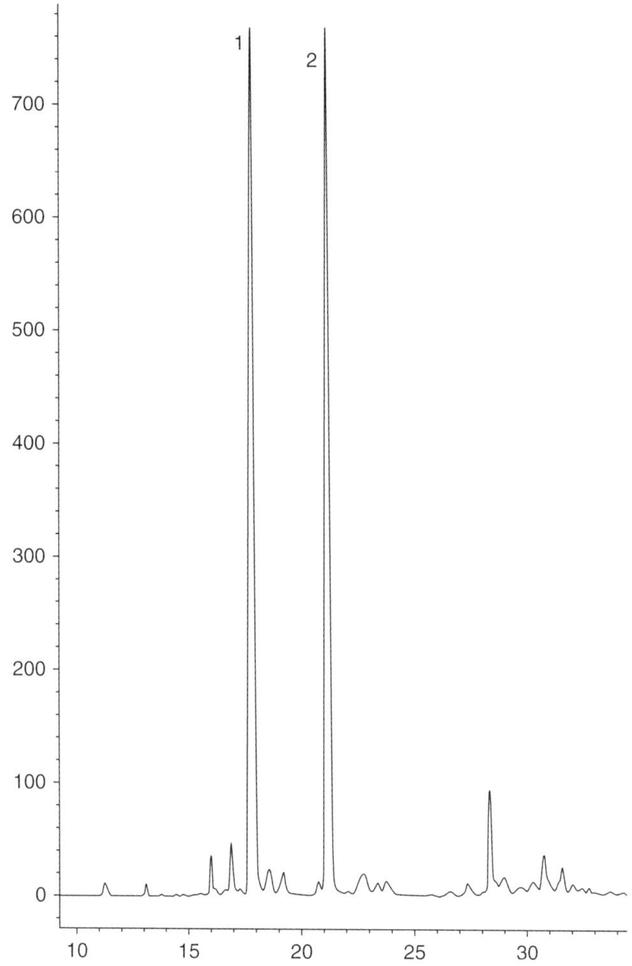

Figure 2.22 Chromatogram relative to analysis of monomer catechins in the seeds extract (fraction diethyl ether from C_{18} cartridge) (sample volume injected 10 μL). 1. (+)-catechin, 2. (−)-epicatechin

Figure 2.24 shows the chromatograms of analysis of proanthocyanidins cleavage products from *Chardonnay* grape seeds (A) and skins (B) extracts.

To perform analysis of proanthocyanidins after thiolysis with benzyl mercaptan (see paragraph 2.3.2) a C_{18} column similar to that described above operating at temperature of 45 °C, can be used. Before injection, the methanol content of the solution is reduced to 20% (v/v) by adding water to minimize peak tailing. As mobile phase, a binary solvent

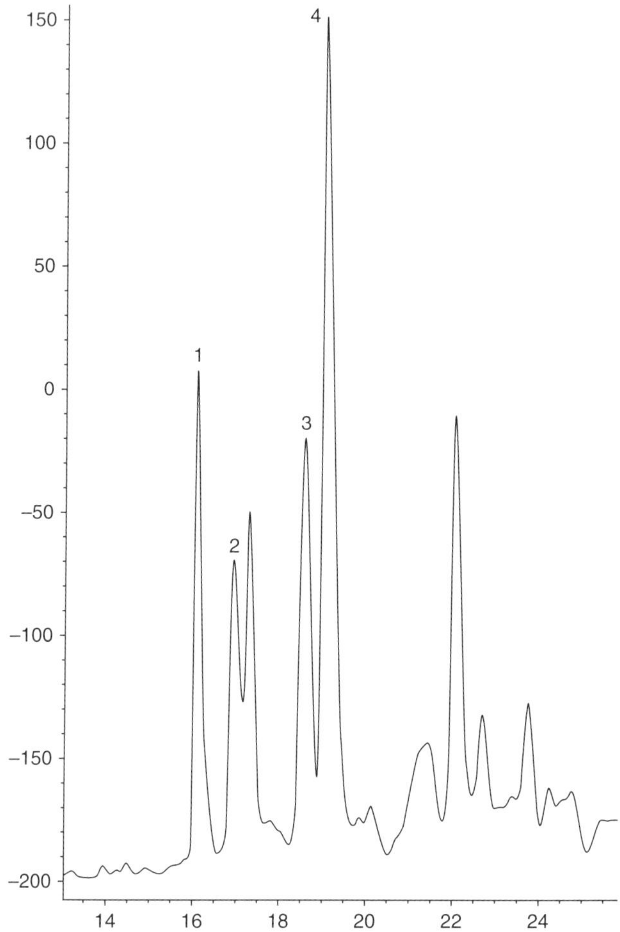

Figure 2.23 Chromatogram relative to analysis of dimer procyanidins in the seeds extract (fraction ethyl acetate from C_{18} cartridge) (sample volume injected 10 μL). 1. procyanidin B1, 2. procyanidin B3, 3. procyanidin B4, 4. procyanidin B2

mixture composed of A) H_3PO_4 200 mM at pH 1.5 and B) methanol is used. The elution gradient program is reported in Table 2.11 (Kennedy *et al.*, 2002).

2.4.4. Free and Glycoside Hydroxystilbenes in Grape

Twenty millilitres of skins extract, must or wine sample are extracted 3 × 10 mL with ethyl acetate. Extracts are combined and organic solvent is evaporated to dryness under vacuum, then the residue is redissolved

Table 2.10 HPLC gradient program used for analysis of phloroglucinol adducts in the determination of the mean polymerization degree of proanthocyanidins performing analysis by a C_{18} (250 × 4.6 mm, 5 μm) column (flow rate 1.0 mL/min) (Kennedy and Jones, 2001).

Time (min)	1% aqueous acetic acid %	methanol %
0	95	5
10	95	5
30	80	20
55	60	40
56	10	90
66	10	90
67	95	5
72	95	5

with 2 mL of 50% aqueous methanol and filtered through a 0.45 μm membrane.

For determination of free and glucoside *trans*-resveratrol in grape, 20 berries are separated from seeds and extracted for 1 h by 30 mL of methanol. After homogenization and centrifugation, the supernatant is separated from the solid parts and methanol is removed under vacuum. Two hundred microlitres of the aqueous residue is added to 100 μL of a 100 mg/L 4′, 5, 7-trihydroxyflavanone methanolic solution as internal standard and of 0.5 g NaCl. The volume is adjusted to 5 mL using distilled water, the solution is transferred in a separation funnel and extracted 3 × 3 mL with ethyl acetate. Extracts are combined and the resulting solution is evaporated to dryness under vacuum. The residue is redissolved with 2 mL of methanol/0.5% aqueous formic acid 3:7 (v/v) solution and filtered through a 0.22 μm membrane before analysis (Bavaresco *et al.*, 1997; 2002).

HPLC analysis of free and glucoside *cis*- and *trans*-resveratrol and of free and glucoside piceatannol is performed using a C_{18} (250 × 4.6 mm, 5 μm) column at 25 °C and elution with a binary solvent mixture composed of A) phosphate buffer 0.02 M pH 3.0 and B) acetonitrile using the gradient program reported in Table 2.12 (Grippi *et al.*, 2007). Detection of free and glucoside *trans*-resveratrol and of free and glucoside piceatannol are performed at wavelength 325 nm, free and glucoside *cis*-resveratrol are determined at wavelength 285 nm (sample volume injected 20 μL).

Figures 2.25 and 2.26 show the chromatograms relative to analysis of hydroxystilbenes in a grape skins extract with the chromatographic conditions reported in Table 2.12 recorded at wavelengths 325 and 285,

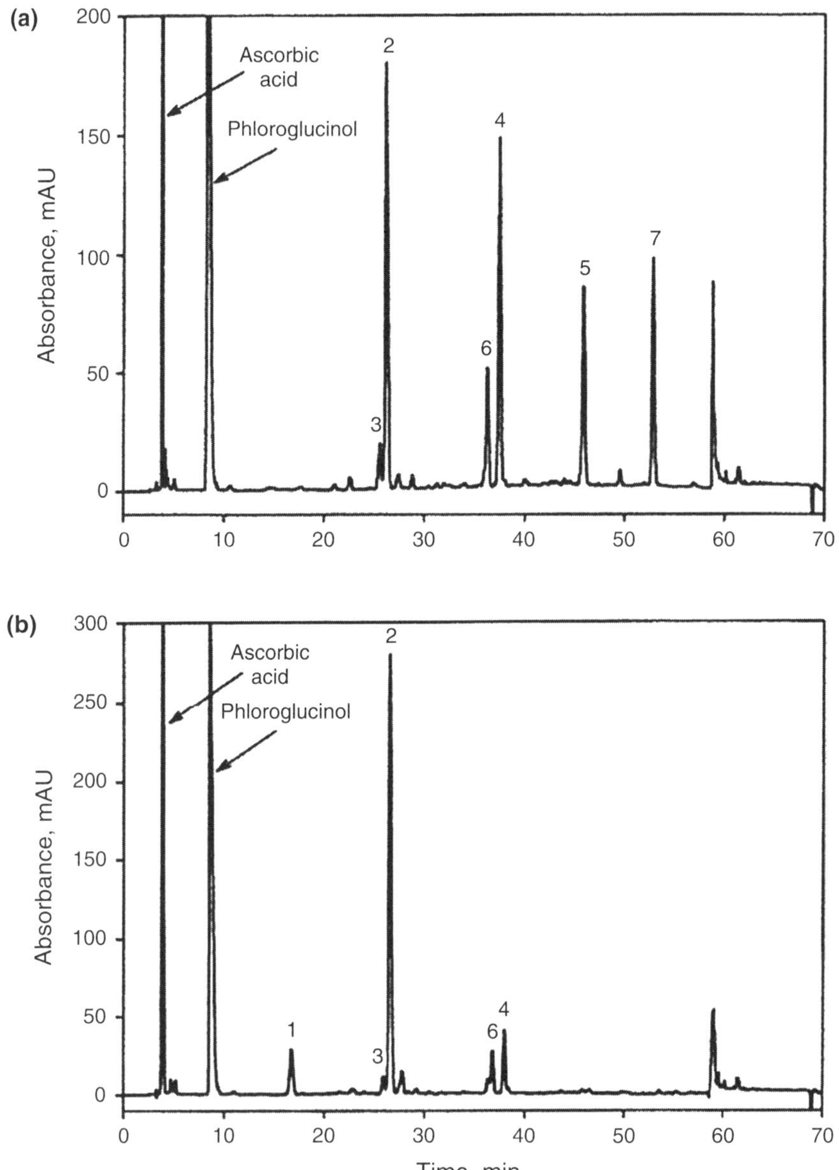

Figure 2.24 HPLC chromatogram of proanthocyanidins cleavage products from *Chardonnay* grape seeds (A) and skins (B) extracts following acid-catalysis in the presence of phloroglucinol. (**1**) epigallocatechin-(4β → 2)-phloroglucinol, (**2**) epicatechin-(4β → 2)-phloroglucinol, (**3**) catechin-(4α → 2)-phloroglucinol, (**4**) epicatechin-3-O-gallate-(4β → 2)-phloroglucinol, (**5**) (−)-epicatechin, (**6**) (+)-catechin, (**7**) (−)-epicatechin-3-O-gallate. (Reprinted from *J. Agric. Food Chem.* (2001) **49**, 1740–1746, Kennedy and Jones. Copyright 2001 with permission from the American Chemical Society)

Table 2.11 HPLC gradient program used for analysis of thiolyzed proanthocyanidins in the determination of the mean polymerization degree of proanthocyanidins performing analysis by a C_{18} (250 × 4.6 mm, 5 μm) column (flow rate 1.0 mL/min) (Kennedy *et al.*, 2002).

Time (min)	H_3PO_4 200 mM pH 1.5 %	methanol %
0	85	15
5	85	15
60	45	55
61	0	100
71	0	100
72	85	15
77	85	15

Table 2.12 HPLC gradient program used for analysis of free and glucoside *cis*- and *trans*-resveratrol and of free and glucoside piceatannol by C_{18} (250 × 4.6 mm, 5 μm) column (flow rate 1.0 mL/min) (Grippi *et al.*, 2007).

Time (min)	phosphate buffer pH 3.0 %	acetonitrile %
0	90	10
5	90	10
8	80	20
10	80	20
13	60	40
25	60	40
27	40	60
30	90	10
35	90	10

respectively. Table 2.13 reports the hydroxystilbenes contents determined in a tartrate buffer skins extract.

In the chromatogram recorded at 320 nm in the HCTA analysis (paragraph 2.4.2) it is also possible to identify the *trans*-resveratrol glucoside peak. The *cis* isomer is detected at 280 nm, but separation from quercetin-3-glucoside and rutin needs optimization of the chromatographic conditions (column and solvent).

Standard of *cis*-resveratrol is not commercially available, but it can be produced by UV-irradiation of the *trans* isomer. Two millilitres of a *trans*-resveratrol (20 mg/L) methanolic solution is transferred in a 2-mL vial and irradiated for 1 h by a UV-lamp operating at wavelength 254 nm at a distance of 2 cm from the sample. Figure 2.27 shows the chromatograms recorded at 307 nm of the sample before (a) and after (b) irradiation. The percentage of transformation of *trans* isomer into

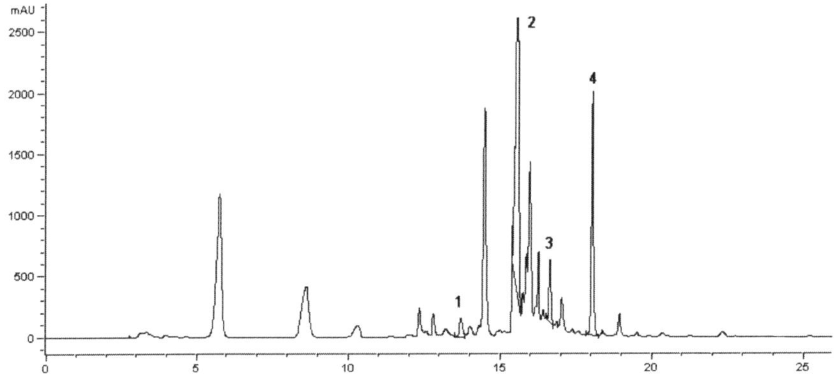

Figure 2.25 Chromatogram relative to the hydroxystilbenes analysis in a grape skins extract recorded at wavelength 325 nm. 1. piceatannol glucoside; 2. *trans*-resveratrol glucoside; 3. piceatannol; 4. *trans*-resveratrol (F. Grippi, personal communication)

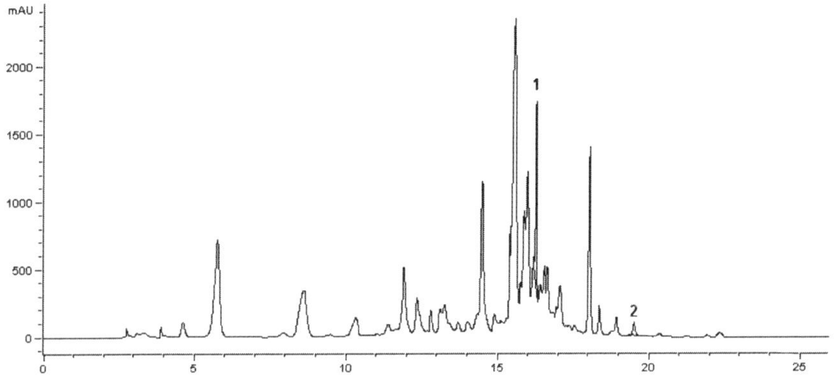

Figure 2.26 Chromatogram relative to the hydroxystilbenes analysis in a grape skins extract recorded at wavelength 285 nm. 1. *cis*-resveratrol glucoside; 2. *cis*-resveratrol (F. Grippi, personal communication)

Table 2.13 Hydroxystilbene contents determined in a tartrate buffer grape skins extract (Grippi, personal communication).

hydroxystilbene	$\mu g\ mL^{-1}$ tartrate buffer extract
piceatannol glucoside	0.35
piceatannol	0.38
trans-resveratrol glucoside	4.51
trans-resveratrol	1.47
cis-resveratrolo glucoside	3.85
cis-resveratrol	0.36

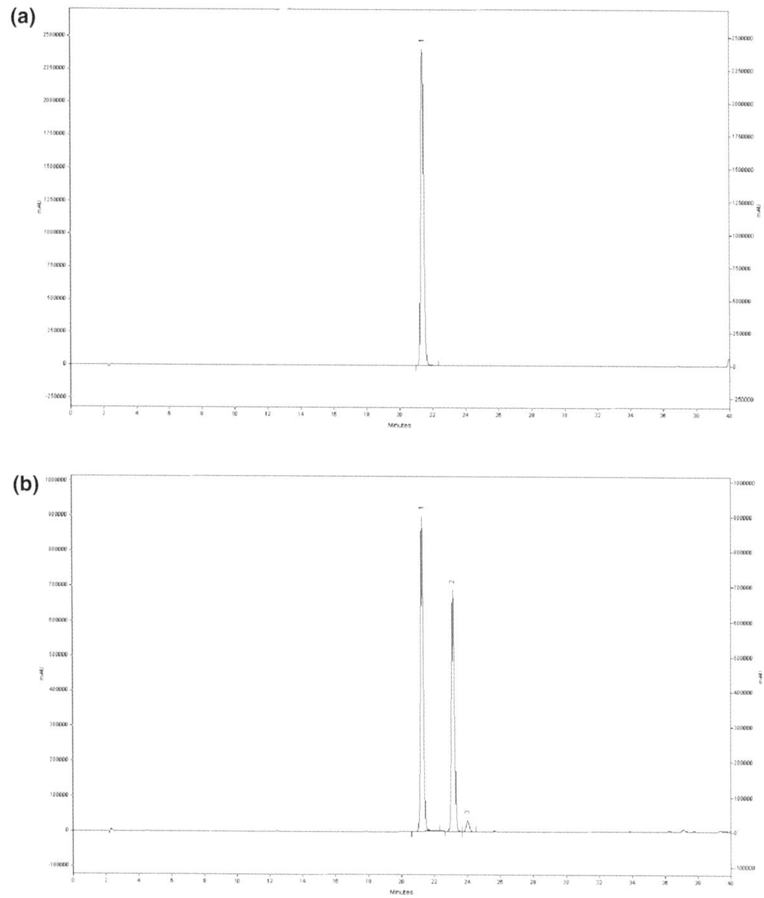

Figure 2.27 Chromatograms recorded at 307 nm before (a) and after (b) UV irradiation at 254 nm of a 16 mg/L *trans*-resveratrol methanolic standard solution. 1. *trans*-resveratrol (max absorbance at 307 nm); 2. *cis*-resveratrol (max absorbance at 285 nm). Percentage of *trans/cis* transformation: 79% (Flamini et coll., unpublished data)

the *cis* one resulted 79% (Romero-Pérez *et al.*, 1996; Flamini et coll., unpublished data).

The total *trans*-resveratrol content can be determined after enzymatic hydrolysis of the glucoside form (La Torre *et al.*, 2003; Wang *et al.* 2002). One millilitre of extract is evaporated to dryness under vacuum, then redissolved in 5 mL of a phosphate buffer solution at pH 6. After addition of 15 mg β-glucosidase 7.55 U/mg enzyme, the sample is incubated overnight at 40 °C. One millilitre of the solution is extracted 3 × 3 mL with ethyl acetate, extracts are combined and the resulting solution is filtered and used for HPLC analysis (Flamini et coll., unpublished data). Performing determination of free *trans*-resveratrol the glucoside form can be calculated as the difference between total and free *trans*-resveratrol.

References

Adams, D.O. (2006) Phenolics and ripening in grape berries, *Am. J. Enol. Vitic.*, **57(3)**, 249–256.

Bavaresco, L. and Fregoni, C. (2001) Physiological role and molecular aspects of grapevine stilbenic compounds, *in Molecular biology & biotechnology of the grapevine*, A.K. Roubelakis-Angelakis (Ed), Kluwer Academic Publishers.

Bavaresco, L., Fregoni, M., Trevisan, M., Mattivi, F., Vrhovsek, U. and Falchetti, R. (2002) The occurence of the stilbene piceatannol in grapes, *Vitis*, **41(3)**, 133–136.

Bavaresco, L., Petegolli, D., Cantù, E., Fregoni, M., Chiusa, G. and Trevisan, M. (1997) Elicitation and accumulation of stilbene phytoalexins in grapevine berries infected by Botrytis cinerea, *Vitis*, **36(2)**, 77–83.

Bavaresco, L., Vezzulli, S., Battilani, P., Giorni, P., Pietri, A. and Bertuzzi, T. (2003) Effect of ochratoxin A-producing Aspergilli on stilbenic phytoalexin synthesis in grapes, *J. Agric. Food Chem.*, **51**, 6151–6157.

Bertelli, A.A., Giovannini, L., Giannessi, D., Migliori, M., Bernini, W., Fregoni, M. and Bertelli, A. (1995) Antiplatelet activity of synthetic and natural resveratrol in red wine, *Int. J. Tissue React.*, **17(1)**, 1–3.

Borsa, D., Carniel, D., Asproudi, A., Monticelli, L., Crespan, M. and Costacurta, A. (2005) Caratterizzazione di uve Malvasia attraverso lo studio dei metaboliti secondari, *Rivista di Viticoltura e di Enologia*, **58 (2–3–4)**, 167–182.

Boss, P.K., Davies, C. and Robinson, S.P. (1996) Expression of anthocyanin biosynthesis pathway genes in red and white grapes, *Plant Mol. Biol.*, **32(3)**, 565–569.

Bourzeix, M., Weyland, D. and Heredia, N. (1986) Étude des catéchines et des procyanidols de la grappe de raisin, du vin et d'autres dérivés de la vigne, *Bull. O.I.V.*, **59(669–670)**, 1171–1254.

Castia, T., Franco, M.A., Mattivi, F., Muggiolu, G., Sferlazzo, G. and Versini, G. (1992) Characterization of grapes cultivated in Sardinia: chemometric methods applied to the anthocyanic fraction, *Sciences des Aliments*, **12**, 239–245.

Cheynier, V., Dueñas-Paton, M., Salas, E., Maury, C., Souquet, J.M., Sarni-Manchado, P. and Fulcrand, H. (2006) Structure and properties of wine pigments and tannins, *Am. J. Enol. Vitic.*, **57(3)**, 298–305.

Cheynier, V. and Rigaud, J. (1986) HPLC separation and characterization of flavonols in the skins of Vitis vinifera var. Cinsault, *Am. J. Enol. Vitic.*, **37(4)**, 248–252.

De Freitas, V.A.P., Glories, Y. and Monique, A. (2000) Developmental changes of procyanidins in grape of red Vitis vinifera varieties and their composition in respective wines. *Am. J. Enol. Vitic.*, **51(4)**, 397–403.

Derdelinckx, G. and Jerumanis, J. (1984) Separation of malt and hop proanthocyanidins on Fractogel TSK HW-40 (S), *J. Chromatogr. A*, **285(1)**, 231–234.

Di Stefano, R. and Cravero, M.C. (1992) The separation of hydroxycinnamates in wine, *Sciences des aliments*, **12**, 139–144.

Downey, M.O., Harvey, J.S. and Robinson, S.P. (2003) Analysis of tannins in seeds and skins of Shiraz grapes throughout berry development, *Aust J. of Grape and Wine Research*, **9(1)**, 15–27.

Favretto, D. and Flamini, R. (2000) Application of electrospray ionization mass spectrometry to the study of grape anthocyanins, *Am. J. Enol. Vitic.*, **51(1)**, 55–64.

Flamini, R. and Dalla Vedova, A. (2004) Fast determination of the total free resveratrol content in wine by direct-exposure-probe, positive-ion-chemical-ionization and collisional-induced-dissociation mass spectrometry (DEP/PICI-MS/MS), *Rapid Commun. Mass Spectrom.* **18(17)**, 1925–1931.

Frankel, E.N., Waterhouse, A.L. and Kinsella, J.E. (1993) Inhibition of human LDL oxidation by resveratrol, *Lancet*, **341(8852)**, 1103–1104.

Fremont, L., Belguendou, L. and Delpal, S. (1999) Antioxidant activity of resveratrol and alcohol-free wine polyphenols related to LDL oxidation and polyunsaturated fatty acids, *Life Sci.*, **64(26)**, 2511–2521.

Fujita, A., Soma, N., Goto-Yamamoto, N., Shindo, H., Kakuta, T., Koizumi, T. and Hashizume, K. (2005) Anthocyanidin reductase gene expression and accumulation of flavan-3-ols in grape berry, *Am. J. Enol. Vitic.*, **56(4)**, 336–342.

Grippi, F., Crosta, L., Giacchino, A., Tolomeo, M., Oliveri, F., Gebbia, N. and Curione, A. (2008) Determination of stilbenes in sicilian pistacho by hygh performance liquid chromathgraphy dios array (HPLC-DAD/FLP) and evaluation of eventually mycotoxin contamination, *Food Chem.* 107(1), 483–488.

Gong, Z., Yamazaki, M., Sugiyama, M., Tanaka, Y. and Saito, K. (1997) Cloning and molecular analysis of structural genes involved in anthocyanin biosynthesis and expressed in a forma-specific manner in Perilla frutescens, *Plant Mol. Biol.*, **35(6)**, 915–927.

Hung, L.M., Chen, J.K., Huang, S.S., Lee, R.S. and Su, M.J. (2000) Cardioprotective effect of resveratrol, a natural antioxidant derived from grapes, *Cardiovasc. Res.*, **47**(3), 549–555.

Jang, M., Cai, L., Udeani, G.O., Slowing, K.V., Thomas, C.F., Beecher, C.W., Fong, H.H., Farnsworth, N.R., Kinghorn, A.D., Mehta, R.G., Moon, R.C. and Pezzuto, J.M. (1997) Cancer chemopreventive activity of resveratrol, a natural product derived from grapes, *Science*, **275**(5297), 218–220.

Jeong, S.T., Goto-Yamamoto, N., Hashizume, K., Kobayashi, S. and Esaka, M. (2006) Expression of *VvmybA1* gene and anthocyanin accumulation in various grape organs, *Am. J. Enol. Vitic.*, **57(4)**, 507–510.

Jeong, S.T., Goto-Yamamoto, N., Kobayashi, S. and Esaka, M. (2004) Effects of plant hormones and shading on the accumulation of anthocyanins and the expression of anthocyanin bioyntetic genes in grape berry skins, *Plant Sci.*, **167(2)**, 247–252.

Jordão, A.M., Ricardo da Silva, J.M. and Laureano, O. (2001) Evolution of catechins and oligomeric procyanidins during grape maturation of Castelão Francês and Touriga Francesa, *Am. J. Enol. Vitic.*, **52(3)**, 230–234.

Kennedy, J.A. and Jones, G.P. (2001) Analysis of proanthocyanidin cleavage products following acid-catalysis in the presence of excess phloroglucinol, *J. Agric. Food Chem.*, **49**(4), 1740–1746.

Kennedy, J.A., Mathews, M.A. and Waterhouse, A.L. (2002) Effect of maturity and vine water status on grape skin an wine flavonoids, *Am. J. Enol. Vitic.*, **53(4)**, 268–274.

Kennedy, J.A., Troup, G.J., Pilbrow, J.R., Hutton, D.R., Hewitt, D., Hunter, C.H., Ristic, R., Iland, P.G. and Jones, G.P. (2000) Development of seeds polyphenols in berries from Vitis vinifera, L. cv. Shiraz, *Aust J. Grape and Wine Research*, **6**(3), 244–254.

Lamuela-Raventós, R.M. and Waterhouse, A.L. (1994) A direct separation of wine phenolics, *Am. J. Enol. Vitic.*, **45**(1), 1–5.

La Torre, G.L., Laganà, G., Bellocco, E., Vialsi, F., Salvo, F. and Dugo, G. (2003) Improvement on enzymatic hydrolysis of resveratrol glucosides in wine, *Food Chem.*, **85**(2), 259–266.

Lea, A.G.H., Bridle, P., Timberlake, C.F. and Singleton, V.L. (1979) The Procyanidins of White Grapes and Wines, *Am. J. Enol. Vitic.*, **30**, 289–300.

Lee, C.Y. and Jaworski, A.W. (1987) Phenolic compounds in white grapes grown in New York, *Am. J. Enol. Vitic.*, **38**(4), 277–281.

Mateus, N., Maqes, S., Gonçalves, A.C., Machado, J.M. and De Freitas, V. (2001) Proanthocyanidin composition of red Vitis vinifera varieties from the Douro valley during ripening: influence of cultivation altitude, *Am. J. Enol. Vitic.*, **52**(2), 151–121.

Mori, K., Saito, H., Goto-Yamamoto, N., Kitayama, M., Kobayashi, S., Sugaya, S., Gemma, H. and Hashizume, K. (2005a) Effect of abscissic acid treatment and night temperature on anthocyanin composition in Pinot noir grapes, *Vitis*, **44**(4), 161–165.

Mori, K., Sugaya, S. and Gemma, H. (2005b) Decreased anthocyanins biosynthesis in grape berries grown under elevated night temperature condition, *Scientia Horticulturae*, **105**(3), 319–330.

Nagel, C.W., Barankowski, J.D., Wulf, L.W. and Powers, J.R. (1979) The Hydroxycinnamic Acid Tartaric Acid Ester Content of Musts and Grape Varieties Grown in the Pacific Northwest, *Am. J. Enol. Vitic.*, **30**, 198–201.

Niketsic, G.K. and Hrazdina, G. (1972) Quantitative analysis of the anthocyanin content in grape juices and wine, *Lebensm. Wiss. Technol.*, **5**, 163–165.

Ong, B.Y. and Nagel, C.W. (1978) High-pressure liquid chromatographic analysis of hydroxycinnamic acid-tartaric acid esters and their clucose esters in Vitis Vinifera, *J. Chromatogr.*, **157**, 345–355.

Oszimianski, J. and Lee, C.Y. (1990) Isolation and HPLC determination of Phenolic compounds in red grapes, *Am. J. Enol. Vitic.*, **41**(3), 204–206.

Oszimianski, J. and Sapis, J.C. (1989) Fractionation and identification of some low molecular weight grape seed phenolics. *J. Agric. Food Chem.* **37**, 1293–1297.

Pace-Asciak, C.R., Hahn, S., Diamandis, E.P., Soleas, G. and Goldberg, D.M. (1995) The red wine phenolics trans-resveratrol and quercetin block human platelet aggregation and eicosanoid synthesis: Implications for protection against coronary heart disease, *Clin. Chim. Acta*, **235**(2), 207–219.

Pastor del Rio, J.L. and Kennedy, J.A. (2006) Development of proanthocyanidins in Vitis vinifera, L. cv. Pinot noir grapes and extraction into wine, *Am. J. Enol. Vitic.*, **57**(2), 125–132.

Peng, Z., Iland, P.G., Oberholster, A., Sefton, M.A. and Waters, E.J. (2002) Analysis of pigmented polymers in red wine by reverse phase HPLC, *Aust J. Grape and Wine Research*, **8**(1), 70–75.

Revilla, E., Ryan, J.M. and Martin-Ortega, G. (1998) Comparison of several procedures used for the extraction of anthocyanins from red grapes, *J. Agric. Food Chem.*, **46**(11) 4592–4597.

Ribéreau-Gayon, P., Glories, Y., Maujean, A. and Dubourdieu, D. (1998) *Traité d'œnologie. 2. Chimie du vin-Stabilisation et traitements*, Dunod, Paris.

Roggero, J.P. and Archier, P. (1989) Mise au point d'une méthode de dosage des phénols simples des vins. Application à des vins d'origines et d'ages différents, *Conn. Vigne Vin*, **23**(**1**), 25–37.

Romero-Pérez, A., Lamuela-Raventós, R.M., Waterhouse, A.L. and de la Torre-Boronat, M.C. (1996) Levels of cis- and trans-resveratrol and their glucosides in white and rosé Vitis vinifera wines from Spain, *J. Agric. Food Chem.*, **44**(8), 2124–2128.

Salagoity-Auguste, M.H. and Bertrand, A. (1984) Wine phenolics. Analysis of low molecular weigh components by high performance liquid chromatography, *J. Sci. Food Agr.*, 35(**11**), 1241–1247.

Souquet, J.M., Cheynier, V., Brossaud, F. and Moutounet, M. (1996) Polymeric proanthocyanidins from grape skins, *Phytochemistry*, **43**(**2**), 509–512.

Spayed, S.E., Tanara, J.M., Mee, D.L. and Ferguson, J.C. (2002) Separation of sunlight and temperature effects on the composition of Vitis vinifera cv. Merlot berries, *Am. J. Enol. Vitic.*, **53**(3), 171–182.

Squadrito, M., Corona, O., Ansaldi, G. and Di Stefano, R. (2007) Relazioni fra i percorsi biosintetici degli HCTA, dei flavonoli e degli antociani nella buccia dell'uva, *Rivista di Viticoltura e di Enologia*, **60**(3), 59–70.

Vidal, S., Cartalade, D., Souquet, J.M., Fulcrand, H. and Cheynier, V. (2002) Changes in proanthocyanidins chain lenght in winelike model solution, *J. Agric. Food Chem.*, 50(8), 2261–2266.

Vuorinen, H., Maatta, K. and Torronen, R. (2000) Content of the flavonols myricetin, quercetin, and kaempferol in Finnish berry wines, *J. Agric. Food Chem.*, **48**(7), 2675–2680.

Wang, Y., Catana, F., Yang, Y., Roderick, R. and van Breemen, R.B. (2002) An LC-MS method for analysing total resveratrol in grape juice, cranberry juice and in wine, *J. Agric. Food Chem.*, **50**(3), 431–435.

Wulf, L.W. and Nagel, C.W. (1976) Analysis of phenolic acids and flavonoids by high-pressure liquid chromatography, *J. Chromatogr.*, **116**, 271–279.

Yamane, T., Jeong, S.T., Goto-Yamamoto, N., Koshita, Y. and Kobayashi, S. (2006) Effects of temperature on anthocyanin biosynthesis in grape berry skins. *Am. J. Enol. Vitic.*, **57**(**1**), 54–59.

3

Polyphenols Analysis by Liquid–Mass Spectrometry

Riccardo Flamini and Mirko De Rosso

3.1. Introduction

Phenols and polyphenols contribute to the colour, flavour and fragrance of the wine. In particular, some phenols with low molecular weight (MW) play an important role in the wine fragrance (e.g. eugenol, vanillin); anthocyanins confer the color to red grape and wines; tannins, by linking proteins of saliva, confer astringency and structure to the wine. Mechanisms of rearrangement and polymerization of polyphenolic compounds contribute in the forming of new molecules in wine. Moreover, polyphenols are the principal compounds related to beneficial effects of wine consumption: grape seeds procyanidins and proanthocyanidins are used as active ingredients in medicinal products for the treatment of circulatory disorders such as capillary fragility, peripheral chronic venous insufficiency and microangiopathy of the retina. It was observed these compounds increase the total antioxidant plasma capacity, decrease the susceptibility of healthy cells to toxic and carcinogenic agents, are associated with reduction of platelet aggregation,

Hyphenated Techniques in Grape and Wine Chemistry Edited by Riccardo Flamini

and have antioxidant activity toward human low density lipoprotein. Quercetin, a flavonol present in grape and wine with different glycoside forms, blocks the aggregation of human platelets by ADP and thrombin and it has gained considerable prominence as an inhibitor of carcinogens and of cancer cell growth in human tumors. Several studies have revealed the beneficial effects of *trans*-resveratrol (a phytoalexin present in grape also as a glucoside called piceid, and as viniferins) on human health, such as anticancer activity, cardioprotection, antioxidant activity, inhibition of platelet aggregation, and anti-inflammatory activity (Flamini, 2003 and references cited herein).

The structure of anthocyanins contains the benzopyrilium ion as base molecule, responsible for the colour of red grape varieties, linked to one or two sugar residues. Characterization of the grape anthocyanin profile is used for chemotaxonomic studies, and allows identification of frauds in countries where production and commercialization of wines from non-*Vitis vinifera* (hybrid) grapes is illegal. These grapes are characterized by peculiar anthocyanin 3,5-O-diglucoside contents which are practically absent in grapes from *Vitis vinifera*. Grape anthocyanins are also used as synthetic colorants in the food and pharmaceutical industries (Hong and Wrolstad, 1990).

Identification and quantification of low molecular weight and volatile phenols is usually performed by Gas Chromatography and Mass Spectrometry (GC-MS). For analysis and structural characterization of more polar compounds such as polyphenols, liquid-phase and Liquid Chromatography Mass Spectrometry (LC-MS) and Multiple Mass Spectrometry (MS/MS and MS^n) techniques are used (Niessen and Tinke, 1995; de Hoffmann, 1996; Abian, 1999; Flamini *et al.*, 2007).

Due to the soft ionization conditions and minor sample purification required, liquid MS methods are the most effective techniques to characterize anthocyanins and polyphenols in grape extracts and wine. Differently from the LC methods coupled with spectrophotometric detection that require hydrolysis or thiolysis for compounds identification, LC-MS is reasonably the most effective tool in the study of structure of compounds correlated to the colour-changing of red wines during ageing formed by reactions of anthocyanins with other compounds. This technique also allows the characterization of grape procyanidins, proanthocyanidins, prodelphinidins and tannins. The MS/MS approach is a very powerful tool, in particular in the study of glycoside compounds, allowing the aglycone and sugar moiety characterization.

3.2. Non-Anthocyanin Polyphenols and Procyanidins in Grape and Wine

Phenols and polyphenols in wine are transferred from the grape in the winemaking and are released from the wood barrels used for the wine ageing. The principal low MW phenols in wine are reported in Figure 3.1.

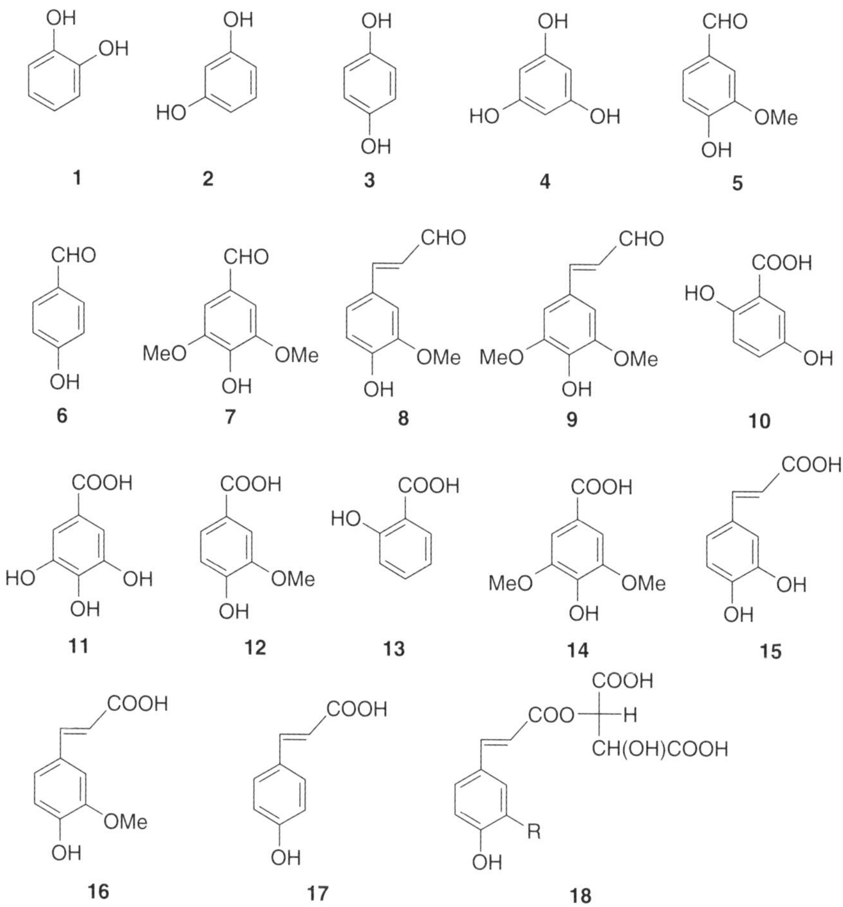

Figure 3.1 Structures of principal low molecular weight phenols in wine: (**1**) pyrocatechol (**2**) resorcinol (**3**) hydroquinone (**4**) phloroglucinol (**5**) vanillin (**6**) *p*-hydroxybenzaldehyde (**7**) syringic aldehyde (**8**) coniferyl aldehyde (**9**) sinapinaldehyde (**10**) gentisic acid (**11**) gallic acid (**12**) vanillic acid (**13**) salicylic acid (**14**) syringic acid (**15**) caffeic acid (**16**) ferulic acid (**17**) *p*-coumaric acid (**18**) hydroxycinnamoyltartaric acids ($R = H$, OH, OCH_3)

The principal non-anthocyanin grape polyphenols are some flavonols present as monoglycosides (see Figure 2.6) and flavan-3-ols (catechins, epicatechins).

Procyanidins and proanthocyanidins (condensed or nonhydrolyzable tannins) are B-type and A-type oligomers and polymers of flavan-3-ols, respectively (see Figures 2.3, 2.4 and 2.5). They are present in grape skin and seeds and are extracted and transferred to the wine in the winemaking. Flavonols and the oxidation-products of flavonols and tannins are the main elements responsible for the colour of white grapes and wines. The sensory perception of astringency and bitterness in wine changes in response to the degree of galloylation (DG) and degree of polymerization (DP) of the flavan-3-ols (Cheynier & Rigaud, 1986; Vidal *et al.*, 2003).

3.2.1. Analysis of Low-Molecular-Weight Phenols and Polyphenols

A method of sample preparation for analysis of grape flavonols is to perform extraction of berries with acidified methanol (0.01% HCl 12 N). The extract is filtered, the solvent removed under vacuum and the residue dissolved in 0.1 M citric acid buffer at pH 3.5. Polyphenolics are fractionated on the basis of their affinity to a C_{18} cartridge (e.g. Sep-Pak 5 g) then on a Sephadex LH-20 (e.g. 3 g) cartridge (crosslinked dextran-based stationary phase used for gel permeation, normal-phase partition and adsorption chromatography). The fractions are eluted by ethyl acetate and methanol. Their composition is reported in the flow-diagram in Figure 3.2 (Lee *et al.*, 2005).

As Figure 3.2 shows, the methanol fraction is evaporated, dissolved in the pH 3.5 buffer and partitioned on a Sephadex LH-20 cartridge in two stages. To perform analysis of isolate II, ethyl acetate is evaporated and the residue is redissolved in the pH 3.5 buffer. The isolates can be analyzed by Liquid-Chromatography-Electrospray-Ionization-Mass Spectrometry (LC/ESI-MS), Collisional-Induced-Dissociation (CID) and MS/MS experiments which allow confirmation of the compounds' identification. By connecting the LC/ESI-MS system to the probe of the mass spectrometer *via* the UV cell outlet, LC-UV chromatograms and spectra can also be recorded. Examples of chromatograms recorded at wavelength 280 and 360 nm of isolate II are reported in Figure 3.3. The

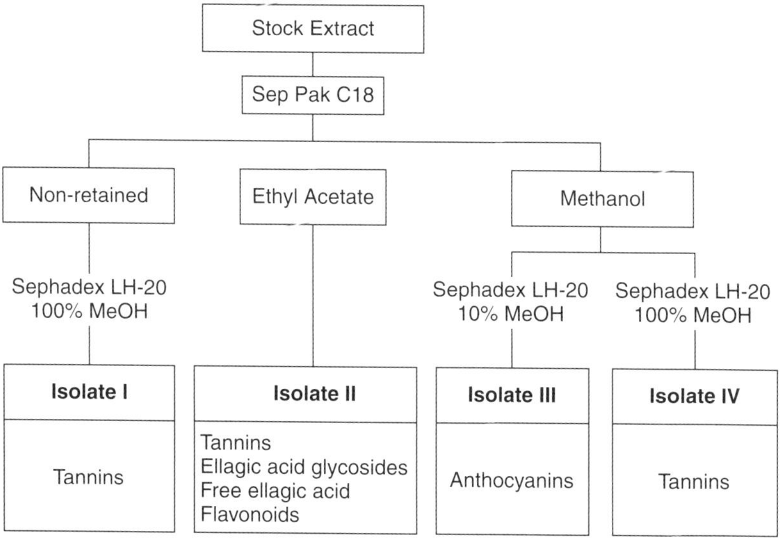

Figure 3.2 Flow diagram of grape polyphenolic fractionation procedure. (Reproduced from *J. Agric. Food Chem.*, 2005, 53, 6003–6010, Lee *et al.*, with permission from the American Chemical Society)

compounds identified in the chromatograms are reported in Table 3.1 with the characteristic fragments originated from MS/MS experiments. As may be seen from the table, the three flavonols myricetin, quercetin and kaempferol in the rhamnoside form, and ellagic acid derivatives, are collected in the isolate II.

Compounds such as vanillin and acid phenols present in wine can be detected by LC/ESI-MS analysis. A list of these compounds is reported in Table 3.2 with some MS conditions used. ESI is also an effective technique for analysis of flavan-3-ols operating in both positive- and negative-ion mode. In the latter mode, using a cone voltage of 60 V, high abundance of the $[M-H]^-$ species of phenols, is observed. Increasing the cone voltage up to 120 V, a reduction of the molecular species intensity, and the most abundant fragments originating from losses of carboxyl $[M-H-45]^-$, hydroxyl $[M-H-17]^-$ or/and formaldehyde $[M-H-30]^-$ groups, are observed (Table 3.3). The positive ionization mode allows a better sensitivity in the flavan-3-ols analysis but, because several acid compounds are present in wine which are not detected in this working mode, the negative ion mode is preferred. Figure 3.4 provides an example of a chromatogram relative to a wine flavan-3-ols negative-ion mode analysis.

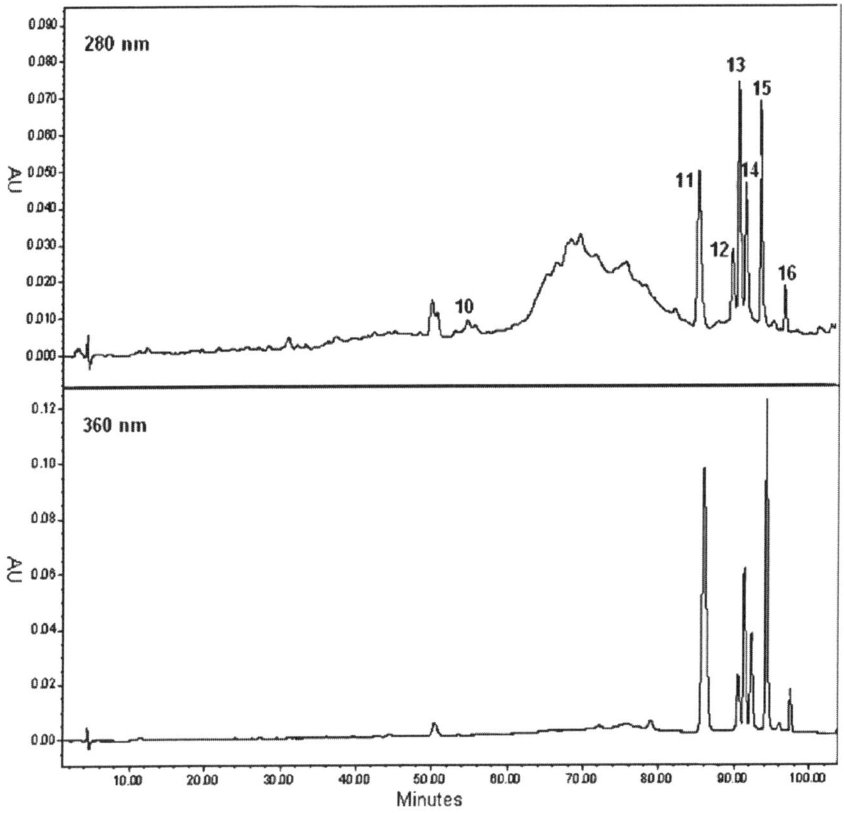

Figure 3.3 LC chromatograms recorded at wavelength 280 and 360 nm of isolate II in Figure 3.2. Analytical conditions: C_{18} 80 Å column (2 × 150 mm, 4 μm); binary solvent composed of A) 0.5% formic acid containing 5 mM ammonium formate and B) 0.5% formic acid in methanol; gradient program: from 5 to 30% of B in 5 min, 30–65% B in 70 min, 65–95% B in 30 min, 95% B isocratic for 20 min (flow rate 0.15 mL/min). (Reproduced from *J. Agric. Food Chem.*, 2005, 53, 6003–6010, Lee *et al.*, with permission from the American Chemical Society)

A method of wine sample preparation is to perform liquid–liquid extraction of 50 mL of wine previously concentrated to 15 mL under vacuum at 30 °C to eliminate ethanol. Two extractions with diethyl ether (3 × 5 mL) and ethyl acetate (3 × 15 mL) are performed. The organic phases are combined, dried with anhydrous Na_2SO_4, the solvent is then removed under vacuum; finally, the residue is dissolved in 2 mL of methanol/H_2O (1:1) and filtrated 0.45 μm (Monagas *et al.*, 2005). In this analysis, several classes of nonanthocyanic phenols are detected: flavan-3-ols and flavonols, hydroxycinnamoyltartaric acids, resveratrol

Table 3.1 Characteristic fragments originated from MS/MS experiments of compounds reported in the chromatograms of Figure 3.3. LC/ESI-MS conditions: positive and negative ion-mode; sheath gas N_2 60 units/min; auxiliary gas N_2 5 units/min; spray voltage 3.3 kV; capillary temperature 250 °C; capillary voltage 1.5 V; tube lens offset 0 V. (Lee *et al.*, 2005).

peak Fig. 3.3	RT (min)	UV (nm)	Compound	MW	ESI	BP (m/z); ID	MS/MS (m/z)	MS^3 (m/z)
10	58–60	**261, 280sh**	ellagitannins	800	(−)	799; $[M-H]^-$	781, 763, 745, 735, 495, 481, 451, 317, 301, 273	763, 745, 735, 719, 479, 461, 301, 275, 247
					(+)	818; $[M+NH_4]^+$	801, 783, 447, 429, 385, 357, 337, 303, 277, 259, 231	429, 411, 385, 357, 303, 277
				814	(−)	813; $[M-H]^-$	781, 763, 753, 735, 301	763, 745, 419, 317, 301, 273, 229
					(+)	832; $[M+NH_4]^+$	797, 779, 461, 447, 443, 397, 335, 317, 303, 277, 259, 241	427, 411, 385, 357, 335, 303, 277
11	86.0	352	myricetin rhamnoside	464	(−)	463; $[M-H]^-$	359, 337, 317	287, 271, 179, 151
					(+)	465; $[M+H]^+$	447, 429, 361, 319	301, 290, 283, 273, 263, 255, 245, 165, 163, 137
12	90.5	360	ellagic acid xyloside	434	(−)	433; $[M-H]^-$	301	257, 229
					(+)	435; $[M+H]^+$	303	285, 275, 257, 247, 229, 165, 153, 137
13	91.3	361	ellagic acid rhamnoside	448	(−)	447; $[M-H]^-$	300, 301	272, 257, 244, 229
					(+)	449; $[M+H]^+$	303	285, 275, 259
14	92.3	366	ellagic acid	302	(−)	301; $[M-H]^-$	301, 284, 257, 229, 185	NA
					(+)	ND		
15	94.2	351	quercetin rhamnoside	448	(−)	447; $[M-H]^-$	301	283, 271, 255, 179, 169, 151, 121, 107
					(+)	449; $[M+H]^+$	431, 413, 303	303, 285, 275, 257, 247, 229, 165, 153, 137
16	97.5	344	kaempferol rhamnoside	432	(−)	431; $[M-H]^-$	327, 299, 285, 256	267, 257, 255, 241, 229, 213, 197, 163
					(+)	433; $[M+H]^+$	415, 397, 375, 287	287, 269, 241, 231, 213, 197, 183, 165, 153

Table 3.2 Negative-ion ESI mass spectra of low MW phenols in wine (Pérez-Margariño *et al.*, 1999).

Compound	MW	Main ions observed (*m/z*)	
		Fragmentor 60 V	Fragmentor 120 V
protocatechuic acid	154	153 (109)	109 (153)
protocatechuic aldehyde	138	137	137 (108)
p-hydroxybenzoic acid	138	137 (93)	93 (137)
p-hydroxybenzoic aldehyde	122	121	121 (92)
vanillic acid	168	167	108 (167, 123, 91)
vanillin	152	151 (136)	136 (151)
ferulic acid	194	193 (134, 149)	134 (193)
syringic aldehyde	182	181 (166)	181 (166, 151)
p-coumaric acid	164	147 (103)	147 (103)
caffeic acid	180	179 (135)	135 (179)
esculetin	178	177	177 (133)
gallic acid	170	169 (125)	125 (169)

Table 3.3 ESI fragment ions of flavan-3-ols generated in negative and positive mode (Pérez-Margariño *et al.*, 1999).

Compound	MW	Main ions observed (*m/z*)		
		API +	API−	
		Fragm. 60 V	Fragm. 60 V	Fragm. 120 V
(+)-catechin	290	291 (139)	289	289 (245)
(−)-epicatechin	290	291 (139, 150)	289	289 (245)
(−)-epigallocatechin	306	307 (139)	305	305
(−)-epicatechin-3-O-gallate	442	443 (123, 273)	441	441 (289, 169)
epigallocatechin-3-O-gallate	458	459 (139, 289)	457	169 (457)

isomers and piceid, dimeric (B1, B3, B4 and B5) and trimeric (C1, T2 and T3) procyanidins, phenolic acids (see Table 3.4).

Alternatively, a binary solvent composed of A) aqueous formic acid 0.5% (v/v) and B) formic acid/acetonitrile/H_2O 5:400:595 (v/v/v) can be used (gradient program: from 0 to 20% B in 15 min; 20% B isocratic 10 min; 20–70% B in 45 min; 70% B isocratic for 5 min; 70–100% B in 10 min; 100% B isocratic for 5 min; flow rate 0.7 mL/min) (Bravo *et al.*, 2006).

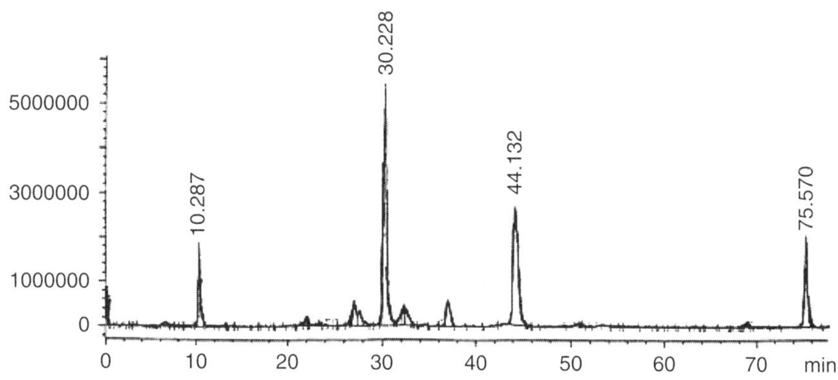

Figure 3.4 Total ion current (TIC) chromatogram of a wine sample LC/ESI-MS analysis in negative-ion mode. Analytical conditions: C_{18} column (250 × 4.6 mm, 3 μm); binary solvent composed of A) 4.5% formic acid in water and B) solvent A/acetonitrile 90:10; gradient program: from 0 to 50% of B in 25 min, 50–80% B in 35 min, 80% B isocratic for 20 min (flow rate 0.7 mL/min). At 10.287 min gallic acid; 30.228 min catechin; 44.132 min epicatechin; 75.57 min epicatechin gallate. (Reprinted from *Journal of Chromatography A*, **847**, Magariño *et al.*, Various applications of liquid chromatography-mass spectrometry to the analysis of phenolic compounds, p. 80, Copyright 1999, with permission from Elsevier)

3.2.2. Analysis of Grape Procyanidins and Proanthocyanidins

Sample extraction and fractionation

Grape seeds are the main source of wine procyanidins (PCs), proanthocyanidins (PAs) and tannins. A method for extraction of PAs from grape seeds is to grind the dry seeds to obtain a homogeneous powder and to perform three consecutive extractions with aqueous 75%-methanol solution under stirring for 15 min at room temperature. The use of ultrasound helps to promote a stronger extraction. The methanolic extracts are combined and concentrated under vacuum at 30 °C to eliminate methanol. The aqueous residue is washed with hexane in order to eliminate liposoluble substances, then fractionated on a Sephadex LH-20 column. The first fractions are eluted with aqueous 90% ethanol solution, followed by acetone aqueous solutions (Gonzáles-Manzano *et al.*, 2006; Gabetta *et al.*, 2000). Ten 500-mL fractions are obtained, their compositions are similar to those reported in Table 3.5.

Another method for extraction and purification of tannins from grape seeds and skins was proposed by Vidal *et al.* (2003). For seeds

Table 3.4 Nonanthocyanin phenolic compounds identified by LC/ESI-MS in wines made with different *Vitis vinifera* grape varieties (*Tempranillo, Garciano, Cabernet Sauvignon, Merlot*). Chromatographic conditions: C_{18} column (300 × 3.9 mm, 4 μm); binary solvent composed of A) H_2O/acetic acid (98:2 v/v) and B) H_2O/acetonitrile/acetic acid (78:20:2 v/v/v); gradient program: from 0 % to 80% of B in 55 min, 80–90% B in 2 min, 90% B isocratic for 3 min, 90–95% B in 10 min, 95–100% B in 10 min (flow rate 0.7 mL/min). ESI/MS condition: negative-ion mode, drying gas N_2 flow 10 L/min, temperature 350 °C; nebulizer pressure 55 psi; capillary voltage 4000 V; fragmentation program: *m/z* 0–200 at 100 V, *m/z* 200–3000 at 200 V.(s) shoulder (Monagas *et al.*, 2005).

RT (min)	λ_{max} (nm)	$[M-H]^-$ (*m/z*)	Fragments (*m/z*)	Compound
7.6	272	169	125	gallic acid
14.0	294,260	153	109	protocatechuic acid
15.7	280	153		dihydroxyphenylethanol
16.7	330,298(s)	311	179	*trans*-caffeyltartaric acid
19.2	310,280	211		2,3-dihydroxy-1-(4-hydroxy-3-methoxyphenyl)-propan-1-one
19.5	272	183	169, 125	methyl gallate
20.5	275	137		tyrosol
22.3	276	593	425	(epi)gallocatechin-(epi)catechin dimer
22.7	280	577	425, 289	procyanidin B3
23.0	280	577	425, 289	procyanidin B1
24.2	313	295	163	*trans*-coumaroyltartaric acid
27.3	279	289		(+)-catechin
28.0	280	865	713, 577, 289	procyanidin T2
28.3	329,301(s)	325	193	*trans*-feruryltartaric acid
28.7	nd	329	167	hexose ester of vanillic acid
29.0	280	865	713, 577, 289	procyanidin T3
29.7	289,262	167		vanillic acid
30.5	283	577	425, 289	procyanidin B4
31.3	323	179	135	*trans*-caffeic acid
33.0	311	325	163, 145	hexose ester of trans-*p*-coumaric acid
33.6	280	577	425, 289	procyanidin B2
34.6	277	197		syringic acid
37.0	312	325	163, 145	hexose ester of trans-*p*-coumaric acid
38.7	279	289		(−)-epicatechin
40.2	280	865	713, 577, 289	trimeric procyanidin
41.3	273	197	169, 125	ethyl gallate
42.2	282	865	713, 577, 289	procyanidin C1
43.2	309	163	119	*trans*-*p*-coumaric acid
43.5	280	865	713, 577, 289	trimeric procyanidin

43.8	278	729	577	procyanidin dimer gallate
44.0	280	577	425, 289	procyanidin B5
48.5	349,300(s),261	493	317	myricetin-3-O-glucuronide
50.1	349,300(s),261	479	317	myricetin-3-O-glucoside
51.3	277	441	289, 169	epicatechin-3-O-gallate
53.2	306(s),319	389	227	*trans*-resveratrol-3-O-glucoside
55.2	368	301		ellagic acid
57.2	354,300(s),256	463	301	quercetin-3-O-galactoside
57.7	354,300(s),256	477	301	quercetin-3-O-glucuronide
58.2	288	449	303	astilbin
58.7	354,300(s),256	463	301	quercetin-3-O-glucoside
60.6	279	160		tryptophol
62.4	346,300(s),265	447	285	kaempferol-3-O-glucoside
67.6	371,300(s),254	317		myricetin
68.9	285	389	227	*cis*-resveratrol-3-O-glucoside
71.9	306,319(s)	227		*trans*-resveratrol
92.7	369,300(s),255	301		quercetin
95.7	284	227		*cis*-resveratrol

Table 3.5 Composition of fractions after separation of methanolic grape seeds extract on Sephadex LH-20 (160 g resin, 50 × 4.5 cm i.d.) column. Elution with A) 90% ethanol and B) 20% C) 40% and D) 70% acetone aqueous solutions (flow rate 16 mL/min) (Gabetta *et al.*, 2000).

Fraction	Main constituents
1	(+)-catechin; (−)-epicatechin
2	(−)-epicatechin-3-O-gallate; dimer; dimer gallate
3	dimer; dimer gallate; trimer
4	dimer; dimer digallate; trimer; trimer gallate; tetramer
5	trimer; trimer gallate; tetramer; tetramer gallate; pentamer
6	trimer gallate; trimer digallate; tetramer; tetramer gallate; tetramer digallate; pentamer; pentamer gallate
7	tetramer gallate; tetramer digallate; pentamer; pentamer gallate; pentamer digallate; hexamer
8	pentamer gallate; tetramer trigallate; pentamer digallate; hexamer; hexamer gallate
9	pentamer digallate; pentamer trigallate; hexamer gallate; hexamer digallate
10	pentamer digallate; pentamer trigallate; hexamer gallate; hexamer digallate; heptamer; heptamer gallate

extraction, 360 g of seeds are frozen with liquid nitrogen and ground with a blender. The resulting powder is extracted twice with 1.5 L acetone/water (60:40 v/v) and the solutions pooled. After centrifugation, the supernatant is concentrated under vacuum and lipophilic compounds are removed by washing the solution with hexane (250 mL). The crude seed extract is used for further purification of PAs: the aqueous acetone extract is chromatographed on a methacrylic size-exclusion resin Toyopearl TSK HW-50 (F) column (18 × 35 cm). Fractions eluting with ethanol/H_2O/trifluoroacetic acid (TFA) 55:45:0.02 (v/v/v) (three bed volumes) and acetone/H_2O (30:70 v/v) (one bed volume) are pooled, concentrated under vacuum and purified further using a DVB-PS resin. The polymeric PAs fraction with degree of polymerization (DP) of 10 units, is desorbed from the column with acetone/H_2O (60:40 v/v). Purification of fractions from the Toyopearl column is performed on DVB-PS resin by dissolution of the residue in water (6 g in 200 mL) and fractionation on a 25 × 50 cm column. After washing with water and ether to eliminate flavan-3-ol monomers, PAs DP 3 are recovered with methanol. Prior to extraction of tannins from the skin, simple phenolic compounds (mainly anthocyanins) are removed by immersing the skins in 12% (v/v) ethanol solution for 72 h at 4 °C; this solution is then discarded. The skins are then ground in methanol and, after 2 h at 4 °C, the methanolic solution is recovered by filtration. Finally, skins are extracted with acetone/H_2O (60:40 v/v) overnight at 4 °C. Methanolic and aqueous acetone extracts are separately concentrated under vacuum and used for further fractionation of PAs the two extracts are separately chromatographed on a Toyopearl TSK HW-50(F) column. Sugars, phenolic acids and anthocyanins are eliminated by passage through the column of ethanol/H_2O/TFA (55:45:0.02 v/v/v) solution followed by acetone/H_2O (30:70 v/v). PAs with DP 12–20 are desorbed from the column with acetone/H_2O (60:40 v/v).

Another method of sample preparation for analysis of PAs from grape seeds and juice is to perform purification by Solid-Phase-Extraction (SPE) using a C_{18} (6–20 mL) cartridge. Seed extracts (250 mg or 50 mL of grape juice) are suspended in 20 mL of water and loaded onto the cartridge previously conditioned by passage of 3 × 5 mL methanol followed by 3 × 5 mL of water. After the sample is passed through, the cartridge is rinsed with 40 mL of water and PAs are eluted with 6–10 mL of acetone/H_2O/HAc (70:29.5:0.5 v/v/v) solution (Lazarus *et al.*, 1999; Núñez *et al.*, 2006).

To study the PAs composition, thiolysis of tannins extract can be performed. The tannin powder is dissolved in methanol (1 mg/mL) and introduced into a glass vial together with an equal volume of a 5% methanolic solution of toluene-α-thiol containing HCl 0.2 M. Reaction is carried on at 90 °C for 2 min and direct analysis of the thiolyzed solution is performed (Fulcrand *et al.*, 1999; Vidal *et al.*, 2003).

LC/MS analysis

LC analysis of PAs is usually performed by reverse phase chromatography. An example of chromatograms relative to analysis of a grape seeds extract, is reported in Figure 3.5.

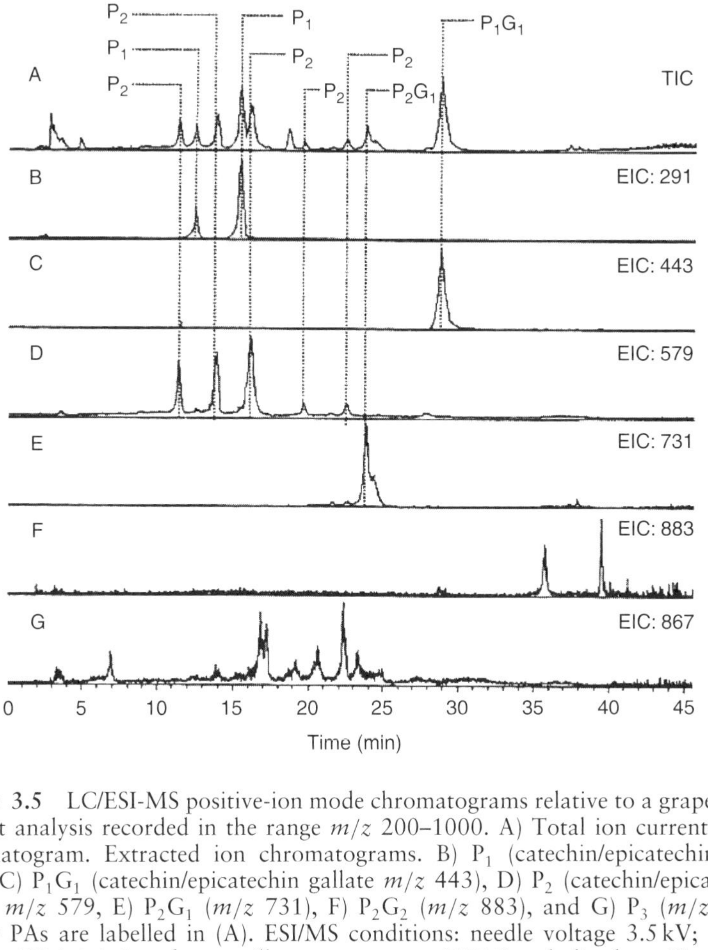

Figure 3.5 LC/ESI-MS positive-ion mode chromatograms relative to a grape seeds extract analysis recorded in the range *m/z* 200–1000. A) Total ion current (TIC) chromatogram. Extracted ion chromatograms. B) P_1 (catechin/epicatechin *m/z* 291), C) P_1G_1 (catechin/epicatechin gallate *m/z* 443), D) P_2 (catechin/epicatechin dimer *m/z* 579, E) P_2G_1 (*m/z* 731), F) P_2G_2 (*m/z* 883), and G) P_3 (*m/z* 867). Major PAs are labelled in (A). ESI/MS conditions: needle voltage 3.5 kV; drying gas N_2 (8 L/min); interface capillary temperature 325 °C; nebulized gas He 40 psi. Chromatographic conditions: column C_{18} (250 × 4.6 mm, 5 μm) at 25 °C; binary solvent composed of A) 0.1% formic acid (v/v) and B) 0.1% formic acid (v/v) in acetonitrile; gradient program: from 10–20% of B in 20 min, isocratic at 20% B for 10 min from 20–50% B in 10 min, isocratic 50% B for 10 min (flow rate 1.0 mL/min, 1/4 of eluent split into mass spectrometer). (Reprinted from *Rapid Communications in Mass Spectrometry* **19**, Wu *et al.*, Determination of proanthocyanidins in fresh grapes and grape products using liquid chromatography with mass spectrometric detection, p. 2065, Copyright 2005, with permission from John Wiley & Sons Ltd)

PAs analysis can be also performed by normal phase chromatography. It is reported that it is a good method to achieve a satisfactory separation of PAs oligomers on the basis of their MW. Analysis is performed by using silica columns (e.g. 250 × 4.6 mm, 5 μm) operating at 37 °C (Lazarus *et al.*, 1999; Núñez *et al.*, 2006). Experimental conditions of two methods used for LC/ESI-MS analysis in normal phase are reported in Table 3.6.

Table 3.6 Experimental conditions of two different methods used for LC/ESI-MS analysis of proanthocyanidins by normal-phase chromatography.

	Method 1 (Lazarus *et al.*, 1999)	Method 2 (Núñez *et al.*, 2006)
Solvent	(A) methylene chloride; (B) methanol; (C) acetic acid/H_2O 1:1 (v/v)	(A) methylene chloride/methanol/H_2O/acetic acid 82:14:2:2 (v/v/v/v) (B) methanol/H_2O/acetic acid 96:2:2 (v/v/v)
Column	silica (250 × 4.6 mm; 5 μm); 37 °C	silica (250 × 2 mm; 5 μm)
Flow rate	1 mL min^{-1}	0.2 mL min^{-1}
Gradient program	linear gradient of B in A with a constant 4% C: elution starting with 14% B in A; 14–28.4% B in A in 30 min; 28.4–50% B in A in 30 min; 50–86% B in A in 5 min; 5 min isocratic	linear gradient of B in A: 0–18% B in A in 30 min; 18–31% B in A in 15 min; 31–88% B in A in 5 min
LC/ESI-MS conditions negative ion mode	buffering reagent 0.75 M ammonium hydroxide in the eluant stream at a flow rate of 0.04 mL min^{-1}; capillary voltage 3 kV, cone voltage 75 V, drying gas temperature at 350 °C	buffering reagent 0.01 M ammonium acetate in the eluant stream at a flow rate of 0.03 mL min^{-1}; capillary voltage 3.2 kV, cone voltage 30 V, source temperature 150 °C, drying gas temperature at 300 °C

Due to their weak acidic nature, LC/MS analysis of PAs is usually performed in negative-ion mode. Hayasaka *et al.* (2003) reported that the use of methanol/acetonitrile (1:1) in negative mode gives the highest intensity of ions, including multiply charged ions through to higher mass region. In addition, analysis in negative ion mode shows simpler mass spectra due to the absence of intense adduct ion species and to the production of more multiply charged ions than in the positive ion mode.

To characterize extracts direct infusion ESI-MS analysis can be used. The sample is dissolved in methanol/acetonitrile (1:1) solution and mass spectra similar to those shown in Figures 3.6(a) and (b) of PAs with DP 3 and 9, are recorded. The $[M–H]^-$ and $[M-2H]^{2-}$ ions of PAs identified, are reported in Table 3.7.

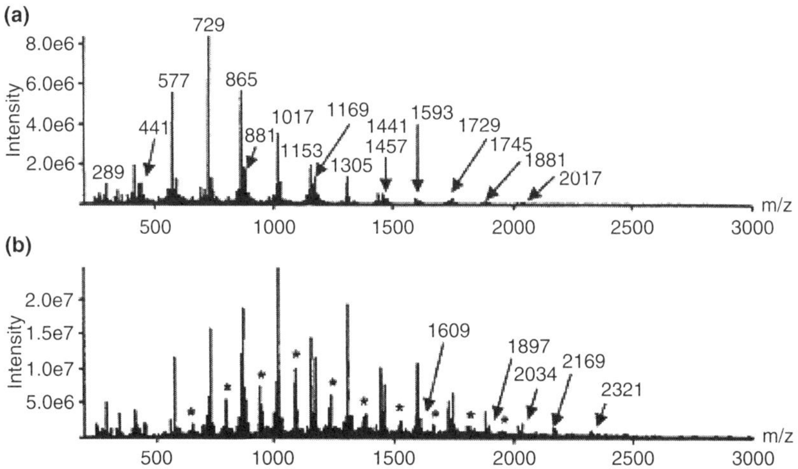

Figure 3.6 ESI mass spectra of PAs with DP (a) 3 and (b) 9 obtained by signal accumulation of 20 consecutive scans. m/z values: major singly charged ions of PAs. *Doubly charged ions of PAs monogallates. Analytical conditions: ESI needle, orifice and ring potentials −4500, −60 and −350 V, respectively; curtain gas N_2, nebulizer gas: air. (Reprinted from *Rapid Communications in Mass Spectrometry* **17**, Hayasaka *et al.*, Characterization of proanthocyanidins in grape seeds using electrospray mass spectrometry, p. 11, Copyright 2003, with permission from John Wiley & Sons Ltd)

Abundant $[M–H]^-$ singly charged ions separated by 288 Da in the range from m/z 289 to 2017 and from m/z 441 to 1881 correspond to the molecular masses of PCs with DP from 1 to 7, and PC monogallates (PC1Gs) with DP l-6, respectively. PAs with DP 9 show the additional larger $[M–H]^-$ ions derived from PC1G with DP 7, PC2Gs (PC digallates) with DP 6 and DP 7, and PC3Gs (PC trigallates) with DP 4 and DP 5 [Figure 3.6(b), Table 3.7]. The ESI mass spectrum in Figure 3.7(a) shows the ions resulting from the product ion spectra obtained from m/z 865 [PC with DP 3, Figure 3.7(b)] and m/z 1017 [PC1G with DP 3, Figure 3.7(c)]). The intensities of these ions relative to the $[M–H]^-$ ions, increase with the orifice potential. Fragmentation pathways of the PAs $[M\text{-}H]^-$ and $[M\text{-}3H]^-$ ions (m/z 577, 575, 729, 727 and 441) could be derived by the cleavage of the interflavanic bond, $[M\text{-}H\text{-}152]^-$ (m/z 713, 425, 865, and 577) and $[M\text{-}H\text{-}152\text{-}H_2O]^-$ (m/z 695, 407, 847 and 559) ions with retro-Diels-Alder fission on the C ring followed by the elimination of water, and $[M\text{-}H\text{–}126]^-$ (m/z 739 and 451) ions after elimination of phloroglucinol. The ion at m/z 881 corresponds to dimer of epicatechin-gallate or epicatechin-epicatechin-epigallocatechin trimer (isobaric compounds).

Table 3.7 [M–H]$^-$ and [M–2H]$^{2-}$ ions of proanthocyanidins with DP 3 and 9. DP, degree of polymerization; na, not assigned/not detected (Hayasaka *et al.*, 2003).

DP	PCs (procyanidins)			PC1Gs (monogallates)			PC2Gs (digallates)		
	Sdp3	Sdp9		Sdp3	Sdp9		Sdp3	Sdp9	
	[M-H]-	[M-H]-	[M-2H]2-	[M-H]-	[M-H]-	[M-2H]2-	[M-H]-	[M-H]-	[M-2H]2-
1	289.2	289.4	/	441.2	441.4	/	/	/	/
2	577.4	577.4	/	729.4	729.4	/	881.4	881.4	/
3	865.4	865.4	/	1017.6	1017.4	/	1169.8	1169.8	/
4	1153.4	1153.4	/	1305.8	1305.8	652.4	1457.6	1457.4	/
5	1441.8	1441.6	720.4	1593.4	1593.4	796.4	1745.4	1745.2	872.6
6	1729.8	1729.2	na	1881.8	1881.6	940.8	/	2034.0	na
7	2017.2	2017.2	1009.2	/	2169.8	1084.2	/	2322.2	1161.0
8	/	/	na	/	/	1228.6	/	/	na
9	/	/	1296.6	/	/	1373.2	/	/	1448.6
10	/	/	na	/	/	1516.8	/	/	na
11	/	/	1584.0	/	/	1661.0	/	/	1737.0
12	/	/	/	/	/	1805.4	/	/	na
13	/	/	/	/	/	1948.8	/	/	2025.2

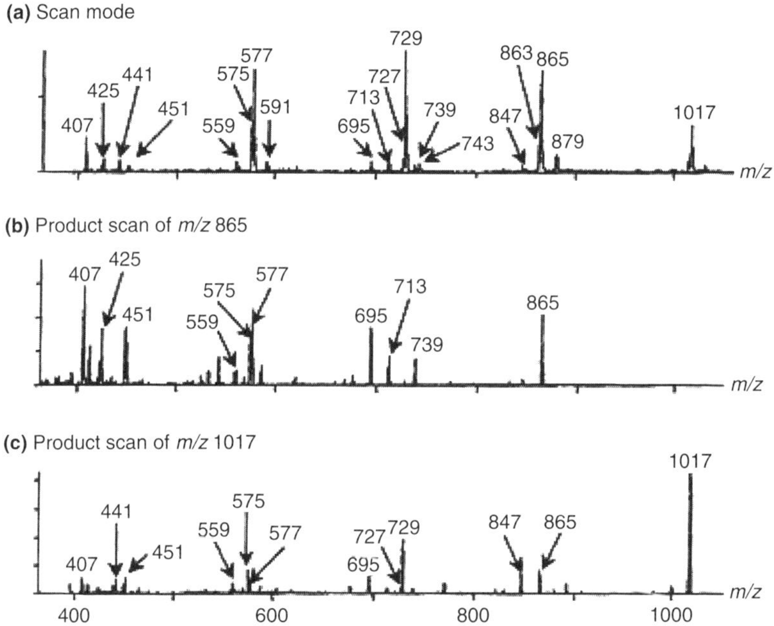

Figure 3.7 MS/MS experiments of PAs with collision gas N_2 (2 units); collision energy potential 40–60 V (Reprinted from *Rapid Communications in Mass Spectrometry* 17, Hayasaka *et al.*, Characterization of proanthocyanidins in grape seeds using electrospray mass spectrometry, p. 13, Copyright 2003, with permission from John Wiley & Sons Ltd)

Table 3.7 (Continued).

PC3Gs (trigallates)			PC4Gs (tetragallates)			PC5Gs (pentagallates)		
Sdp3	Sdp9		Sdp3	Sdp9		Sdp3	Sdp9	
[M-H]-	[M-H]-	[M-2H]2-	[M-H]-	[M-H]-	[M-2H]2-	[M-H]-	[M-H]-	[M-2H]2-
/	/	/	/	/	/	/	/	/
/	/	/	/	/	/	/	/	/
/	/	/	/	/	/	/	/	/
/	1609.2	804.4	/	/	/	/	/	/
/	1897.8	948.8	/	/	1024.8	/	/	1100.2
/	/	1092.6	/	/	na	/	/	1243.8
/	/	1236.8	/	/	1312.6	/	/	1388.4
/	/	1380.6	/	/	na	/	/	1532.2
/	/	1524.6	/	/	1601.2	/	/	1676.4
/	/	1669.2	/	/	na	/	/	1821.8
/	/	1813.2	/	/	1889.2	/	/	/
/	/	/	/	/	/	/	/	/
/	/	/	/	/	/	/	/	/

Doubly charged ions show a series of abundant ions separated by 144 Da from *m/z* 652.4 to 1948.8 (indicated with asterisk in Figure 3.6(b)), which correspond to the $[M-2H]^{2-}$ ions of PC1Gs with DP 4–13.

Increasing the orifice voltage, two different fragmentation patterns of trimeric species are observed. From ion at *m/z* 863 (A-type) two different ions at *m/z* 575 and 573 form, by Retro-Diels-Alder (RDA)

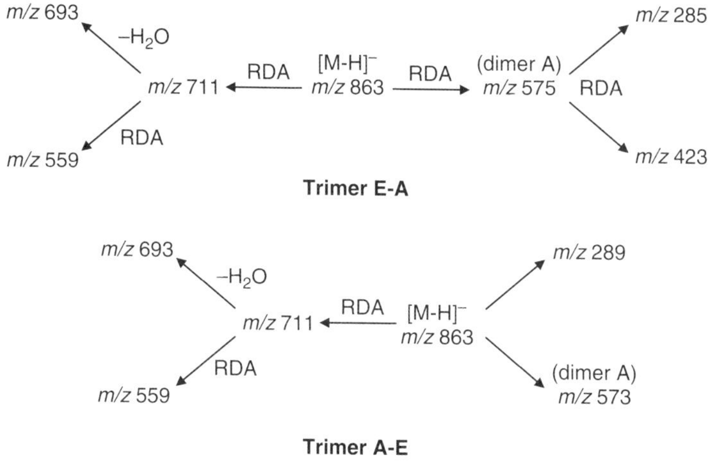

Figure 3.8 Fragmentation patterns of A-type procyanidin trimers by LC/ESI-MS in negative-ion mode

fragmentation the ions at *m/z* 711 is observed. By the loss of a neutral fragment of 152 Da (3,4-dihydroxy-α-hydroxystyrene), two fragments at *m/z* 285 and 289, generated by cleavage of the A-type interflavanic linkage, are formed. Fragmentation patterns are shown in Figure 3.8 (Cheynier *et al.* 1997).

MS analysis of PCs and PAs can also be performed operating in positive-ion mode, the compounds are identified on the *m/z* values of their protonated molecules. Signals of $[M+H]^+$ ions of catechin dimers, trimers and tetramers (*m/z* 579, 867, 1155), their mono- and di- galloyl derivatives (*m/z* 731, 1019, 1307, 883, 1171, 1459) and the trigalloyl derivatives of trimers and tetramers (*m/z* 1323 and 1611) are evident, and $[M+H]^+$ ions of flavan-3-ols pentamers, hexamers and heptamers (*m/z* 1443, 1731, 2019), their monogalloyl derivatives (*m/z* 1595, 1883, 2171), pentamers and hexamers digalloyl derivatives (*m/z* 1747 and 2035), pentamers and hexamers trigalloyl derivatives (*m/z* 1899 and 2187), are also identified (Gabetta *et al.*, 2000). Figure 3.9 shows the fragmentation patterns proposed for trimeric PCs in positive-ion mode: ions with *m/z* ratios corresponding to the most intense peaks of MS spectra are isolated and fragmented by ion trap, MS^3 spectra are acquired by fragmentation of principal ions in the MS^2 spectra (Pati *et al.*, 2006).

An example of positive ion mode mass spectrum of a grape seeds extract analysis, and the full scan spectra of ESI-MS^2 positive-ion mode analysis of PCs, is reported in Figure 3.10.

3.2.3. Analysis of Procyanidins and Proanthocyanidins in Wine

Two different methods of sample preparation by size-exclusion and reverse phase chromatography for LC analysis of PCs and PAs in wine, have been proposed. To perform sample preparation by size exclusion, 5 mL of dealcoholized wine is deposited on a Fractogel Toyopearl TSK gel HW-50 (F) (12 × 120 mm) column. After washing with 25 mL water, simple polyphenols are eluted with 50 mL of ethanol/H_2O/TFA (55:45:0.005 v/v/v) solution. The polymeric fraction is recovered with 50 mL of acetone/H_2O (60:40 v/v). Figure 3.11 shows the extracted ion chromatograms of dimers and trimers from the TIC of a wine LC/ESI-MS analysis (Fulcrand *et al.*, 1999).

Sample preparation by reverse phase chromatography is performed using a C_{18} (e.g. 20-mL) SPE cartridge: a volume of 30 mL of

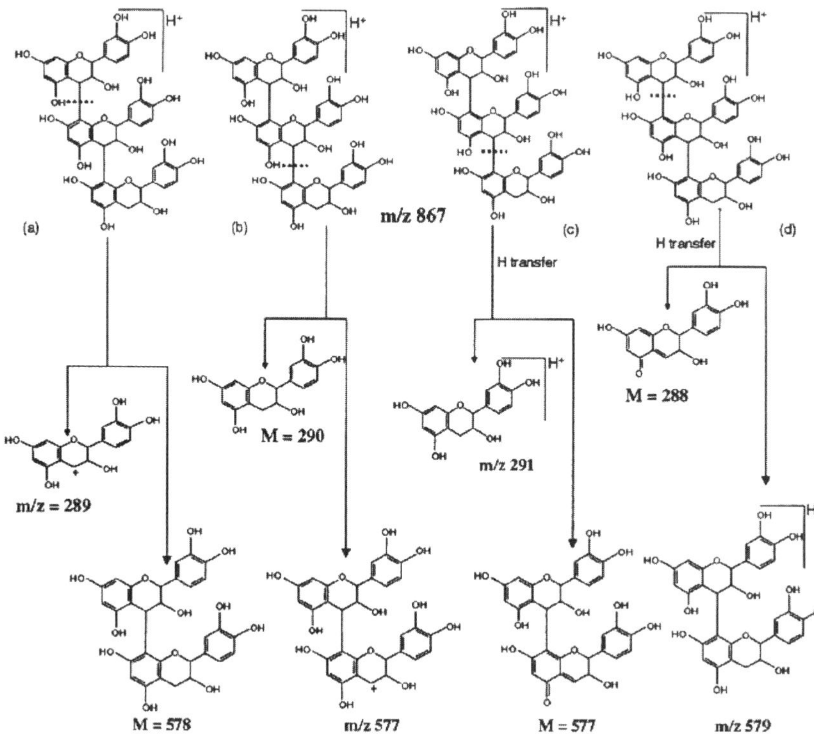

Figure 3.9 Fragmentation patterns in positive-ion mode of PCs. ESI-MS conditions: spray voltage 4.5 kV; sheath gas nitrogen 0.9 L/min; capillary voltage 35 V; capillary temperature 200 °C; tube lens offset voltage 15 V. (Reprinted from Pati *et al.*, 2006. Simultaneous separation and identification of oligomeric procyanidins and anthochyanidins-derived pigments in raw red wine by HPLC-UV-ESI-MS^n, *Journal of Mass Spectrometry* **41**, p. 869, with permission from John Wiley & Sons Ltd)

dealcoholized wine is loaded on the cartridge; after rinsing the stationary phase with 40 mL of water, PAs are recovered with 10 mL of acetone/water/acetic acid (70:29.5:0.5 v/v/v) solution (Lazarus *et al.*, 1999). A scheme of wine PAs fractionation is showed in the flow-diagram in Figure 3.12 (Sun *et al.*, 2006).

Fractions 1–4 can be analyzed by ESI-direct infusion recording mass spectra in negative ionization mode. Due to the higher complexity of fractions 8–10, for these samples direct ESI-MS analysis does not provide useful information, and for their characterization LC/ESI-MS and MS^n analyses have to be performed. Compounds identified are reported in Table 3.8 with the analytical conditions used.

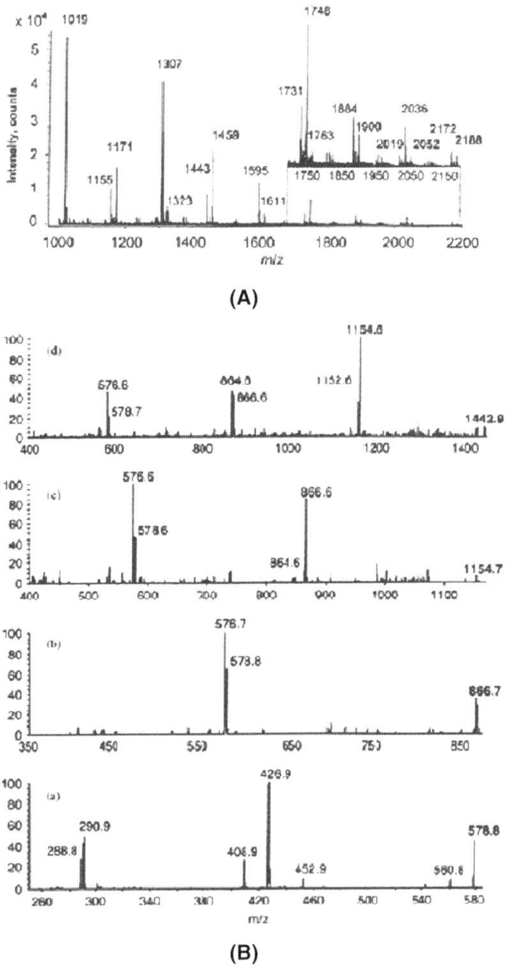

Figure 3.10 (A): ESI/MS positive-ion mode analysis of a grape seeds extract: $[M+H]^+$ ions of high-polymerized PAs from P_3G_1 to P_7G_1 in the composite mass spectrum with expansion of the higher mass range (needle voltage 3.5 kV; drying gas N_2; interface capillary temperature 325 °C; nebulizer gas He 40 psi. (Reprinted from *Rapid Communications in Mass Spectrometry* **19**, Wu *et al.*, Determination of proanthocyanidins in fresh grapes and grape products using liquid chromatography with mass spectrometric detection, p. 2065, Copyright 2005, with permission from John Wiley & Sons Ltd). (B): ESI-MS^2 full scan spectra of (a) dimeric (*m/z* 579), (b) trimeric (*m/z* 867), (c) tetrameric (*m/z* 1155), and (d) pentameric (*m/z* 1449) PCs (Reprinted from *Journal of Mass Spectrometry* **41**, Pati *et al.*, Simultaneous separation and identification of oligomeric procyanidins and anthochyanidins-derived pigments in raw red wine by HPLC-UV-ESI-MS^n, p. 868, Copyright 2006, with permission from John Wiley & Sons Ltd)

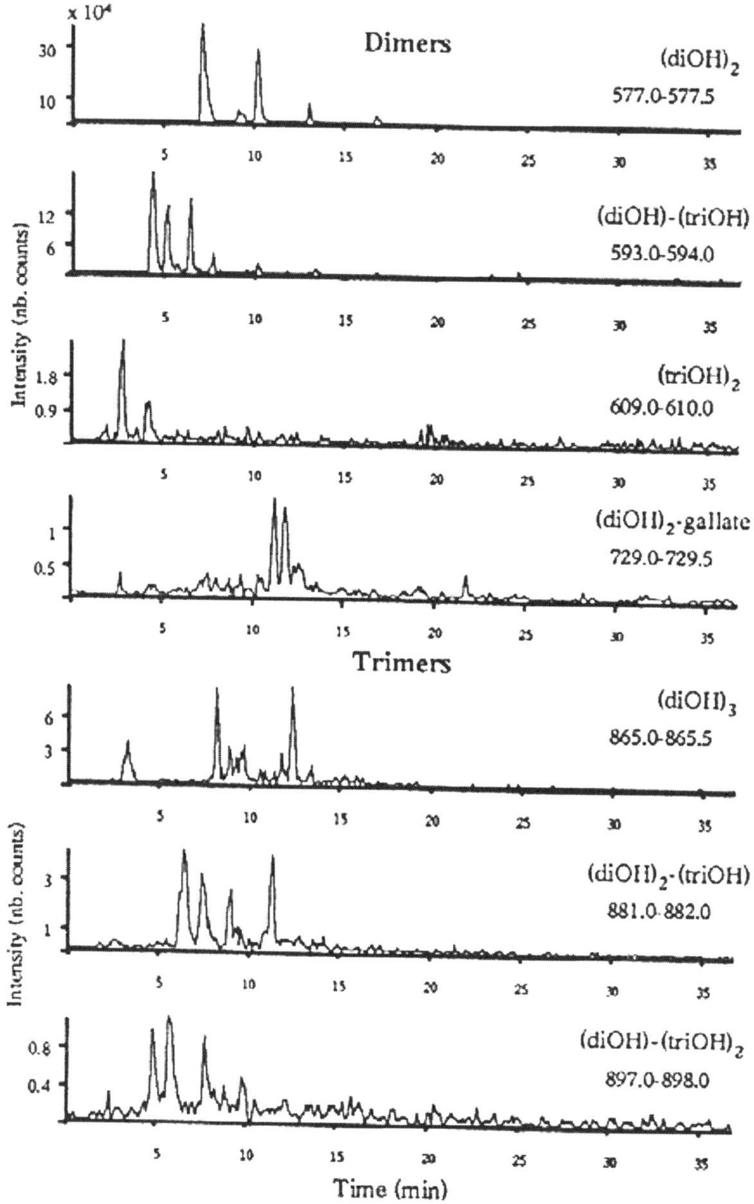

Figure 3.11 Chromatograms extracted from the TIC recorded in negative-mode of PCs and PAs dimers and trimers of a wine LC/ESI-MS analysis. Analytical conditions: C_{18} (125 × 2 mm i.d., 3 μm) narrow-bre column; ion spray voltage −4000 V, orifice voltage −60 V. Binary solvent composed of A) aqueous 2% formic acid and B) acetonitrile/H_2O/formic acid (80:18:2 v/v/v). Gradient program: from 5% to 30% of B in 20 min, 30–50% B in 10 min (flow rate 200 μL/ min, flow rate in the ESI source 50 μL/ min). (Reproduced from *J. Agric. Food Chem.*, 1999, 47, 1023–1028, Fulcrand *et al.*, with permission of American Chemical Society)

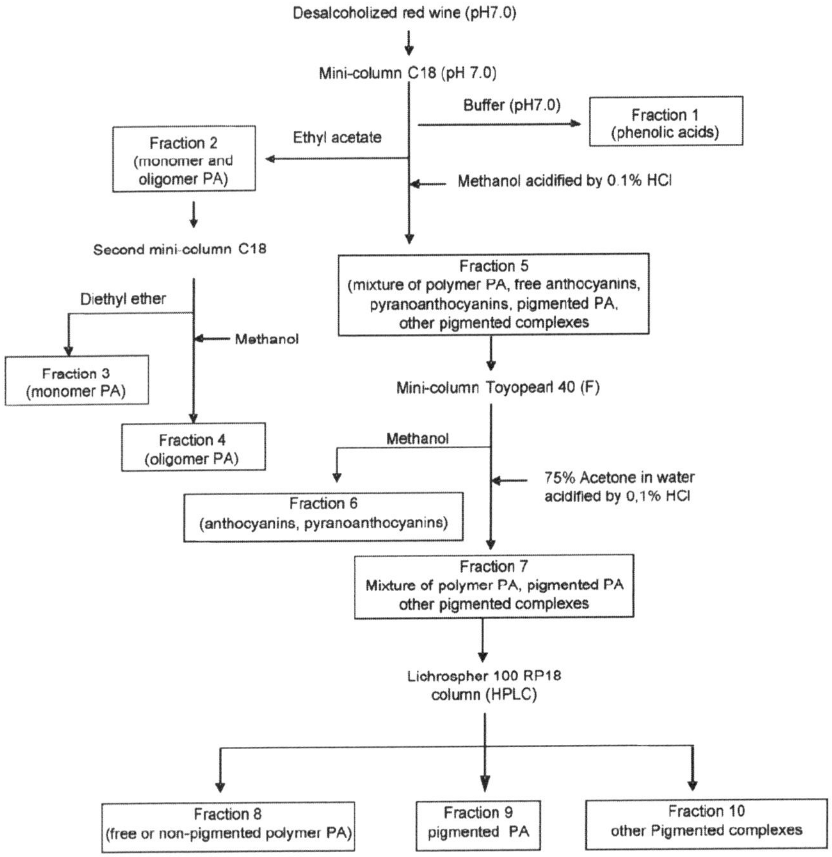

Figure 3.12 Flow-diagram of fractionation polyphenols in red wine. PA, proanthocyanidins (Reprinted from *Journal of Chromatography* A **1128**, Sun *et al.*, Fractionation of red wine polyphenols by solid phase extraction and liquid chromatography, p. 29, Copyright 2006, with permission from Elsevier)

3.3. Anthocyanins of Grape and Wine

The five anthocyanins in the *Vitis vinifera* grape are delphinidin (Dp), cyanidin (Cy), petunidin (Pt), peonidin (Pn) and malvidin (Mv) (see Figure 2.7). These compounds can be present as 3-O-monoglucosides, 3-O-acetylmonoglucosides and 3-O-(6-O-*p*-coumaroyl)monoglucosides; Mv-3-O-(6-O-caffeoyl)monoglucoside can be also present. In the non-*Vitis vinifera* (hybrid) grape, anthocyanins with a second glucose linked to the C-5 hydroxyl group (diglucosides) are often present.

Colour changes of red wines occurring during ageing are mainly due to chemical reaction and polymerization of anthocyanins with other

Table 3.8 LC/ESI-MS and MS^n analysis of fractions 8 and 9 in Figure 3.12. Analytical conditions: C_{18} column (250×4 mm, 5 μm) at 30 °C. Binary solvent A) acetonitrile/H_2O/formic acid (2.5/97.4/0.1 v/v/v) and B) acetonitrile/H_2O/formic acid (30/69.9/0.1 v/v/v). Gradient program for fraction 8: from 0% to 30% of B in 20 min, 30–50% B in 20 min, 50–100% B in 20 min. Gradient program for fraction 9: from 0% to 15% of B in 20 min, 15–40% B in 20 min, 40–100% B in 40 min (flow-rate 0.7 mL/min). Skimmer and the capillary voltages for negative ion mode −40 and +4000 V, respectively (+40 and −4000 V for positive ion mode); nebulizer gas N_2 10 psi; drying gas N_2 10 L/min; dry temperature 250 °C (Sun *et al.*, 2006).

peak number	phenolic compounds	$[M-H]^-$ (m/z)	$[M+H]^+$ (m/z)	$[M]^+$ (m/z)	$[M-2H]^{2-}$ (m/z)	MS^2 (m/z)	MS^3 (m/z)	MS^4 (m/z)
1	PC1PD1	593				441,289	315, 289, 153	
2	PC3	865				847, 577, 289	451, 289	
3	PC4	1153				863, 577	451, 289	
4	PC6-4G or PC6PD1-2G				1168	1017, 729	575	
5	PC7PD1-6G or PC7PD2-4G or PC7PD3-2G				1616	1084, 1641		
6	PC2PD1		883			847, 731, 579	427, 289	
7	PC3	865	867			577, 291	289	
8	PC4	1153	1155			863, 577	451, 289	
9	PC5		1443			1425, 1291, 1155, 865, 577	847, 577	
10	PC2-Mv-3-glu			1069		781	619	601, 331
11	Mv-3-glc-PC3	1357				1067, 779	617	437

compounds in wine. As the wine becomes older, the percentage of orange compounds increases inducing a colour change. During barrel ageing, new compounds such as B-type vitisins, vinylphenol, vinylcatechol and vinylguaiacol anthocyanin adducts, form (Alcade-Eon *et al.*, 2006). More than 100 different compounds were identified in aged wines belonging to the pigment families of anthocyanins, pyranoanthocyanins, direct flavanol-anthocyanin condensation products (structures summarized in Figure 3.13) and acetaldehyde-mediated flavanol-anthocyanin

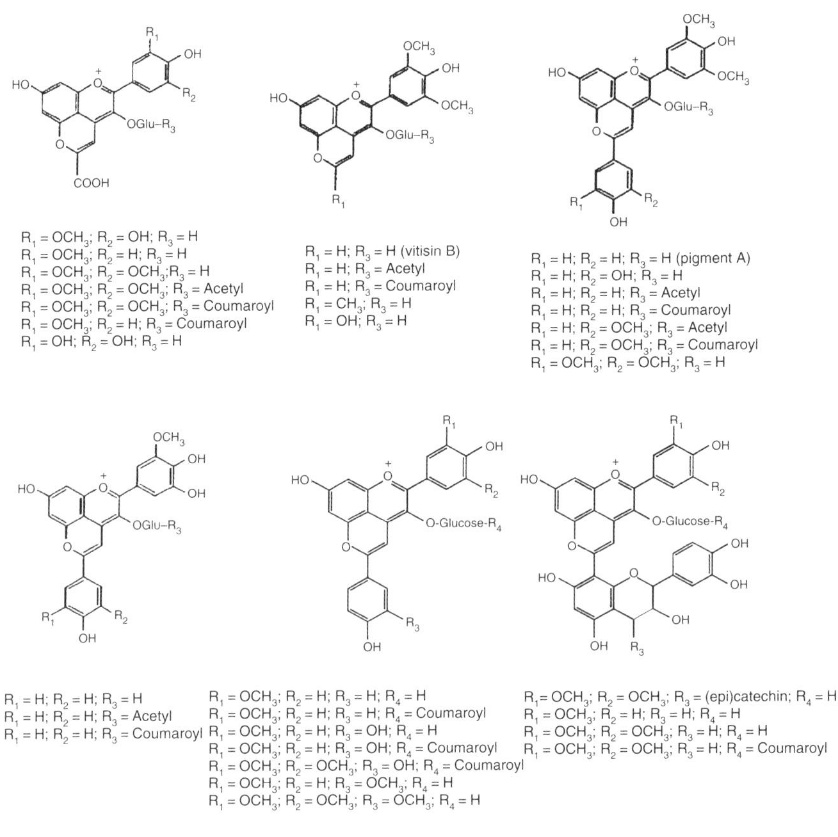

Figure 3.13 Structures of C-4 substituted anthocyanins identified in aged red wines: A-type vitisins and pyranoanthocyanins originate from reaction between anthocyanins and vinylphenol, vinylcatechol, vinylguaiacol, or vinyl(epi)catechin. (Reprinted from *Analytica Chimica Acta* **513**, Alcade-Eon *et al.*, Separation of pyranoanthocyanins from red wine by column chromatography, p. 311, Copyright 2004, with permission from Elsevier and from *Analytica Chimica Acta* **570**, Gómez-Ariza *et al.*, Anthocyanins profile as fingerprint of wines using atmospheric pressure photoionisation coupled to quadrupole time-of-flight mass spectrometry, p. 104, Copyright 2006, with permission from Elsevier)

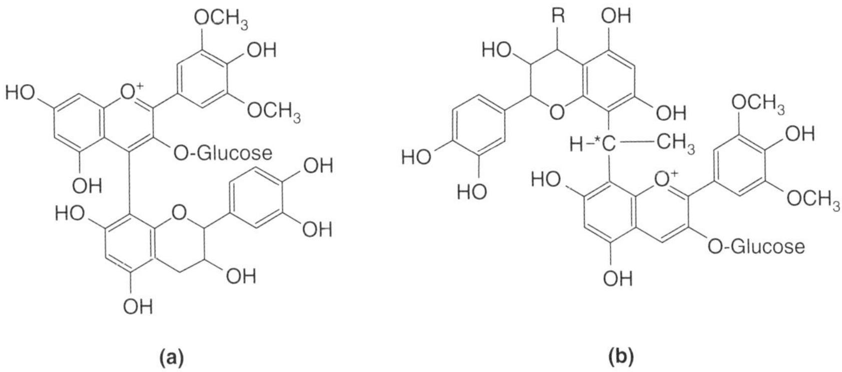

Figure 3.14 (a) Structure formed by direct polymerization of anthocyanin and a flavan-3-ol proposed by Somers, 1971; (b) structure formed with ethyl bridge proposed by Timberlake and Bridle, 1976

condensation products with anthocyanin linked directly or by an ethyl bridge to a flavan-3-ol (Figure 3.14).

3.3.1. Analysis of Anthocyanins in Grape

Reverse Phase LC is usually used for analysis of anthocyanins in grape extracts by recording the UV-Vis anthocyanin profile at wavelength 520 nm. An example of analysis of a hybrid grape extract is reported in Figure 3.15 with the chromatographic conditions used, compounds identified are reported in Table 3.9. Using this approach, the peak assignment is mainly based on the sequence elution of compounds from the column, using mass spectrometry, confident structural data are also achieved.

To reduce the risk of hydrolysis of acetylated compounds non-acidified methanol is the best solvent to perform extraction of anthocyanins from grape: the skins of 20 berries are extracted with 50 mL methanol for 12 h at room temperature (Revilla *et al.*, 1998). The extract is then reduced to half volume under vacuum at 30 °C, the volume adjusted to 100 mL by water addition and 10 mL of the resulting solution is diluted to 50 mL with water in order to reduce the percentage methanol content and used for analysis. The solution is purified by passage through to a C_{18} cartridge (1 g) previously activated by passage of 3 mL methanol and 5 mL of water. After the solution is passed through, the cartridge is washed with 5 mL of water to remove sugars and more polar compounds. The non-anthocyanin phenols are eluted

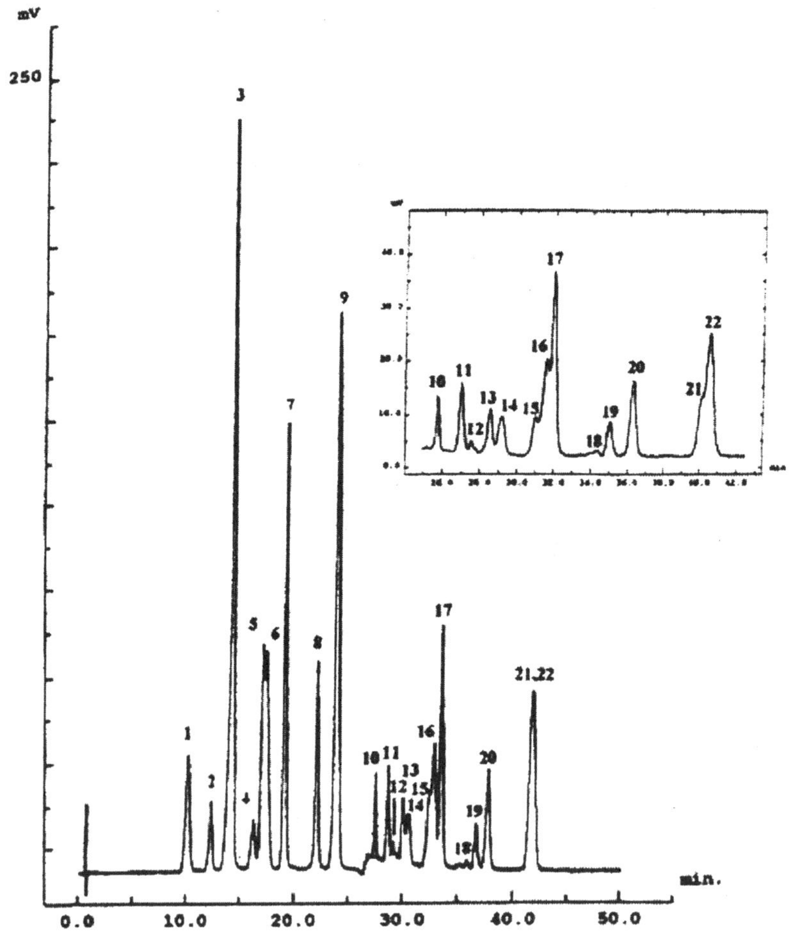

Figure 3.15 LC-UV anthocyanin profile recorded at wavelength 520 nm of a hybrid grape crude extract (Clinton, *Vitis labrusca* × *Vitis riparia*). Analytical conditions: column C_{18} (250 × 4.6 mm, 5 μm), binary solvent composed of A) H_2O/formic acid (90:10 v/v) and B methanol/H_2O/formic acid (50:40:10 v/v/v); gradient program: from 15% to 45% of B in 15 min, 45–70% B in 30 min, 70–90% B in 10 min, 90–99% B in 5 min, 99–15% B in 5 min (flow rate 1 mL/min). (Reprinted from *American Journal of Enology and Viticulture* **51**, Favretto and Flamini, Copyright 2000)

from the cartridge using 3 mL ethyl acetate and anthocyanins retained are recovered with 3 mL of methanol.

To achieve structural and semi-quantitative information on the anthocyanins ESI-MS/MS direct-injection analysis of extract can be performed. An example of an ESI-MS spectrum is reported in Figure 3.16.

Where the M^+ ions of anthocyanins are evident. Structural identification of compounds is achieved by performing multiple mass

Table 3.9 Retention times and molecular masses of anthocyanins (by off-line ESI-MS) separated in the chromatogram Figure 3.15 (Favretto and Flamini, 2000).

peak Fig. 3.18	RT (min)	MW	Compound
1	10.50	627	Delphinidin-3,5-O-diglucoside (1)
2	12.50	611	Cyanidin-3,5-O-diglucoside (2)
3	14.22	641 465	Petunidin-3,5-O-diglucoside (3A)+ Delphinidin-3-O-monoglucoside (3B)
4	16.08	625	Peonidin-3,5-O-diglucoside (4)
5	16.67	655	Malvidin-3,5-O-diglucoside (5)
6	17.14	449	Cyanidin-3-O-monoglucoside (6)
7	18.59	479	Petunidin-3-O-monoglucoside (7)
8	21.22	463	Peonidin-3-O-monoglucoside (8)
9	22.88	493	Malvidin-3-O-monoglucoside (9)
10	25.78	507	Delphinidin-3-O-acetylmonoglucoside (10)
11	27.01	773	Delphinidin-3-(6-O-p-coumaroyl), 5-O-diglucoside (11)
12	27.52	757 491	Cyanidin-3-(6-O-p-coumaroyl), 5-O-diglucoside (12A)+ Cyanidin-3-O-acetylmonoglucoside (12B)
13	28.53	787	Petunidin-3-(6-O-p-coumaroyl), 5-O-diglucoside (13)
14	29.15	521 801	Petunidin-3-O-acetylmonoglucoside (14A)+ Malvidin-3-(6-O-p-coumaroyl), 5-O-diglucoside (14B)
15	30.98	505	Peonidin-3-O-acetylmonoglucoside (15)
16	31.58	535	Malvidin-3-O-acetylmonoglucoside (16)
17	31.97	611	Delphinidin-3-(6-O-p-coumaroyl) monoglucoside (17)
18	34.37	655	Malvidin-3-(6-O-caffeoyl) monoglucoside (18)
19	35.02	595	Cyanidin-3-(6-O-p-coumaroyl) monoglucoside (19)
20	36.29	625	Petunidin-3-(6-O-p-coumaroyl) monoglucoside (20)
21	40.11	609	Peonidin-3-(6-O-p-coumaroyl) monoglucoside (21)
22	40.50	639	Malvidin-3-(6-O-p-coumaroyl) monoglucoside (22)

spectrometry and collisional induced dissociation (CID) experiments (supplementary radio frequency field in the range 1–15 V is applied to the endcaps of the ion trap in order to make selected ions collide with helium). Identification of compounds on the basis of their fragments is reported in Table 3.10.

In general, MS^n is also highly effective for differentiation of isobaric compounds. The detection of fragment ions $[M\text{-}162]^+$ ($Y_0^{\ +}$), $[M\text{-}324]^+$ (due to consecutive losses of two sugar residues), $[M\text{-}204]^+$, $[M\text{-}308]^+$

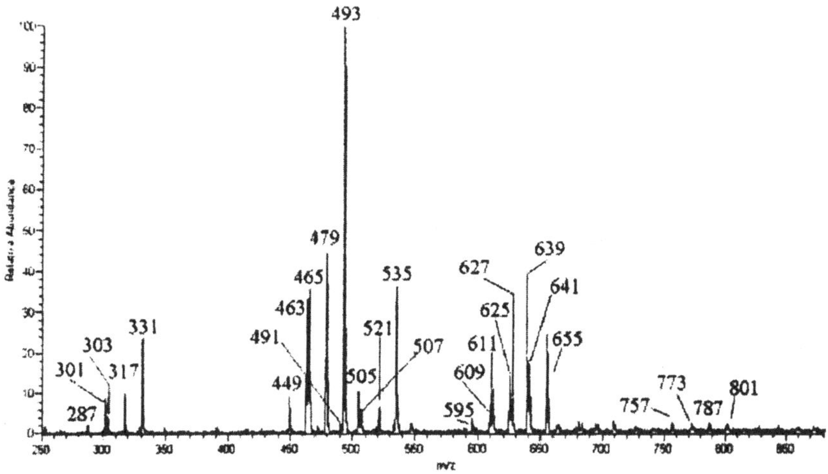

Figure 3.16 ESI-MS anthocyanin profile of Clinton extract recorded by direct injection. ESI conditions: source voltage 4.2 kV in positive ion mode, capillary voltage 3.14 V, capillary temperature 220 °C, sheath gas flow rate 40 (arbitrary units), flow rate injection into the ESI source 3 μL/min. (Reprinted from *American Journal of Enology and Viticulture* **51**, Favretto and Flamini, Copyright 2000)

and $[M\text{-}470]^+$ (due to consecutive losses of *p*-coumaroylglucose and sugar residues) allows the characterization of the five monoglucoside and diglucoside anthocyanins. Of course, the collision energy affects the relative abundance of diagnostic fragments. In the particular case of Mv diglucoside and Mv caffeoylmonoglucoside, due to the identical molecular mass and aglycone moiety, MS^n experiments are not effective in the differentiation between two compounds. They can be distinguished by deuterium exchange experiments: samples are dissolved in deuterated water and different mass shifts, in agreement with the different number of exchangeable, acidic protons present in the molecules, are observed (Figure 3.17).

Quantitative ESI/MS data of anthocyanins present in the extract can be achieved by calibration curve using Mv-3-O-glucoside (M^+ species at m/z 493) as standard compound for monoglucosides and Mv-3,5-O-diglucoside (M^+ species at m/z 655) for diglucosides (both compounds are commercially available). A 40-ppm solution of Mv-3-O glucoside in water/acetonitrile (95:5 v/v) is infused in the ESI interface to optimize the ESI parameters in order to maximize the signal (e.g. experimental conditions: spray voltage, 4.5 kV; sheath gas nitrogen 0.9 L/min, capillary voltage 35 V; capillary temperature 200 °C; tube lens offset voltage 15 V) (Favretto and Flamini, 2000; Pati *et al.*, 2006).

Table 3.10 Fragmentation of M^+ ions of anthocyanins present in the Clinton extract (chromatogram Figure 3.15) by direct infusion ESI multiple step mass spectrometry (MS^n). (Reprinted from *American Journal of Enology and Viticulture* **51**, Favretto and Flamini, Copyright 2000).

Compound (#)	M^+	$[M\text{-}C_6H_{10}O_5]^+$		
Malvidin-3-O-monogluc. (**9**)	493	331		
Petunidin-3-O-monogluc. (**7**)	479	317		
Delphinidin-3-O-monogluc. (**3B**)	465	303		
Peonidin-3-O-monogluc. (**8**)	463	301		
Cyanidin-3-O-monogluc. (**6**)	449	287		
	M^+	$[M\text{-}C_8H_{12}O_6]^+$		
Malvidin-3-O-(6-O-acetyl) monogluc. (**16**)	535	331		
Petunidin-3-O-(6-O-acetyl) monogluc. (**14A**)	521	317		
Delphinidin-3-O-(6-O-acetyl) monogluc. (**10**)	507	303		
Peonidin-3-O-(6-O-acetyl) monogluc. (**15**)	505	301		
Cyanidin-3-O-(6-O-acetyl) monogluc. (**12B**)	491	287		
	M^+	$[M\text{-}C_{15}H_{16}O_7]^+$		
Malvidin-3-O-(6-O-*p*-coumaroyl) monogluc. (**22**)	639	331		
Petunidin-3-O-(6-O-*p*-cou maroyl) monogluc. (**20**)	625	317		
Delphinidin-3-O-(6-O-*p*-cou maroyl) monogluc. (**17**)	611	303		
Peonidin-3-O-(6-O-*p*-cou maroyl) monogluc. (**21**)	609	301		
Cyanidin-3-O-(6-O-*p*-cou maroyl) monoglu. (**19**)	595	287		
	M^+	$[M\text{-}C_6H_{10}O_5]^+$	$[M\text{-}2(C_8H_{10}O_5)]^+$	
Malvidin-3,5-O-diglucoside (**5**)	655	493	331	
Petunidin-3,5-O-diglucoside (**3A**)	641	479	317	
Delphinidin-3,5-O-diglucoside (**1**)	627	465	303	
Peonidin-3,5-O-diglucoside (**4**)	625	463	301	
Cyanidin-3,5-O-diglucoside (**2**)	611	449	287	
	M^+	$[M\text{-}C_6H_{10}O_5]^+$	$[M\text{-}C_{15}H_{18}O_7]^+$	$[M\text{-}C_{18}H_{16}O_7\text{-}C_6H_{10}O_5]^+$
Malvidin-3-(6-O-*p*-coumaroyl)-5-O-diglucoside (**14B**)	801	639	493	331
Petunidin-3-(6-O-*p*-coumaroyl)-5-O-diglucoside (**13**)	787	625	479	317
Delphinidin-3-(6-O-*p*-coumaroyl)-5-O-diglucoside (**11**)	773	611	465	303
Cyanidin-3-(6-O-*p*-coumaroyl)5-O-diglucoside (**12A**)	757	595	449	287
	M^+	$[M\text{-}C_{15}H_{16}O_8]^+$		
Malvidin-3-O-(6-O-caffeoyl) monogluc. (**18**)	655	331		

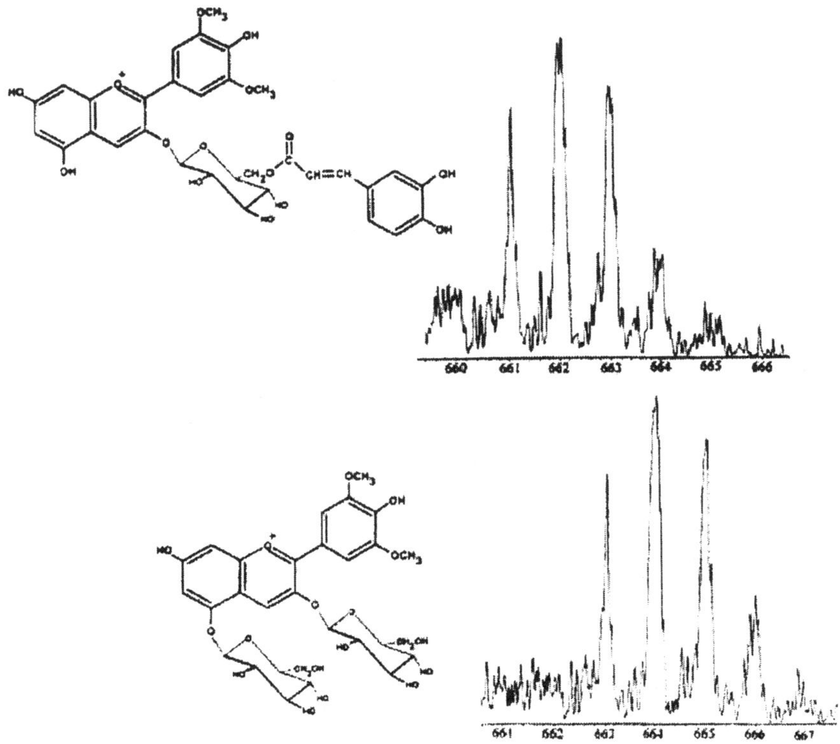

Figure 3.17 Differentiation of two isobaric compounds with the same aglycone moiety by deuterium exchange experiment: positive-ion ESI mass spectrum of Mv-3-O(6-O-caffeoyl)monoglucoside (above) and of Mv-3,5-O-diglucoside (below) dissolved in D_2O. (Reprinted from *American Journal of Enology and Viticulture* 51, Favretto and Flamini, Copyright 2000)

Several oligomeric anthocyanins and Mv-4-vinyl-polycatechins were identified in grape skin extracts (Table 3.11) (Asenstorfer *et al.*, 2001; Vidal *et al.*, 2004). An example of identification of two dimeric anthocyanins by direct infusion ESI-MS/MS analysis is reported in Figure 3.18.

3.3.2. Analysis of Anthocyanin-Derivatives in Wine

LC/MS analysis of anthocyanins and their derivatives in wine can be performed by direct injection of sample without a prior sample preparation. Several methods using different chromatographic conditions have been proposed, four of which are reported in Table 3.12.

Table 3.11 Oligomeric anthocyanins identified in Shiraz grape skin. F, fragment ion; M molecular ion. Dp, delphinidin; Cy, cyanidin; Pt, petunidin; Pn, peonidin; Mv, malvidin; G, glucose, pCG, *p*-coumaroyl glucoside (Vidal *et al.*, 2004).

m/z	mass assignment	*m/z*	mass assignment	*m/z*	mass assignment
287^{F}	Cy	795^{F}	MvDp + G	1315^{F}	MvMvMv + 2G
301^{F}	Pn	809^{F}	MvPt + G	1417^{F}	MvMvCy + G · pCG
303^{F}	Dp	823^{F}	MvMv + G	1431^{F}	MvMvPn + G · pCG
317^{F}	Pt	941^{M}	MvCy + 2G	1433^{F}	MvMvDp + G · pCG
331^{F}	Mv	955^{M}	MvPn + 2G	1433^{M}	MvMvCy + 3G
449^{M}	Cy + G	957^{M}	MvDp + 2G	1447^{F}	MvMvPt + G · pCG
463^{M}	Pn + G	971^{M}	MvPt + 2G	1447^{M}	MvMvPn + 3G
465^{M}	Dp + G	985^{M}	MvMv + 2G	1449^{M}	MvMvDp + 3G
479^{M}	Pt + G	1087^{M}	MvCy + G · pCG	1461^{F}	MvMvMv + G · pCG
493^{M}	Mv + G	1101^{M}	MvPn + G · pCG	1463^{M}	MvMvPt + 3G
617^{F}	MvCy	1103^{M}	MvDp + G · pCG	1477^{M}	MvMvMv + 3G
631^{F}	MvPn	1117^{M}	MvPt + G · pCG	1579^{M}	MvMvCy + 2G · pCG
633^{F}	MvDp	1131^{M}	MvMv + G · pCG	1593^{M}	MvMvPn + 2G · pCG
647^{F}	MvPt	1271^{F}	MvMvCy + 2G	1595^{M}	MvMvDp + 2G · pCG
661^{F}	MvMv	1285^{F}	MvMvPn + 2G	1609^{M}	MvMvPt + 2G · pCG
779^{F}	MvCy + G	1287^{F}	MvMvDp + 2G	1623^{M}	MvMvMv + 2G · pCG
793^{F}	MvPn + G	1301^{F}	MvMvPt + 2G		

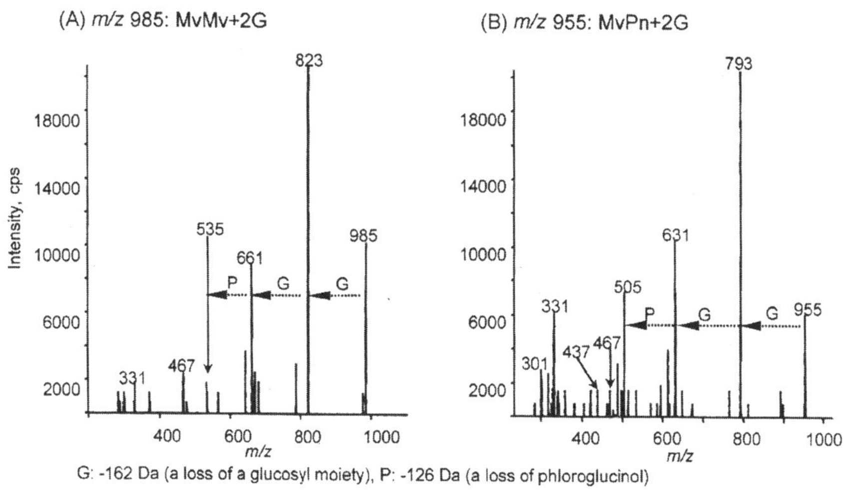

Figure 3.18 Direct-infusion ESI-MS/MS product ion spectra of anthocyanin dimers composed of Mv-glucoside with (A) Mv-3-glucoside (*m/z* 985, MvMv + 2G) and (B) Pn-3-glucoside (*m/z* 955, MvPn + 2G) identified in grape skin (ESI needle, orifice, and ring potentials at 5000, 150, and 250 V, respectively; collision gas N_2; collision energy 30–60 V). (Reproduced from *J. Agric. Food Chem.*, 2004, 52, 7144–7151, Vidal *et al.*, with permission of American Chemical Society)

Table 3.12 Chromatographic conditions of four different methods used for LC/ESI-MS analysis of anthocyanins and anthocyanin derivatives in wine by direct injection of the sample.

	Program A (Monagas *et al.*, 2003)	Program B (Košir *et al.*, 2004)	Program C (Salas *et al.*, 2004)	Program D (Pati *et al.*, 2006)
Column	C_{18}(150 × 3.9 mm, 4 μm)	C_{18} (250 × 4.6 mm, 5 μm)	C_{18}(250 × 2 mm, 5 μm)	C_{18}(150 × 2 mm, 5 μm)
Temperature	room temperature	30 °C	30 °C	room temperature
Binary solvent	(A) H_2O/formic acid 90:10 (v/v) (B) H_2O/methanol/formic acid 45:45:10 (v/v/v)	(A) 0.3% $HClO_4$ in H_2O (B) 96% ethanol in H_2O	(A) H_2O/formic acid 95:5 (v/v) (B) acetonitrile/solvent A 80:20 (v/v)	(A) H_2O/acetonitrile 95:5 (v/v) +0.1% (v/v) TFA (B) H_2O/acetonitrile 10:90 (v/v) +0.1% (v/v) TFA
Flow rate	0.8 mL · min^{-1}	0.8 mL · min^{-1}	0.25 mL · min^{-1}	0.2 mL · min^{-1}
Gradient program	15–80% B, 0–30 min; isocratic 80% B, 30–43 min	71.5 – 54.5% B, 0–32 min; 54.5 – 31.5 B, 32–47 min; 31.5–100% B, 47–49 min; isocratic 100% B, 49–52 min	isocratic 2% B, 0–7 min; 2–20% B, 7–22 min; 20–30% B, 22–30 min; 30–40% B, 30–40 min; 40–50% B; 40–45 min; 50–80% B, 45–50 min	isocratic 2% B, 0–2 min; 2–10% B, 2–8 min; 10–13% B; 8–30 min; 13–20% B, 30–50 min; 20–30% B, 50–75 min

An example of compounds identified in three different wines by performing analysis with program A is reported in Table 3.13. In program D, the TFA percentage of mobile phase is kept low to limit the formation of ionic pairs that may decrease the detection sensibility of MS. In Table 3.14, compounds detected in a Primitivo wine sample by performing analysis with program D and characteristic fragment ions recorded in the MS/MS and MS^3 experiments, are reported.

A previous purification of wine sample can be performed by reverse-phase SPE. A volume of 5 mL of wine is added of 15 mL of water and passed through a C_{18} cartridge previously activated by passage of methanol and water. After washing the cartridge with 6 mL of 0.3% formic acid aqueous solution and 4 mL of water, anthocyanin compounds are recovered with 5 mL of methanol. The solution is dried and the residue re-dissolved in the LC mobile phase (Košir *et al.*, 2004).

A method for isolation and fractionation of oligomeric pigments by using cation-exchange chromatography in the presence of bisulfite buffer, was proposed (Asenstorfer *et al.*, 2001). Bisulfite excess is used for the resistance of the 4-substituted anthocyanins to form anionic bisulfite adduct allowing separation of these compounds from the others. The procedure is shown in the flow-diagram in Figure 3.19.

Before performing the sample preparation wine is concentrated under vacuum to remove ethanol. The methanolic solution from the C_{18} 50 g column (**B**) is concentrated under vacuum, the volume is adjusted to 200 mL by water addition and approximately 50 mL of the resulting solution is loaded onto the sulfoxyethyl cellulose 40 × 200 mm column. Vitisin is present in the neutral/anionic fraction i) eluted with 2 L of 10% methanol. Fraction ii) is eluted with a solution of 2 M NaCl in 50% (v/v) aqueous methanol; non-acidified methanol is used to avoid acid hydrolysis of pigments. To remove NaCl the solution is passed through a C_{18} column and the stationary phase is washed with water. Fraction iii), eluted from the column with the 0.1 M potassium metabisulfite solution, contains for the most part anthocyanins existing primarily as their anionic bisulfite addition products. The pigments retained on the column are eluted with a NaCl 2 M in 50% methanol solution (fraction iv). This fraction is further purified by preparative TLC using a silica gel plate and 70% (v/v) aqueous propanol as eluent. The separated red band (Rf 0.8) is extracted with a 10% (v/v) aqueous methanol solution.

Analyses can be performed either by direct injection of the sample into the mass spectrometer or by performing LC separation. ESI/MS analysis

Table 3.13 Anthocyanins and their derivatives identified by LC/ESI-MS in Graciano, Tempranillo and Cabernet Sauvignon wines by performing analysis with the program A in Table 3.12. ESI-MS parameters: positive ion mode; drying gas N_2; temperature 350 °C; nebulizer pressure 380 Pa (55 psi); capillary voltage 4 kV; fragmentator voltage: 100 V from 0 to 17 min, 120 V from 17 to 55 min (Monagas *et al.*, 2003).

RT (min)	λ_{max} (nm)	$[M]^+$ (*m/z*)	Fragments (*m/z*)	Compound
4.5	530	781		Malvidin-3-O-glucoside-(epi)catechin
6.9	524	465	303	Delphinidin-3-O-glucoside
8.6	515	449	287	Cyanidin-3-O-glucoside
9.8	526	479	317	Petunidin-3-O-glucoside
11.2	516	463	301	Peonidin-3-O-glucoside
12.0	520	493	331	Malvidin-3-O-glucoside
13.4	509	531	369	Peonidin-3-O-glucoside pyruvate
14.0	533	507	303	Delphinidin-3-(6-O-acetylglucoside)
14.3	513	561	399	Malvidin-3-O-glucoside pyruvate
15.2	518	603	399	Malvidin-3-(6-O-acetylglucoside) pyruvate
15.8	516	491	287	Cyanidin-3-(6-O-acetylglucoside)
16.1	543	809		Malvidin-3-O-glucoside-8-ethyl-(epi)catechin
16.2	532	521	317	Petunidin-3-(6-O-acetylglucoside)
18.0	513	707	399	Malvidin-3-(6-O-p-coumaroylglucoside) pyruvate
18.7	520	505	301	Peonidin-3-(6-O-acetylglucoside)
19.0	532	611	303	Delphinidin-3-(6-O-p-coumaroylglucoside)
19.3	530	535	331	Malvidin-3-(6-O-acetylglucoside)
20.1	524	625	301	Peonidin-3-(6-O-caffeoylglucoside)
20.6	536	655	331	Malvidin-3-(6-O-caffeoylglucoside)
21.1	527	595	287	Cyanidin-3-(6-O-p-coumaroylglucoside)
21.2	537	639	331	Malvidin-3-(6-O-p-coumaroylglucoside)-cis isomer
21.9	532	625	317	Petunidin-3-(6-O-p-coumaroylglucoside)
22.2	503	805		Malvidin-3-O-glucoside-4-vinyl-catechin
22.3	540	955		Malvidin-3-(6-O-p-coumaroylglucoside)-8-ethyl-(epi)catechin
23.6	508	847		Malvidin-3-(6-O-acetylglucoside)-4-vinyl-catechin
24.1	524	609	301	Peonidin-3-(6-O-p-coumaroylglucoside)
24.4	535	639	331	Malvidin-3-(6-O-p-coumaroylglucoside)-trans isomer
25.3	514	625	463	Malvidin-3-O-glucoside-4-vinylcatechol
26.0	514	847		Malvidin-3-(6-O-acetylglucoside)-4-vinyl-epicatechin
26.3	508	805		Malvidin-3-O-glucoside-4-vinyl-epicatechin
27.8	504	609	447	Malvidin-3-O-glucoside-4-vinylphenol
28.6	504	639	477	Malvidin-3-O-glucoside-4-vinylguaiacol
29.7	509	651	447	Malvidin-3-(6-O-acetylglucoside)-4-vinylphenol
34.7	504	755	447	Malvidin-3-(6-O-p-coumaroylglucoside)-4-vinylphenol

Table 3.14 Anthocyanins and their derivatives identified in a Primitivo wine by LC/ESI-MS analysis performing program D in Table 3.12. The characteristic fragment ions from the MS/MS and MS^3 experiments of the most intense *m/z* signals of mass spectra, are reported. MS conditions: positive ion mode; spray voltage 4.5 kV; sheath gas N_2 0.9 L/min; capillary voltage 35 V; capillary temperature 200 °C; tube lens offset voltage, 15 V (Pati *et al.*, 2006).

RT (min)	*m/z* of the base peak	*m/z* of main MS^2 fragments	*m/z* of main MS^3 fragments	Compound
16.8	751	589	571,437,463	(epi)catechin-peonidin-3-O-glucoside
18.3	781	619	601,493,467	(epi)catechin-malvidin-3-O-glucoside
20.0	465	303	/	Delphinidin-3-O-glucoside
20.3	1069	907,781,619	/	di(epi)catechin-malvidin-3-O-glucoside
22.3	449	287	/	Cyanidin-3-O-glucoside
23.4	781	619	601,493,467	(epi)catechin-malvidin-3-O-glucoside
23.5	1069	907,781,619	/	di(epi)catechin-malvidin-3-O-glucoside
24.0	479	317	/	Petunidin-3-O-glucoside
26.7	463	301	/	Peonidin-3-O-glucoside
28.3	493	331	/	Malvidin-3-O-glucoside
29.5	561	399	/	Malvidin-3-O-glucoside pyruvate
33.5	517	355	/	Malvidin-3-O-glucoside acetaldehyde
38.4	809	647,519,357	/	Malvidin-3-O-glucoside-8-ethyl-(epi)catechin
40.7	809	647,519,357	/	Malvidin-3-O-glucoside-8-ethyl-(epi)catechin
43.5	1093	931	641	Malvidin-3-O-glucoside-4-vinyl-di(epi)catechin
		803	/	
44.4	809	647,519,357	/	Malvidin-3-O-glucoside-8-ethyl-(epi)catechin
45.5	1093	931	641	Malvidin-3-O-glucoside-4-vinyl-di(epi)catechin
		803	/	
46.4	809	647,519,357	/	Malvidin-3-O-glucoside-8-ethyl-(epi)catechin
47.3	927	619	601,493,467	Malvidin-3-(6-O-p-coumaroylglucoside)-(epi)catechin
48.7	707	399	/	Malvidin-3-(6-O-p-coumaroylglucoside) pyruvate
48.7	505	301	/	Peonidin-3-(6-O-acetylglucoside)
50.4	535	331	/	Malvidin-3-(6-O-acetylglucoside)
52.1	611	303	/	Delphinidin-3-(6-O-p-coumaroylglucoside)
54.3	847	643	491	Malvidin-3-(6-O-acetylglucoside)-4-vinyl-(epi)catechin

Table 3.14 (Continued).

RT (min)	*m/z* of the base peak	*m/z* of main MS^2 fragments	*m/z* of main MS^3 fragments	Compound
56.1	655	331	/	Malvidin-3-(6-O-caffeoylglucoside)
56.8	595	287	/	Cyanidin-3-(6-O-p-coumaroylglucoside)
58.3	625	317	/	Petunidin-3-(6-O-p-coumaroylglucoside)
58.5	639	331	/	Malvidin-3-(6-O-p-coumaroylglucoside)
59.4	1239	931	641	Malvidin-3-(6-O-p-coumaroylglucoside)-4-vinyl-di(epi)catechin
		641	/	
59.7	805	643	491	Malvidin-3-O-glucoside-4-vinyl-(epi)catechin
60.5	925	635,617,327	/	Peonidin-3-(6-O-p-coumaroylglucoside)-8-ethyl-(epi)catechin
61.5	955	665,357	/	Malvidin-3-(6-O-p-coumaroylglucoside)-8-ethyl-(epi)catechin
62.8	609	301	/	Peonidin-3-(6-O-p-coumaroylglucoside)
63.7	639	331	/	Malvidin-3-(6-O-p-coumaroylglucoside)
63.9	625	463	/	Malvidin-3-O-glucoside-4-vinylcatechol
65.5	805	643	491	Malvidin-3-O-glucoside-4-vinyl-(epi)catechin
65.7	951	643	491	Malvidin-3-(6-O-p-coumaroylglucoside)-4-vinyl-(epi)catechin
67.5	951	643	491	Malvidin-3-(6-O-p-coumaroylglucoside)-4-vinyl-(epi)catechin
67.9	609	447	/	Malvidin-3-O-glucoside-4-vinylphenol
69.8	639	477	/	Malvidin-3-O-glucoside-4-vinylguaiacol
71.2	955	665,357	/	Malvidin-3-(6-O-p-coumaroylglucoside)-8-ethyl-(epi)catechin

of fraction iv) from a red wine and a grape marc extract allowed determination of the anthocyanidin C-4 substituted compounds reported in Table 3.15.

In addition to this, a sample preparation method for LC/MS analysis of pyranoanthocyanidins and other anthocyanin-derivatives in wine was proposed and is reported in the flow-diagram Figure 3.20.

As shown in the figure, eluates 1, 2, 3 and 4 are polled and analysis is performed on the resulting solution (fraction A). By connecting the

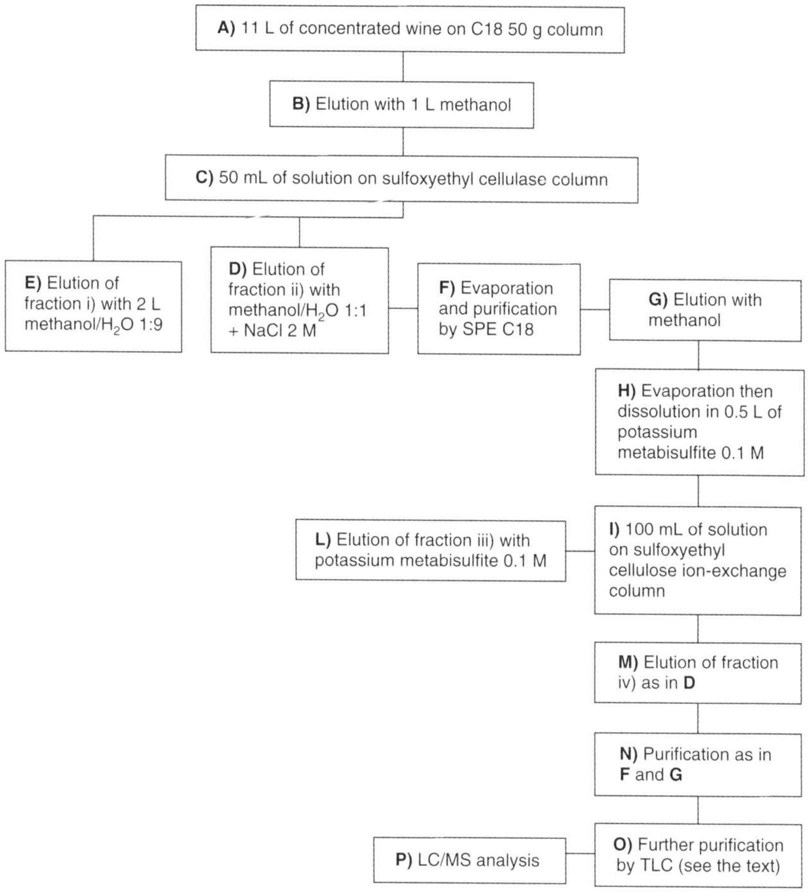

Figure 3.19 A method for fractionation of oligomeric pigments in wine (Asenstorfer *et al.*, 2001)

LC/ESI-MS system to the probe of the mass spectrometer *via* the UV cell outlet, the chromatograms in Figure 3.21 (relative to fraction A analysis of a wine at different aging stages), can be recorded.

A list of anthocyanins and derivatives identified in 4–23 months aged wines is reported in Table 3.16. As may be seen, a number of compounds are identified: simple anthocyanins from grape, ethyl-bridge derivatives, pyranoanthocyanins and pigments formed by anthocyanin-flavanol linkage. Some compounds are present in the wine in the first stages of ageing and disappear in time, others form during ageing.

MS/MS experiments are used to confirm identification of compounds, in particular anthocyanin-flavan-3-ols derivatives. Examples of

Table 3.15 ESI/MS spectral data of pigments isolated in fraction iv) (scheme of Figure 3.19) from Shiraz grape marc extract and wine. nd: not detected. Analytical conditions: ion source and orifice potentials 5.5 kV and 30 V respectively, positive ion mode. Curtain gas N_2 8 units; nebulizer gas air 10 units; injected solution 50% acetonitrile acidified with 2.5% acetic acid (rate 5 μL/min) (Asenstorfer *et al.*, 2001).

Marc $[M]^+$ (m/z)	Wine $[M]^+$ (m/z)	compound
609.4	609.4	pigment A
nd	639.4	3″-O-methyl- pigment A
651.4	651.4	(acetyl)pigment A
707.2	707.2	(*p*-coumaryl)vitisin A
nd	755.6	(*p*-coumaryl)pigment A
805.4	805.4	malvidin-3-glucose-4-vinyl-catechin
847.4	nd	malvidin-3-(acetyl)glucose-4-vinyl-catechin
951.4	951.4	malvidin-3-(p-coumaryl)glucose-4-vinyl-catechin
1093.4	1093.4	malvidin-3-glucose-4-vinyl-dicatechin
1135.4	nd	malvidin-3-(acetyl)glucose-4-vinyl-dicatechin
1239.6	nd	malvidin-3-(p-coumaryl)glucose-4-vinyl-dicatechin
1381.6	nd	malvidin-3-glucose-4-vinyl-tricatechin
1423.4	nd	malvidin-3-(acetyl)glucose-4-vinyl-tricatechin
1527.6	nd	malvidin-3-(p-coumaryl)glucose-4-vinyl-tricatechin
1669.4	nd	malvidin-3-glucose-4-vinyl-tetracatechin

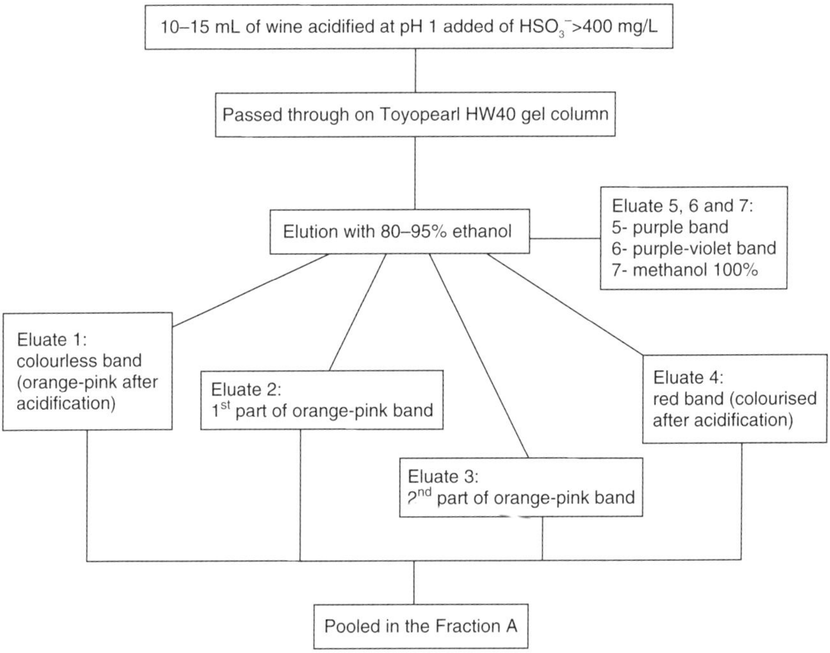

Figure 3.20 A method of sample preparation for analysis of pyranoanthocyanidins and anthocyanin-derivatives in wine (Alcade-Eon *et al.*, 2004)

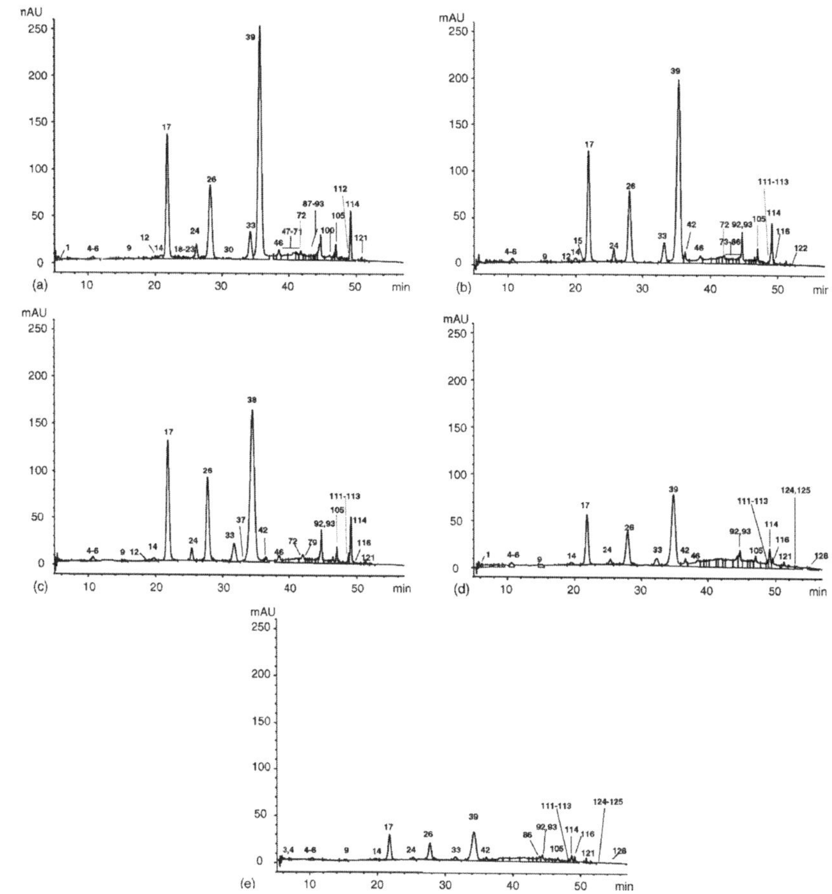

Figure 3.21 Chromatograms recorded at 520 nm of a wine sample aged (a) 4 months, (b) 8 months; (c) 13 months; (d) 16 months; (e) 23 months. Compounds corresponding to the numbered peaks are reported in Table 3.16. Chromatographic conditions: Column C_{18} (150×4.6 mm, 5 μm) at 35 °C; binary solvent A) aqueous solution TFA 0.1 % and B) acetonitrile; gradient program: isocratic 10% B for 5 min, from 10% to 15% B in 15 min, isocratic 15% B for 5 min, from 15% to 18% B in 5 min, from 18% to 35% B in 20 min (flow rate 0.5 mL/min). (Reprinted from *Analytica Chimica Acta* **563**, Alcade-Eon *et al.*, Changes in the detailed pigment composition of red wine during maturity and ageing. A comprehensive study, p. 240, Copyright 2006, with permission from Elsevier)

fragmentation spectra recorded for (epi)catechin-Mv-3-glu (M^+ at m/z 781 with fragmentation scheme), (epi)catechin-Pn-3-glu (M^+ at m/z 751), Mv-3-glu-8-ethyl-(epi)catechin (M^+ at m/z 809) and Mv-3-glu-8-vinyl(epi)catechin (M^+ at m/z 805), are reported in Figure 3.22.

Table 3.16 Anthocyanins and derivatives identified in 4–23 months old wines. Dp, delphinidin; Cy, cyanidin; Pt, petunidin; Pn, peonidin; Mv, malvidin; glc, glucose; p-coumglc, *p*-coumaroylglucoside; cafglc, caffeoylglucoside; acetylglc, acetylglucoside; catechin; GC, gallocatechin; EC, epicatechin; ECG, epigallocatechin; (*) detected; n.d., not detected. Chromatographic conditions are reported in the caption of Figure 3.21. MS conditions: positive ion mode; sheath gas and auxiliary gas is a mixture of N_2 and He; sheath gas flow 1.2 L/min; auxiliary gas flow 6 L/min; capillary voltage 4 V; capillary temperature 195 °C; normalized energy of collision 45% (Alcade-Eon *et al.*, 2006).

peak number Figure 3.21	RT (min)	$[M]^+$ (*m/z*)	MS^2 fragments (*m/z*)	MS^3 fragments (*m/z*)	λ_{max} (nm)	Compound	aged wine (months)				
							4	8	13	16	23
17	21.7	465	303	303	277, 342, 524	Dp-3-glc	*	*	*	*	*
24	26.1	449	287	287	279, 516	Cy-3-glc	*	*	*	*	*
26	28.1	479	317	317	277, 347, 525	Pt-3-glc	*	*	*	*	*
33	34.1	463	301	301	280, 517	Pn-3-glc	*	*	*	*	*
39	35.5	493	331	331	277, 348, 527	Mv-3-glc	*	*	*	*	*
46	38.3	507	303	303	276, 346, 527	Dp-3-acetylglc	*	*	*	*	*
63	41	491	287	287	280, 523	Cy-3-acetylglc	*	*	*	*	*
72	41.6	521	317	317	270, 529	Pt-3-acetylglc	*	*	*	*	*
87	43.6	505	301	301	280, 522	Pn-3-acetylglc	*	*	*	*	*
92	44.3	535	331	331	278, 350, 530	Mv-3-acetylglc	*	*	*	*	*
84	43.1	611	303	303	280, 301, 534	Dp-3-p-coumglc *cis*	*	*	*	*	*
93	44.3	611	303	303	282, 313, 531	Dp-3-p-coumglc *trans*	*	*	*	*	*
96	45.1	595	287	287	280, 301, 533	Cy-3-p-coumglc *cis*	*	*	*	*	*
103	46.3	595	287	287	284, 314, 524	Cy-3-p-coumglc *trans*	*	*	*	*	*
97	45.3	625	317	317	281, 301, 536	Pt-3-p-coumglc *cis*	*	*	*	*	*
105	46.6	625	317	317	282, 313, 532	Pt-3-p-coumglc *trans*	*	*	*	*	*
107	47.5	609	301	301	283, 300, 535	Pn-3-p-coumglc *cis*	*	*	*	*	*
112	48.6	609	301	301	283, 313, 526	Pn-3-p-coumglc *trans*	*	*	*	*	*
108	47.5	639	331	331	280, 301, 535	Mv-3-p-coumglc *cis*	*	*	*	*	*
114	48.7	639	331	331	282, 313, 532	Mv-3-p-coumglc *trans*	*	*	*	*	*
68	41.1	627	303	303	283, 331, 532	Dp-3-cafglc *trans*	*	n.d.	n.d.	n.d.	n.d.
88	43.6	641	317	317	283, 328, 531	Pt-3-cafglc *trans*	*	*	n.d.	n.d.	n.d.
99	45.6	625	301	301	283, 328, 525	Pn-3-cafglc *trans*	*	n.d.	n.d.	n.d.	n.d.
94	44.8	655	331	331		Mv-3-cafglc *cis*	*	*	n.d.	n.d.	n.d.
100	45.7	655	331	331	282, 328, 534	Mv-3-cafglc *trans*	*	*	*	*	*

10	16.7	627	303	303	279, 523	Dp-3,7-diglc	*	n.d.	n.d.	n.d.	n.d.
13	20.4	641	317	317	275, 521	Pt-3,5-diglc	*	n.d.	n.d.	n.d.	n.d.
23	24.6	641	317	317	275, 349, 522	Pt-3,7-diglc	*	*	*	*	*
28	28.7	625	301	301		Pn-3,7-diglc	*	*	n.d.	n.d.	n.d.
19	23.5	655	331	331	275, 524	Mv-3,5-diglc	*	*	n.d.	n.d.	n.d.
30	30.8	655	331	331	278, 350, 526	Mv-3,7-diglc	*	*	*	*	*
34	34.7	537				Dp-3-glc + L(+)lactic acid	*	*	*	*	*
44	37.5	551				Pt-3-glc + D(−)lactic acid	*	*	*	*	n.d.
52	39.2	551	317	317	278, 526	Pt-3-glc + L(+)lactic acid	*	*	*	*	*
57	40.5	535				Pn-3-glc + D(−)lactic acid	*	*	n.d.	n.d.	n.d.
73	41.7	535	301	301	281, 525	Pn-3-glc + L(+)lactic acid	*	*	*	*	*
61	40.8	565	331	331	278, 350, 530	Mv-3-glc + D(−)lactic acid	*	*	*	*	*
78	42.1	565	331	331	278, 348, 531	Mv-3-glc + L(+)lactic acid	*	*	*	*	*
1	5.7	769	607	439	531	Dp-3-glc-GC	*	*	*	*	*
2	7.1	753	591	453	282, 524	Cy-3-glc-GC	*	*	*	*	*
3	7.2	783	621	453	279, 532	Pt-3-glc-GC	*	*	*	*	*
5	10.8	767	605	437		Pn-3-glc-GC	*	*	*	*	*
4	10.6	797	635	467	281, 531	Mv-3-glc-GC	*	*	*	*	*
18	22.3	797	635	467		Mv-3-glc-EGC	*	*	*	*	*
21	24.4	839				Mv-3-acetylglc-GC	n.d.	*	*	*	n.d.
31	30.9	915	607	439		Dp-3-p-coumglc-GC	n.d.	*	*	*	n.d.
37	35.1	899				Cy-3-p-coumglc-GC	n.d.	*	*	n.d.	n.d.
38	35.4	929				Pt-3-p-coumglc-GC	n.d.	*	*	*	n.d.
48	38.5	913	605	437		Pn-3-p-coumglc-GC	n.d.	*	*	*	*
45	38.3	943	635	467		Mv-3-p-coumglc-GC	n.d.	*	*	*	*
66	41.1	943				Mv-3-p-coumglc-EGC	n.d.	*	*	*	*
6	10.8	753	591	439	282, 534	Dp-3-glc-C	*	*	*	*	*
7	14.8	753	591	439		Dp-3-glc-EC	*	*	*	*	*
8	14.9	737	575	423	286, 526	Cy-3-glc-C	*	*	*	*	*
11	18	737	575	423		Cy-3-glc-EC	*	*	*	*	*
9	16.2	767	605	453	279, 532	Pt-3-glc-C	*	*	*	*	*
16	21.6	767	605	453		Pt-3-glc-EC	*	*	*	*	*
12	20.3	751	589	437	283, 524	Pn-3-glc-C	*	*	*	*	*
20	24.3	751	589	437		Pn-3-glc-EC	*	*	*	*	*
14	21	781	619	467	280, 532	Mv-3-glc-C	*	*	*	*	*

Table 3.16 (Continued).

peak number Figure 3.21	RT (min)	$[M]^+$ (m/z)	MS^2 fragments (m/z)	MS^3 fragments (m/z)	λ_{max} (nm)	Compound	aged wine (months)				
							4	8	13	16	23
29	29.9	781	619	467	279, 533	Mv-3-glc-EC	*	*	*	*	*
41	35.9	823	619	467		Mv-3-acetylglc-C	n.d.	*	*	*	*
51	39	899	591	439		Dp-3-p-coumglc-C	n.d.	*	*	*	*
58	40.5	899				Dp-3-p-coumglc-EC	n.d.	*	n.d.	n.d.	n.d.
54	39.5	883				Cy-3-p-coumglc-C	n.d.	*	*	*	n.d.
70	41.4	883				Cy-3-p-coumglc-EC	n.d.	*	n.d.	n.d.	n.d.
64	41	913	605	453		Pt-3-p-coumglc-C	n.d.	*	*	*	*
81	42.6	913				Pt-3-p-coumglc-EC	n.d.	*	n.d.	n.d.	n.d.
75	41.8	897	589	437		Pn-3-p-coumglc-C	n.d.	*	*	*	*
90	43.8	897				Pn-3-p-coumglc-EC	n.d.	*	n.d.	n.d.	n.d.
85	43.4	927	619	467	290, 538	Mv-3-p-coumglc-C	*	*	*	*	*
102	46	927				Mv-3-p-coumglc-EC	n.d.	*	n.d.	n.d.	n.d.
40	35.8	781	329	329		Dp-3-glc-ethyl-C	*	n.d.	n.d.	*	n.d.
43	36.7	781				Dp-3-glc-ethyl-EC	*	n.d.	n.d.	*	n.d.
53	39.5	765				Cy-3-glc-ethyl-C	*	n.d.	n.d.	n.d.	n.d.
56	39.6	795	343	343		Pt-3-glc-ethyl-C	*	n.d.	n.d.	*	*
76	42	779	327	327		Pn-3-glc-ethyl-C	*	*	n.d.	*	*
82	43.1	779				Pn-3-glc-ethyl-EC	*	n.d.	n.d.	*	*
65	41.1	809	357	357		Mv-3-glc-ethyl-C	*	*	*	*	*
79	42.2	809	357	357	282, 539	Mv-3-glc-ethyl-C	*	*	*	*	*
83	43.1	809	357	357	276, 537	Mv-3-glc-ethyl-C	*	*	*	*	*
101	45.8	851	357	357		Mv-3-acetylglc-ethyl-C	*	*	n.d.	n.d.	n.d.
106	47.4	955	357	357		Mv-3-p-coumglc-ethyl-C	*	*	*	*	n.d.
35	34.7	797	329	329		Dp-3-glc-ethyl-GC	*	n.d.	n.d.	n.d.	*
49	33.6	781				Cy-3-glc-ethyl-GC	*	n.d.	n.d.	n.d.	n.d.
50	38.7	811	343	343		Pt-3-glc-ethyl-GC	*	n.d.	n.d.	*	*
62	40.9	795				Pn-3-glc-ethyl-GC	*	n.d.	n.d.	n.d.	n.d.
59	40.6	825	357	357		Mv-3-glc-ethyl-GC	*	n.d.	*	*	*
69	41.1	825	357	357	539	Mv-3-glc-ethyl-GC	*	*	*	*	*
74	41.8	825	357	357		Mv-3-glc-ethyl-GC	*	n.d.	*	*	*
95	45	867				Mv-3-acetylglc-ethyl-GC	*	n.d.	n.d.	n.d.	n.d.

15	21	533	371	371	297, 368, 507	A-type vitisin of Dp-3-glc	n.d.	*	*	*	*
25	27	517				A-type vitisin of Cy-3-glc	n.d.	n.d.	n.d.	*	n.d.
27	28.7	547	385	385	299, 371, 508	A-type vitisin of Pt-3-glc	n.d.	*	*	*	*
36	35	531	369	369	503	A-type vitisin of Pn-3-glc	n.d.	n.d.	n.d.	*	*
42	36	561	399	399	299, 372, 510	Vitisin A	n.d.	*	*	*	*
60	40.7	693				A-type vitisin of Pt-3-p-coumglc	n.d.	n.d.	*	*	n.d.
89	43.8	677	369	369	284, 508	A-type vitisin of Pn-3-p-coumglc	n.d.	*	*	*	*
91	44.1	707	399	399	271, 514	A-type vitisin of Mv-3-p-coumglc	n.d.	*	*	*	*
22	24.4	489	327	327		B-type vitisin of Dp-3-glc	n.d.	*	*	*	*
32	33.5	503	341	341	492	B-type vitisin of Pt-3-glc	n.d.	*	*	*	*
47	38.5	487	325	325		B-type vitisin of Pn-3-glc	*	*	*	*	n.d.
55	39.5	517	355	355	294, 358, 490	Vitisin B	*	*	*	*	*
71	41.4	529	325	325		B-type vitisin of Pn-3-acetylglc	*	*	n.d.	n.d.	n.d.
80	42.4	559	355	355	298, 361, 494	B-type vitisin of Mv-3-acetylglc	*	*	*	*	n.d.
67	41.1	501	339	339	475	Acetone derivative of Pn-3-glc	n.d.	*	*	*	n.d.
77	42.1	531	369	369	480	Acetone derivative of Mv-3-glc	n.d.	*	*	*	*
98	45.5	581	419	419	264, 412, 503	Dp-3-glc 4-vinylphenol adduct	*	*	*	*	*
109	47.5	565				Cy-3-glc 4-vinylphenol adduct	n.d.	*	*	*	*
111	48.3	595	433	433	264, 413, 502	Pt-3-glc 4-vinylphenol adduct	*	*	*	*	*
119	50.5	579	417	417	278, 406, 500	Pn-3-glc 4-vinylphenol adduct	*	*	*	*	*
121	51	609	447	447	263, 412, 504	Mv-3-glc 4-vinylphenol adduct	*	*	*	*	*
125	53.2	651	447	447	298, 416, 505	Mv-3-acetylglc 4-vinylphenol adduct	*	*	*	*	*
117	49.9	727	419	419		Dp-3-p-coumglc 4-vinylphenol adduct	n.d.	n.d.	n.d.	*	*
123	52.4	741	433	433	314, 504	Pt-3-p-coumglc 4-vinylphenol adduct	n.d.	n.d.	n.d.	*	*

Table 3.16 (Continued).

peak number Figure 3.21	RT (min)	$[M]^+$ (m/z)	MS^2 fragments (m/z)	MS^3 fragments (m/z)	λ_{max} (nm)	Compound	aged wine (months)				
							4	8	13	16	23
127	54.6	725	417	417	314, 501 264, 313, 416,	Pn-3-p-coumglc 4-vinylphenol adduct	n.d.	n.d.	n.d.	*	*
128	55.2	755	447	447	505	Mv-3-p-coumglc 4-vinylphenol adduct	n.d.	*	*	*	*
86	43.5	597	435	435	509	Dp-3-glc 4-vinylcatechol adduct	*	*	*	*	*
104	46.5	611	449	449	510	Pt-3-glc 4-vinylcatechol adduct	*	*	*	*	*
113	48.6	595	433	433	506	Pn-3-glc 4-vinylcatechol adduct	*	*	*	*	*
116	49.2	625	463	463	510	Mv-3-glc 4-vinylcatechol adduct	*	*	*	*	*
120	50.9	667	463	463	513	Mv-3-acetylglc 4-vinylcatechol adduct	n.d.	n.d.	*	*	*
110	47.6	743	435	435		Dp-3-p-coumglc 4-vinylcatechol adduct	*	*	*	*	*
118	50.5	757	449	449		Pt-3-p-coumglc 4-vinylcatechol adduct	n.d.	*	*	*	*
124	53.2	771	463	463	312, 511	Mv-3-p-coumglc 4-vinylcatechol adduct	*	*	*	*	*
122	52	639	477	477	511	Mv-3-glc 4-vinylguaiacol adduct	*	*	*	*	*
126	54	681	477	477	514	Mv-3-acetylglc 4-vinylguaiacol adduct	n.d.	*	*	*	*
129	55.7	785	477	477	514	Mv-3-p-coumglc 4-vinylguaiacol adduct	n.d.	*	*	*	*
115	49	805				Mv-3-glc 4-vinylepi-catechin adduct	*	*	*	*	*

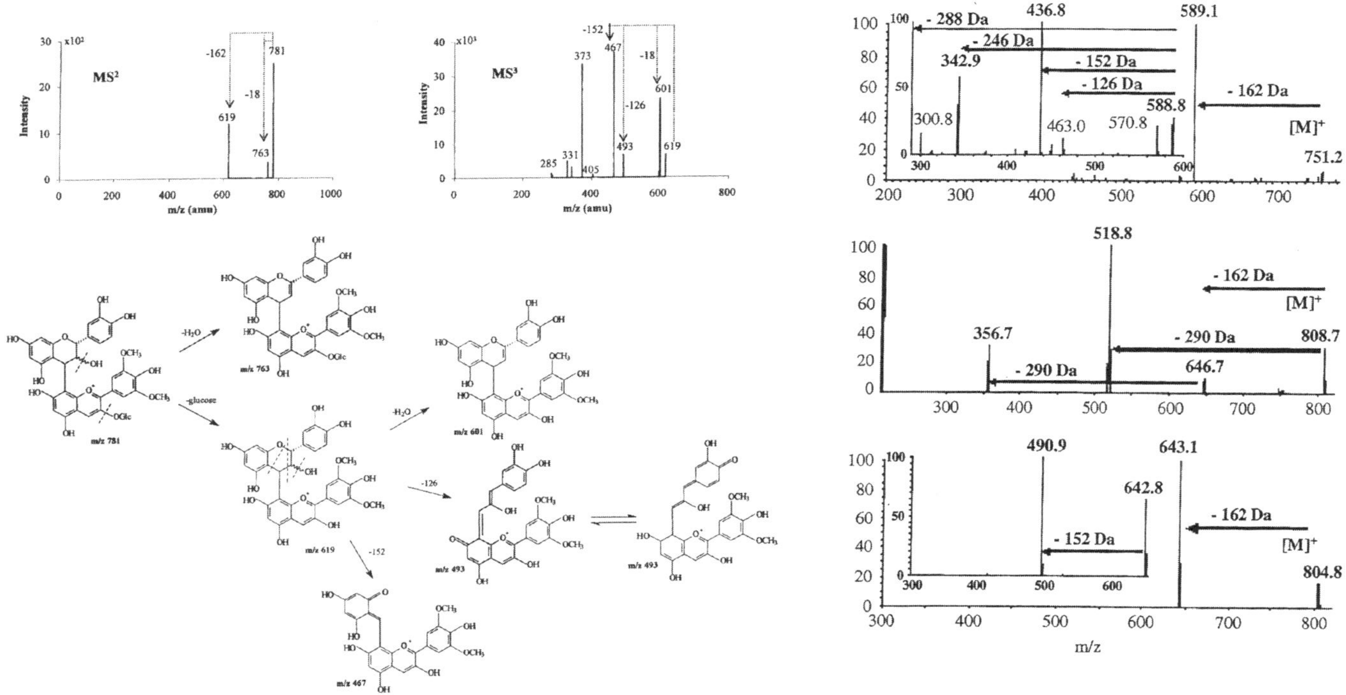

Figure 3.22 On the left: MS^2 and MS^3 fragmentation spectra of (epi)catechin-Mv-3-glu (M^+ at m/z 781) with a proposed fragmentation scheme. MS conditions: positive ion mode; source voltage 4.5 kV; capillary voltage 23.5 V; capillary temperature 250 °C; collision energy fragmentation 25% for MS^2, 30% for MS^3 and 35% for MS^4. (Reprinted from *Analytica Chimica Acta* **513**, Salas *et al.*, Demonstration of the occurrence of flavanols-anthocyanin adducts in wine and in model solutions, p. 328, Copyright 2004, with permission from Elsevier). On the right: fragmentation spectra of (epi)catechin-Pn-3-glu (M^+ at m/z 751), Mv-3-glu-8-ethyl-(epi)catechin (M^+ at m/z 809) and Mv-3-glu-8-vinyl(epi)catechin (M^+ at m/z 805). (Reprinted from *Journal of Mass Spectrometry* **41**, Pati *et al.*, Simultaneous separation and identification of oligomeric procyanidins and anthochyanidins-derived pigments in raw red wine by HPLC-UV-ESI-MS^n, p. 867, Copyright 2006, with permission from John Wiley & Sons Ltd)

References

Abian, J. (1999) The coupling of gas and liquid chromatography with mass spectrometry, *J. Mass Spectrom.*, **34**(3), 157–168.

Alcade-Eon, C., Escribano-Bailón, M.T., Santos-Buelga, C. and Rivas-Gonzalo, J.C. (2004) Separation of pyranoanthocyanins from red wine by column chromatography, *Anal. Chim. Acta*, **513**(1), 305–318.

Alcade-Eon, C., Escribano-Bailón, M.T., Santos-Buelga, C. and Rivas-Gonzalo, J.C. (2006) Changes in the detailed pigment composition of red wine during maturity and ageing. A comprehensive study, *Anal. Chim. Acta*, **563**(**1–2**), 238–254.

Asenstorfer, R.E., Hayasaka, Y. and Jones, G.P. (2001) Isolation and structures of oligomeric wine pigments by bisulfite-mediated ion-exchange chromatography, *J. Agric. Food Chem.*, **49**(**12**), 5957–5963.

Bravo, M.N., Silva, S., Coelho, A.V., Vilas Boas, L. and Bronze, M.R. (2006) Analysis of phenolic compounds in Muscatel wines produced in Portugal, *Anal. Chim. Acta*, **563**(**1–2**), 84–92.

Cheynier, V. and Rigaud, J. (1986) HPLC separation and characterization of flavonols in the skin of Vitis vinifera var. Cinsault, *Am. J. Enol. Vitic.*, **37**(**4**), 248–252.

Cheynier, V., Doco, T., Fulcrand, H., Guyot, S., Le Roux, E., Souquet, J.M., Rigaud, J. and Moutounet, M. (1997) ESI-MS analysis of polyphenolic oligomers and polymers, *Analusis Magazine*, **25**(**8**), 32–37.

de Freitas, V.A.P., Glories, Y., Bourgeois, G. and Virty, C. (1998) Characterisation of oligomeric and polymeric procyanidins from grape seeds by liquid secondary ion mass spectrometry, *Phytochem.*, **49**(5),1435–1441.

de Hoffmann, E. (1996) Tandem Mass Spectrometry: a Primer, *J. Mass. Spectrom.*, **31**(**2**), 125–137.

Favretto, D. and Flamini, R. (2000) Application of electrospray ionization mass spectrometry to the study of grape anthocyanins, *Am. J. Enol. Vitic.*, **51**(**1**), 55–64.

Flamini, R. (2003) Mass spectrometry in grape and wine chemistry. Part I: Polyphenols, *Mass. Spectrom. Rev.*, **22**(**4**), 218–250.

Flamini, R., Dalla Vedova, A., Cancian, D., Panighel, A. and De Rosso, M. (2007) GC/MS-Positive Ion Chemical Ionization and MS/MS study of volatile benzene compounds in five different woods used in barrel-making, *J. Mass Spectrom.*, **42**, 641–646.

Francia-Aricha, E.M., Guerra, M.T., Rivas-Gonzalo, J.C. and Santos-Buelga, C. (1997) New anthocyanin pigments formed after condensation with flavanols, *J. Agric. Food Chem.*, **45**(**6**), 2262–2266.

Fulcrand, H., Remy, S., Souquet, J.M., Cheynier, V. and Moutounet, M. (1999) Study of wine tannin oligomers by on-line liquid chromatography electrospray ionization mass spectrometry, *J. Agric. Food Chem.*, **47**(**3**), 1023–1028.

Gabetta, B., Fuzzati, N., Griffini, A., Lolla, E., Pace, R., Ruffilli, T. and Peterlongo, F. (2000) Characterization of proanthocyanidins from grape seeds, *Fitoterapia*, **71**(**2**), 162–175.

Gómez-Ariza, J.L., García-Barrera, T. and Lorenzo, F. (2006) Anthocyanins profile as fingerprint of wines using atmospheric pressure photoionisation coupled to quadrupole time-of-flight mass spectrometry, *Anal. Chim. Acta*, **570**(**1**), 101–108.

González-Manzano, S., Santos-Buelga, C., Pérez-Alonso, J.J. Rivas-Gonzalo, J.C. and Escribano-Bailón, M.T. (2006) Characterization of the mean degree of polymerization of proanthocyanidins in red wines using liquid chromatography-mass spectrometry (LC-MS), *J. Agric. Food Chem.*, **54**(**12**), 4326–4332.

Hayasaka, Y. and Asenstorfer, R.E. (2002) Screening for potential pigments derived from anthocyanins in red wine using nanoelectrospray tandem mass spectrometry, *J. Agric. Food Chem.*, **50**(**4**), 756–761.

Hayasaka, Y., Waters, E.J., Cheynier, V., Herderich, M.J. and Vidal, S. (2003) Characterization of proanthocyanidins in grape seeds using electrospray mass spectrometry, *Rapid Commun. Mass Sp.*, **17**(**1**), 9–16.

Hong, V. and Wrolstad, R.E. (1990) Characterization of anthocyanin-containing colorants and fruit juices by HPLC/photodiode array detection, *J. Agric. Food Chem.*, **38**(**3**), 698–708.

Košir, I.J., Lapornik, B., Andrenšek, S., Wondra, A.G. Vrhovšek, U. and Kidrič, J. (2004) Identification of anthocyanins in wines by liquid chromatography, liquid chromatography-mass spectrometry and nuclear magnetic resonance, *Anal. Chim. Acta*, **513**(**1**), 277–282.

Lazarus, S.A., Adamson, G.E., Hammerstone, J.F. and Schmitz, H.H. (1999) High-performance liquid chromatography/mass spectrometry analysis of proanthocyanidins in foods and beverages, *J. Agric. Food Chem.*, **47**(**9**), 3693–3701.

Lee, J.H., Johnson, J.V. and Talcott, S.T. (2005) Identification of ellagic acid conjugates and other polyphenolics in Muscadine grapes by HPLC-ESI-MS, *J. Agric. Food Chem.*, **53**(**15**), 6003–6010.

Monagas, M., Núñez, V., Bartolomé, B. and Gómez-Cordovés, C. (2003) Anthocyanin-derived pigments in Graciano, Tempranillo, and Cabernet Souvignon wines produced in Spain, *Am. J. Enol. Vitic.*, **54**(**3**), 163–169.

Monagas, M., Suárez, R., Gómez-Cordovés, C. and Bartolomé, B. (2005) Simultaneous determination of nonanthocyanin phenolic compounds in red wines by HPLC-DAD/ESI-MS, *Am. J. Enol. Vitic.*, **56**(**2**), 139–147.

Niessen, W.M.A. and Tinke, A.P. (1995) Liquid Chromatography-Mass Spectrometry. General principles and instrumentation, *J. Chromatogr. A*, **703**(**1–2**), 37–57.

Núñez, V., Gómez-Cordovés, C., Bartolomé, B. Hong, Y.J. and Mitchell, A.E. (2006) Non-galloylated and galloylated proanthocyanidin oligomers in grape seeds from Vitis vinifera, L. cv. Graciano, Tempranillo and Cabernet Sauvignon, *J. Sci. Food Agr.*, **86**(**6**), 915–921.

Pati, S., Losito, I., Gambacorta, G., La Notte, E., Palmisano, F. and Zambonin, P.G. (2006) Simultaneous separation and identification of oligomeric procyanidins and anthocyanin-derived pigments in raw red wine by HPLC-UV-ESI-MS^n, *J. Mass Spectrom.*, **41**(**7**), 861–871.

Pérez-Magariño, S., Revilla, I., Gonzáles-SanJosé, M.L. and Beltrán, S. (1999) Various applications of liquid chromatography-mass spectrometry to the analysis of phenolic compounds. *J. Chromatogr. A*, **847**(**1–2**), 75–81.

Revilla, E., Ryan, J.M. and Martín-Ortega, G. (1998) Comparison of several procedures used for the extraction of anthocyanins from red grapes, *J. Agric. Food Chem.*, **46**(**11**), 4592–4597.

Salas, E., Atanasova, V., Poncet-Legrand, C., Meudec, E., Mazauric, J.P. and Cheynier, V. (2004) Demonstration of the occurrence of flavanol–anthocyanin adducts in wine and in model solutions, *Anal. Chim. Acta*, **513**(**1**), 325–332.

Somers, T. (1971) The polymeric nature of wine pigments. *Phytochemistry*, **10**(**9**), 2175–2186.

Sun, B., Leandro, M.C. de Freitas, V. and Spranger, M.I. (2006) Fractionation of red wine polyphenols by solid phase extraction and liquid chromatography, *J. Chromatogr. A*, **1128(1–2)**, 27–38.

Timberlake, C.F. and Bridle, P. (1976) Interactions between anthocyanins, phenolic compounds, and acetaldehyde and their significance in red wines. *Am. J Enol. Vitic.*, **27(3)**, 97–105.

Vidal, S., Francis, L., Guyot, S., Marnet, N., Kwiatkowski, M., Gawel, R., Cheynier, V. and Waters, E.J. (2003) The mouth-feel properties of grape and apple proanthocyanidins in a wine-like medium, *J. Sci. Food Agr.*, **83(6)**, 564–573.

Vidal, S., Meudec, E., Cheynier, V., Skouroumounis, G. and Hayasaka, Y. (2004) Mass spectrometric evidence for the existence of oligomeric anthocyanins in grape skins, *J. Agric. Food Chem.*, **52(23)**, 7144–7151.

Wu, Q., Wang, M. and Simon, J.E. (2005) Determination of proanthocyanidins in fresh grapes and grape products using liquid chromatography with mass spectrometric detection, *Rapid Commun. Mass Sp.*, **19(14)**, 2062–2068.

4

Grape and Wine Contaminants: Ochratoxin A, Biogenic Amines, Trichloroanisole and Ethylphenols

Riccardo Flamini and Roberto Larcher

4.1. Introduction

The various contaminations in foods are due to the presence of compounds dangerous for consumer health or that affect organoleptic characteristics of the product. Ochratoxin A (OTA) and biogenic amines are dangerous for human health, and their legal limits are fixed in grape and wine. Due to their negative sensory properties and very low sensory thresholds, 2,4,6-trichloroanisole and ethylphenols have to be practically absent in wine, or present at as low a level as is possible.

Hyphenated Techniques in Grape and Wine Chemistry Edited by Riccardo Flamini

4.1.1. Ochratoxin A in Grape and Wine

The structure of ochratoxin A (R-N-[(5-chloro-3,4-dihydro-8, hydroxy-3-methyl-1-oxo-1H-2-benzopyran-7-yl)carbonyl]phenylalanine), constituted of an isocoumarin-derivative linked to phenylalanine through a carboxyl group, is shown in Figure 4.1. This compound is a secondary metabolite of fungi *Penicillium verrucosum* and *Aspergillus ochraceus*, *A. carbonarius* and *A. niger*. Development of these fungi is promoted through advantageous environmental conditions, in particular during storage of products. Toxicity of OTA is due to renal damage with nephrotoxic effect (Schwerdt *et al.*, 1999), interference with mitochondrial respiratory function and pH homeostasis (Sauvant *et al.*, 1998), and impaired organic anion transport (Eder *et al.*, 2000). Moreover, the toxin has an inhibition effect on tRNA-synthetase, accompanied by the reduced protein synthesis and enhanced lipid peroxidation *via* the generation of free radicals (Hohler, 1998). It can also induce the formation of renal tumours, with cytochrome P450-related reactions and DNA adduct generation as potential contributing mechanisms (Castegnaro *et al.*, 1998; Pfohl-Leszkowicz *et al.*, 1998). The fungi develop on the grape, in particular during the harvest, promoted by conditions of high humidity and high temperature. As a consequence, the toxin is present in raisins and grape, and it is transferred to the wine in the winemaking. Because the fungi develops on the grape surface, for red wines, made by grape skins maceration, and sweet wines, there is a higher risk of contamination.

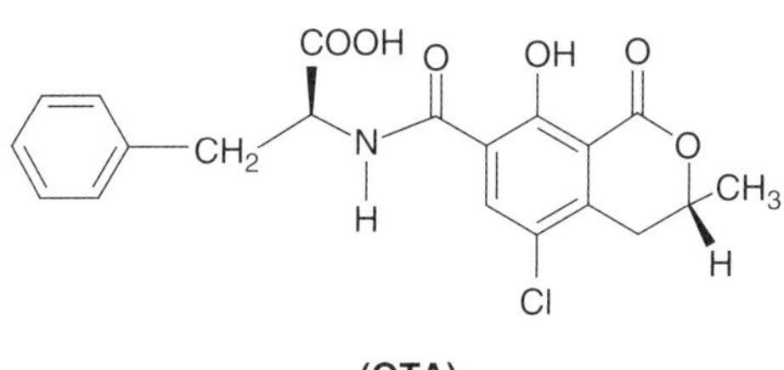

Figure 4.1 Structure of ochratoxin A

The World Health Organization (WHO) has set a provisional tolerable weekly intake level for OTA at 100 ng/Kg of body weight, taking its potentially carcinogenic effect into account (JEFCA, 2001). The tolerated maximum concentrations are based on a tolerable daily intake of 5 ng/(kg of body weight) suggested by the ED Scientific Committee on Food (Commission of the European Communities, 1998). CE

Regulation n° 123/2005 fixed the OTA legal maximum limit in grape, wine and grape juice at 0.002 ppm (Flamini and Panighel, 2006 and references cited herein).

4.1.2. Biogenic Amines in Grape and Wine

Biogenic amines (BAs), as spermine, spermidine and cadaverine, are synthesized in several parts of the vine, such as berries, seeds and leaves, with growth-regulating roles, while histamine, tyramine and 1-methylhistamine can be present just in trace amounts (Adams *et al.*, 1990; Radler and Fath, 1991; Geny *et al.*, 1997; Nicolini *et al.*, 2003). BAs were instead found in significant amounts in fermented foods and beverages, such as cheeses, beer and fish and meat products (Stratton *et al.*, 1991; Shalaby, 1996). They were also found in wine, the most abundant being histamine, tyramine, putrescine and phenylethylamine (Radler and Fath, 1991; Lehtonen, 1996).

BAs are organic bases with low molecular weight (MW) and high biological activity, their structures are reported in Figure 4.2.

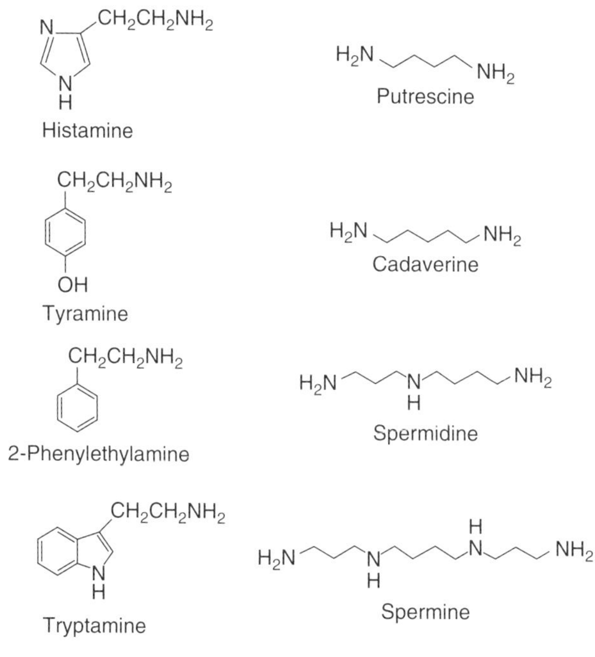

Figure 4.2 Structures of some biogenic amines

Shalaby (1996) reported they are toxic substances with deleterious effects on the human health. Histamine is perhaps the most important amine in terms of physiological effects, in small amounts it normally degrades without problems after uptake from food, but can nevertheless induce a drastic response in an increasing number of sensitive people. Symptoms are redness of the skin, headache, nausea, stomach disorder, and respiratory troubles. This pathology is known as 'Histamine Intolerance' and usually people affected by this disorder cannot convert amines such as histamine into harmless products. The simultaneous consumption of foods with high contents of BAs and alcoholic beverages should be avoided, because ethanol reduces the ability of the human detoxification system to degrade histamine by the diamine oxidase enzyme. Hypertensive crises have been observed in psychiatric patients treated with drugs inhibiting monoamine oxidase following the consumption of alcoholic beverages (Kalač and Křížek, 2003). Tyramine is generally associated with migraines and hypertension.

As the individual toxicity threshold of BAs has not been defined exactly, varying significantly between individuals, histamine levels above 500–1000 mg/kg must be considered of potential risk to human health (Taylor, 1985). In some European Countries tolerance levels for histamine in food and beverages have been defined. The maximum level of tolerance of histamine in wine has been established in Switzerland at 10 mg/L, in France at 8 mg/L, in Belgium at 5–6 mg/L, and in Germany at 2 mg/L, however the level for histamine-free wines should be lower than 0.5 mg/L (Bauza *et al.*, 1995; Lehtonen, 1996).

The presence of BAs in wines is considered as an evident indicator of amino decarboxylase activity by dangerous micro-organisms. It is believed that *lactobacilli* and *pediococci* are the main producers of amines, although some *oenococci* can also produce amines (Delfini, 1989; Farias *et al.*, 1993; Leitao *et al.*, 2000). The ability of wine lactic acid bacteria to decarboxylate histidine to histamine, and tyrosine to tyramine, has been demonstrated in detail (Lonvaud-Funel, 2001). The addition of bacteria starter cultures, followed by a direct microscopic control, makes it possible to effectively prevent BAs production. Some authors have reported that the concentration of aliphatic primary amines in wines produced from botrytized grapes is higher than in those made from intact grape berries (Kiss *et al.*, 2006). In particular, Eder and co-workers established that the content of 2-methyl-butyl amine, 3-methyl-butyl amine and phenylethylamine is higher in products from botrytized grapes (Eder *et al.*, 2002). Important decreases in the histamine content can be obtained by treating wines with bentonite, but this treatment does not affect the levels of putrescine, isoamylamine, phenylethylamine and tyramine (Mayer and Pause, 1985).

4.1.3. 2,4,6-Trichloroanisole in Wine

Due to their physical properties as an excellent seal for liquids, cork stoppers are the principal means of closure for bottled wines. It may occur that some compounds released from the wood used for making barrels for wine ageing or from cork stoppers, confer defects to the wine. The tainted corks may cause off-flavor or 'corkiness' in wine, and a global dollar loss of up to US$ 10 billion was estimated for bottled Champagne wines affected by this defect (Fuller, 1995). Due to its very low odour threshold (1.4–10 ppt), 2,4,6-trichloroanisole (TCA) has been identified as the major cause of corkiness in wine. Corks may be contaminated by chloroanisoles from packaging and shipping containers during transport, microbiological methylation of chlorophenols during the corks' bleaching with hypochlorite, microbial contamination of raisins (Aung *et al.*, 1996). Chloroanisoles may also be present as residues of pesticides and insecticides used in the cork forest (Burttshel *et al.*, 1951; Lee and Simpson, 1993). Irradiation of TCA causes the formation of 2-chloroanisole, 4-chloroanisole, 2,4-dichloroanisole and 2,6-dichloroanisole (Careri *et al.*, 2001; Flamini and Panighel, 2006 and references cited herein). The structure of TCA and its relationship with trichlorophenol are shown in Figure 4.3.

4.1.4. Volatile Phenols in Wine

Volatile phenols (VPs), i.e. 4-ethylphenol (4-EP), 4-ethylguaiacol (4-EG), 4-vinylphenol (4-VP) and 4-vinylguaiacol (4-VG) (Figure 4.4), are one of the most significant problems in modern wine-making. 4-EP was found in wine for the first time in 1967 by Webb and colleagues (Webb, 1967) and its presence, together with the other phenols cited, was confirmed in 1970 by Dubois and Brulé (Dubois and Brulè, 1970).

Ethyl phenols are a result of enzymatic activities linked to the decarboxylation of cinnamic acids and the subsequent reduction in vinyl phenols caused by the *Brettanomyces/Dekkera* yeast genus (Chatonnet *et al.*, 1992), apart from very small quantities produced in peculiar

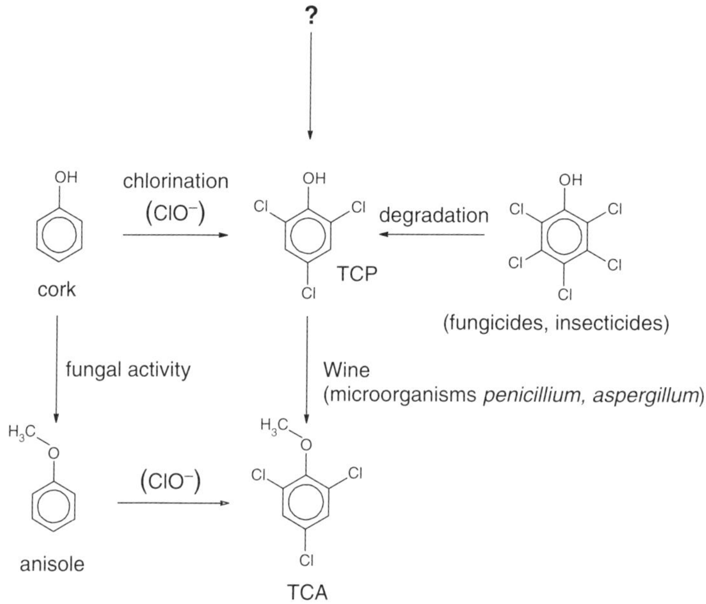

Figure 4.3 Origin of 2,4,6-trichloroanisole and its relationship with 2,4,6-trichlorophenol

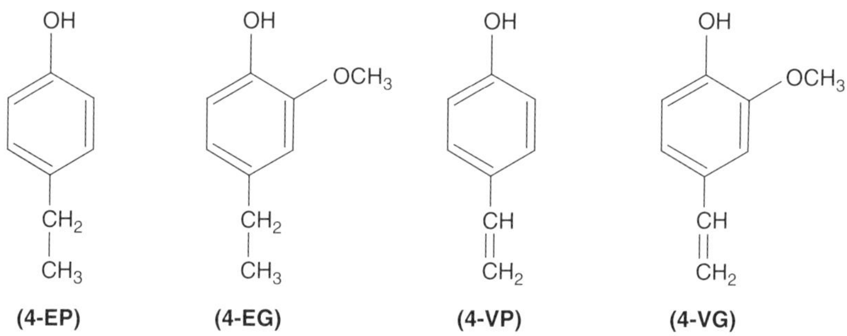

Figure 4.4 Structure of Volatiles Phenols (VPs) in wine

conditions by some yeasts and lactic acid bacteria (Chatonnet *et al.*, 1995; Barata *et al.*, 2006; Couto *et al.*, 2006). In contrast to *S. cerevisiae, Brettanomyces* decarboxylase is not inhibited by proanthocyanidines (Chatonnet *et al.*, 1993). 4-EP is liable to confer an odour defined as stable, horse-sweat or leather-like (Etievant, 1991) and, because of its relatively low limit preference threshold (Chatonnet *et al.*,

1992), it must be considered as the most critical volatile phenol in red wine. 4-EG was also described in beer as sweet (Meilgaard, 1975). Ethylphenols are usually present in wine with a 4-EP/4-EG ratio ranging from 3.5 to 16 (Chatonnet *et al.*, 1992; Pollnitz *et al.*, 2000; Alessandria *et al.*, 2005; Nicolini *et al.*, 2006).

White wines can contain vinylphenols in varying quantities up to several hundred μg/L, but usually lack ethylphenols, while the contrary is true for red wines where ethylphenols can reach a few mg/L (Chatonnet *et al.*, 1992; Chatonnet *et al.*, 1993; Chatonnet, 1993). Vinylphenols are mainly formed during alcoholic fermentation by strains of *S. cerevisiae* yeast called POF+ (Phenolic Off Flavour positive) capable of stereospecific enzymatic decarboxylation of the *trans* forms of *p*-coumaric and ferulic acids (Albagnac, 1975; Grando *et al.*, 1993; Chatonnet *et al.*, 1993). Oligomer proanthocyanidins inhibit cinnamate decarboxylase of *S. cerevisiae* (Chatonnet *et al.*, 1990), justifying the very low amounts of vinylphenols in red wines. They are liable to give sensory characteristics generally classified among the 'off flavours' and described in general as phenolic, medicinal, pharmaceutical, smoky, spicy and clove-like (Montedoro and Bertuccioli, 1986; Rapp and Versini, 1996). 4-VP, even if lower than the sensory threshold, is deemed to negatively affect and mask the fruity scent of white wines (Dubois, 1983; Chatonnet, 1993) conferring odours resembling 'band-aid' and gouache (van Wyk and Rogers, 2000). It can be considered less negatively, contributing, when combined with 4-VG, to the genista-like floral aroma of *Chardonnay* wines (Versini *et al.*, 1992). 4-VG contributes to the spicy note of *Gewürztraminer* wines (Versini, 1985), as well as to the typical characteristics of some beers from Belgium and Bavaria, e.g. Lambic and Weizen (Narziss *et al.*, 1990; Coghe *et al.*, 2004).

4.2. Determination of Ochratoxin A in Grape and Wine

In general, a satisfactory quantitative analytical method has to be characterized by high sensitivity and low limits of detection and quantification (LOD and LOQ), linearity range of at least 3–4 magnitude orders, high precision (to provide high repeatable and reproducible data) and high accuracy (experimental data as closer is possible to the 'true value'). In particular, for methods of analysis of OTA in wine, including both sample preparation and analysis, a quantification limit lower than the legal limit of 2 ppb (at least 0.6 μg/L), detection limit lower than 0.2 μg/L, extraction yield from red and white wines not lower than 84.6% and 88.4% respectively (International Organization of Vine and Wine, 2006), and linearity at least in the 0.1–100 ppb range are required. Sample preparation performed by solid phase extraction (SPE) and analysis by High Performance Liquid Chromatography (HPLC) coupled with fluorescence, or more recently mass spectrometry, are suitable to satisfy the performances required.

4.2.1. Sample Preparation

Grape sample

Heterogeneity of the grape cluster constituted of berries and stalks, and the spot-contamination of fungi in the vineyard, make it difficult to get a representative grape sample for analysis. This problem arises when samples are picked in the vineyard and frozen, and when the sample is taken from the packages for analysis. Because fungi develop on both grape skins and stalks, a representative sample for analysis can be prepared by extracting both the juice, which washes the grape surfaces as it exits the berries, and the solid parts of the cluster. A 100-g sample constituted of berries, stalks and juice is extracted with 100 mL of chloroform for 24 h under stirring. On the assumption that OTA is localized on the cluster surfaces, no homogenization of the sample is performed in order to avoid extraction of interfering substances (Garcia Moruno *et al.*, 2004; Tonus *et al.*, 2005). Ninety millilitres of the organic phase is separated from the solid residue and filtered, the solvent is removed under vacuum at 30 °C and the residue is re-dissolved in the mobile phase used for HPLC analysis.

An alternative method for OTA extraction from grape is the use of a hydrogen carbonate and polyethylene glycol (PEG) solution (1% PEG 8000 and 5% $NaHCO_3$ in water), followed by a clean-up of the extract on an immunoaffinity column (IAC) (Serra *et al.*, 2004). Berries are slightly homogenized and 50 g is made to 150 mL with $NaHCO_3$/PEG solution. The mixture is stirred for 30 min, then centrifuged at 4 °C. The supernatant is filtered through a glass microfibre (porosity 1.5 μm) and 20 mL of the filtrate is finally passed through the IAC. Addition of PEG to the extraction solution seems to reduce the number and intensity of interfering chromatographic peaks. A method of extraction by the use of acidified methanol has been reported: 50 g of homogenized berries is extracted with 50 mL methanol and 5 mL orthophosphoric acid for

2 min, the mixture is filtered thorough a 1.5 μm glass microfibre and diluted to 100 mL with 1% PEG – 5% $NaHCO_3$ solution prior to the IAC purification (MacDonald *et al.*, 1999).

Wine sample

For analysis of OTA in wine the sample is preliminarily centrifuged and degassed by ultrasound; afterwards different methods of sample preparation can be used. The more common method is to perform SPE using an immunoaffinity column (IAC) specific for OTA (Burdaspal and Legarda, 1999; Visconti *et al.*, 1999; Castellari *et al.*, 2000). A volume of 10 mL of wine is added to 10 mL of PEG/$NaHCO_3$ solution (1% PEG 8000 and 5% $NaHCO_3$ in water). The pH of the solution is adjusted to 8.5 by addition of concentrated NaOH, the resulting solution is filtered through a glass microfibre to remove any present solid and 10 mL of the sample is applied to the IAC. After the sample loading the column is washed with 5 mL of a NaCl/$NaHCO_3$ aqueous solution (2.5% NaCl and 0.5% $NaHCO_3$) followed by 5 mL of water. The analyte elution is performed by passing through the cartridge 2 mL of methanol containing 2% (v/v) of acetic acid. The solvent is evaporated under a nitrogen stream at 50 °C, then the residue is redissolved in the HPLC mobile phase (Sàez *et al.*, 2004).

Another SPE method for sample preparation is by C_{18} cartridge (Zöllner *et al.*, 2000). Ten milliliters of wine is applied to a 100-mg cartridge previously activated by the passage of 5 mL methanol followed by 5 mL water. After the sample loading, the column is washed with 2 mL water then the stationary phase is dried. OTA is recovered by 1.25 mL of methanol; alternatively, methanol/acetic acid 99.5:0.5 solution (Sàez *et al.*, 2004) or 3 mL of ethyl acetate/methanol/acetic acid 95:5:0.5 (Chiodini *et al.*, 2006), can be used. The solvent is evaporated and the residue redissolved as reported above.

Liquid-liquid extraction of OTA from the wine can be also performed (Zimmerli and Dick, 1996). A volume of 5 mL wine is mixed with 10 mL of aqueous solution containing 3.4% orthophosphoric acid (85%) and 1.18 g of NaCl, and extraction is performed twice with 5 mL of chloroform each. The organic phases are combined, the solvent is removed under vacuum and the residue is redissolved in 5 mL of phosphate buffer solution (NaCl 120 mM, KCl 2.7 mM, phosphate buffer 10 mM, pH 7.4) prior to purification on IAC cartridge.

4.2.2. OTA Analysis

HPLC and fluorescence (HPLC/FL)

Analysis of OTA in wine is commonly performed by HPLC coupled with fluorimetric detection operating at excitation wavelength 330 nm (λ_{ex}) and emission wavelength 470 nm (λ_{em}) (Burdaspal and Legarda, 1999; Visconti *et al.*, 1999; Castellari *et al.*, 2000). Chromatography is performed using a reverse phase C_{18} column (e.g. 250 × 3.0 mm, 5 μm) and isocratic elution with a binary mobile phase composed of sodium acetate 4 mM/acetic acid (95:5 v/v) at pH 3.50 (solvent A), and acetonitrile (solvent B) prepared by mixing A/B 52:48 (v/v) (Tonus *et al.*, 2005), or a mixture of acetonitrile/H_2O/acetic acid (49.5:49.5:1 v/v/v), at a flow rate 0.5–1.0 mL/min (Visconti *et al.*, 1999). LOD and LOQ in analysis of grape extracts were established at 0.004 and 0.007 μg/kg, respectively (Serra *et al.*, 2004). In some grape extracts and juices, the presence of a compound with similar LC retention time and fluorimetric characteristics to OTA, was reported (Zimmerli and Dick, 1996; Serra *et al.*, 2004). Examples of OTA analysis in wine and grape extract are reported in Figure 4.5. The OTA emission fluorescence spectra is shown above right.

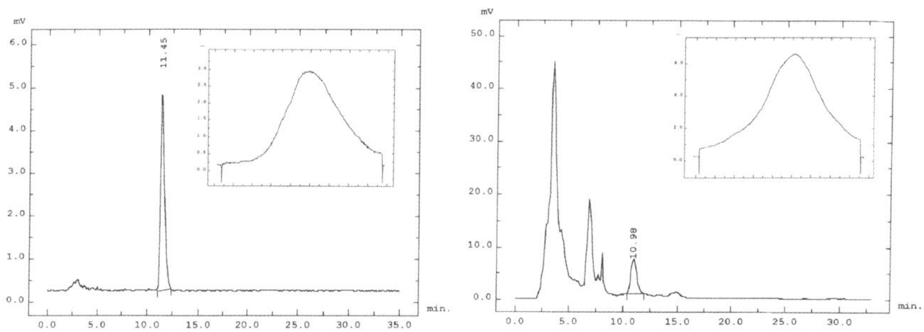

Figure 4.5 Chromatogram of a wine sample containing OTA 4.41 μg/L (left) and a grape extract naturally OTA-contaminated at 4.92 μg/L (right). Emission spectra are recorded in the wavelength range 350–550 nm, the maximum is at 470 nm (Tonus *et al.*, 2005)

Identification of the analyte is based on the presence in the chromatogram of a peak with the same retention time as standard compound. Confirmation of identification has to be performed, it is achieved by recording the fluorescence spectra in the range 350–550 nm and the maximum wavelength emission of OTA is at 470 nm, or by synthesis of methyl (or ethyl) ester. To perform synthesis of methyl ester, the sample used for analysis (dissolved in the HPLC mobile phase) is extracted three times with 1 mL of chloroform each. Extracts are combined, the solvent is removed under vacuum and the residue is re-dissolved in 0.5 mL of boron trifluoride-methanol-complex solution ($2CH_3OH \cdot BF_3$). Reaction is carried out for 20 min at 60 °C, then the solvent is evaporated to dryness by nitrogen flow, the residue is redissolved in the HPLC mobile phase and used for analysis. To achieve a reliable spectrum for confirmation of the OTA identity, a minimum content of 0.1 μg/Kg grape OTA is necessary; 0.46 μg/Kg grape is the minimum content necessary to perform the methyl ester synthesis confirmation (Tonus *et al.*, 2005). In Figure 4.6, chromatograms relative to analysis of an OTA-positive grape extract and a wine sample before (above) and after (below) derivatization, are reported.

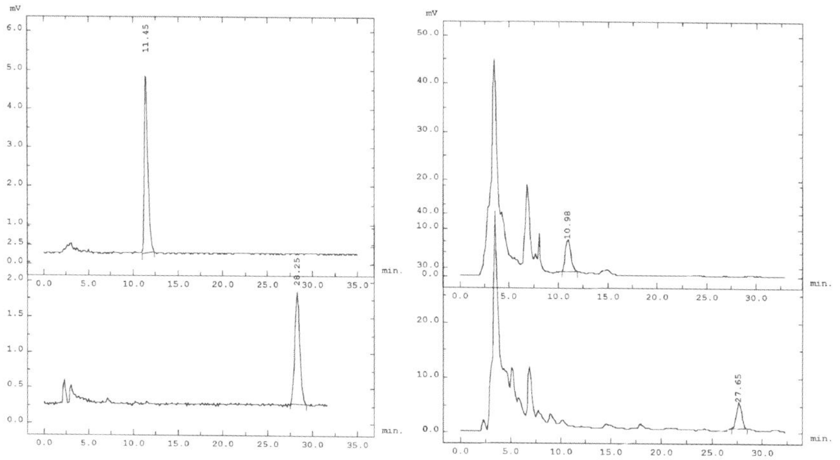

Figure 4.6 OTA analysis of the wine sample (left) and grape extract sample (right) reported in Figure 4.5 before (above) and after (below) synthesis of the methyl ester (Tonus *et al.*, 2005)

HPLC and mass spectrometry (LC/MS)

For unambiguous confirmation in samples containing OTA higher than 0.2 ng/ML, HPLC-electrospray ionization and ion trap mass spectrometry (LC/ESI-MS) were used (Medina *et al.*, 2006). Analysis was performed with a C_{18} HPLC column, confirmation was based on the protonated molecule $[M+H]^+$ at m/z 404 and the most abundant product ion $[(M+H)-HCOOH]^+$ at m/z 358 (ESI source parameters: nebulizer gas N_2 60 psi, dry gas 10 L/min, dry temperature 220 °C, capillary voltage 3 kV).

To perform LC/ESI-MS analysis, the commercially available compound zearalanone (ZAN) can be used as internal standard LOD of method is 0.5 ppb. Additional confirmation can be achieved by performing tandem-mass spectrometry (MS/MS) experiments. The ions selected for multiple reaction monitoring (MRM) are at m/z 404→239 (relative to $[M+H-Phe]^+$ species), m/z 404→257 and m/z 406→241 for OTA, and at m/z 321→123/189 for ZAN (Zöllner *et al.*, 2000). The OTA confirmation (positive-ion mode) can also be performed by collision-induced dissociation (CID)

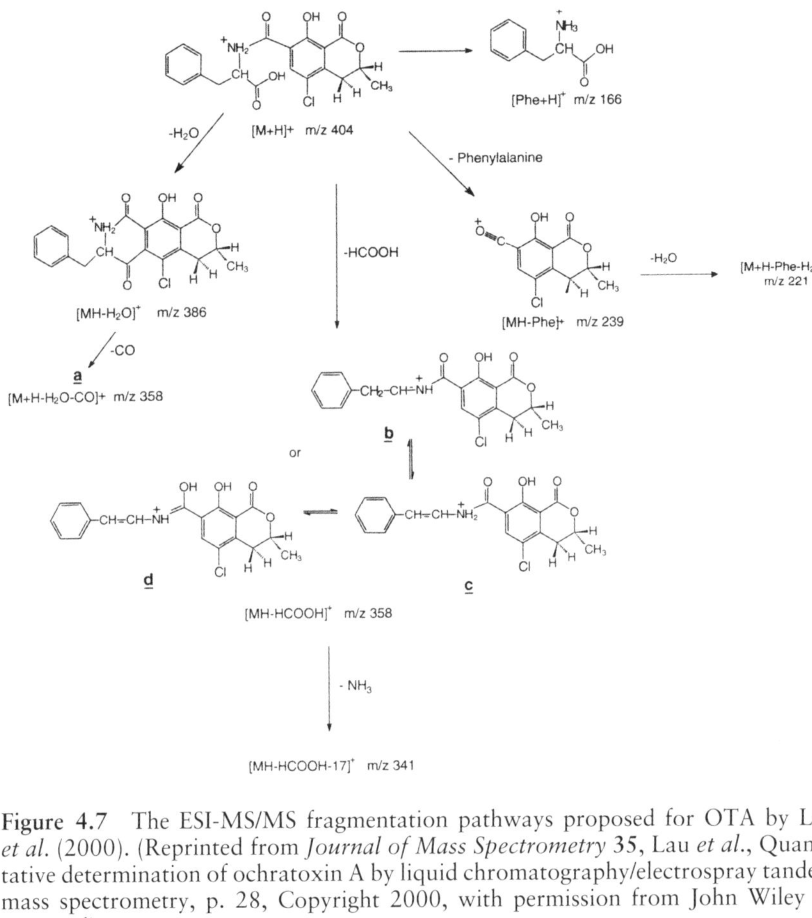

Figure 4.7 The ESI-MS/MS fragmentation pathways proposed for OTA by Lau *et al.* (2000). (Reprinted from *Journal of Mass Spectrometry* 35, Lau *et al.*, Quantitative determination of ochratoxin A by liquid chromatography/electrospray tandem mass spectrometry, p. 28, Copyright 2000, with permission from John Wiley & Sons Ltd)

experiments on the protonated molecular ion $[M+H]^+$ of analyte at *m/z* 404, and selected reaction monitoring (SRM) of the resultant product ions $[M+H-H_2O-CO]^+$ at *m/z* 358 and $[M+H-H_2O]^+$ at *m/z* 386 (Shephard *et al.*, 2003). The ESI-MS/MS fragmentation pathways of OTA are reported in Figure 4.7 (Lau *et al.*, 2000). The LC-MS/MS chromatograms of a sample prepared from a 5 μg/L OTA-spiked wine containing ZAN 100 μg/L and extracted by SPE using a C_{18} cartridge, are shown in Figure 4.8.

Also, stable isotope dilution assay can be applied for quantitative analysis of OTA, using d_5-OTA as internal standard and performing LC/MS analysis in positive-ion mode (Lindenmeier *et al.*, 2004; MacDonald *et al.*, 1999). The ^{35}Cl-containing $[M+H]^+$ ion at *m/z* 404, and the ^{37}Cl-containing analog at *m/z* 406 species, are monitored. As OTA contains a chlorine atom, additional confirmation is achieved by monitoring the two isotopic forms, and comparing the *m/z* 404/406 peak area ratio in the sample with that of standard solutions.

Recently, by performing direct nano-LC/ESI-MS sample injection onto a reverse phase capillary without immunoaffinity clean-up, LOD and LOQ in grape extracts of 1 and 2 pg/g, respectively, were reported (Timperio *et al.*, 2006). HPLC analysis was carried out at flow 200 nL/min and introducing 1 μL sample onto the column. Figure 4.9 shows a nano-LC/ESI-MS chromatogram recorded in positive mode in the analysis of acetonitrile solutions containing various concentrations of OTA.

In general, performing sample preparation using imunoaffinity clean-up coupled with HPLC/FL analysis provides limits of detection and quantification comparable with those achieved by SPE using C_{18} cartridges and LC/ESI-MS/MS analysis. Purification on IAC column followed by HPLC/MS analysis did not show advantages in terms of sensitivity and accuracy (Leitner *et al.*, 2002). One advantage in performing MS analysis is that it offers the possibility of using the cheaper C_{18} cartridges in the sample preparation, which is not possible with HPLC/FL because the matrix interferences have to be eliminated prior to analysis by performing the sample clean up using IAC.

Recently, a LC/MS method to determine OTA in wine and grape extracts by using Surface-Activated-Chemical-Ionization and multi-stage fragmentation mass spectrometry (LC/SACI-MS^3), was proposed (Flamini *et al.*, 2007). SACI is an ionization source developed to analyze medium-high polar compounds in a wide range of *m/z* ratios (100–4000 Da). This device, used in determination of compounds such as

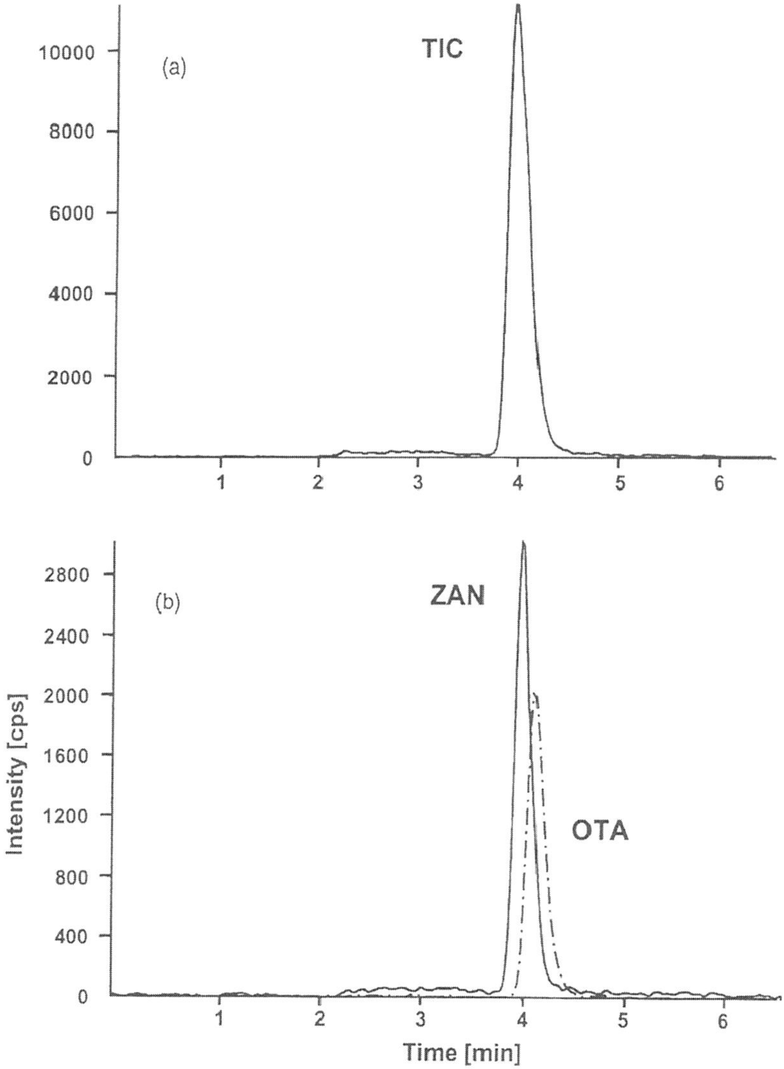

Figure 4.8 LC/ESI-MS/MS chromatograms of a wine sample spiked with OTA 5 μg/L and containing ZAN 100 μg/L: (a) total ion current (TIC) chromatogram, (b) individual multiple reaction monitoring (MRM) chromatograms of ZAN (solid line signal at *m/z* 321.0-123.0) and OTA (broken line signal at *m/z* 404.1-239.0). Analytical conditions: C_{18} (125 × 3 mm) column at room temperature; eluent methanol/H_2O/acetic acid (70:30:2.5 v/v/v) at flow rate 0.5 mL/min split 1:50 into the ion source; electrospray voltage +5600 V, multiple-reaction monitoring (MRM) mode; collision energy 32.5 eV; collision gas N_2 (LOD 0.05 μg/L). (Reprinted from *Analytica Chimica Acta* 453, Leitner *et al*., Comparison of methods for the determination of ochratoxin A in wine, p. 38, Copyright 2002, with permission from Elsevier)

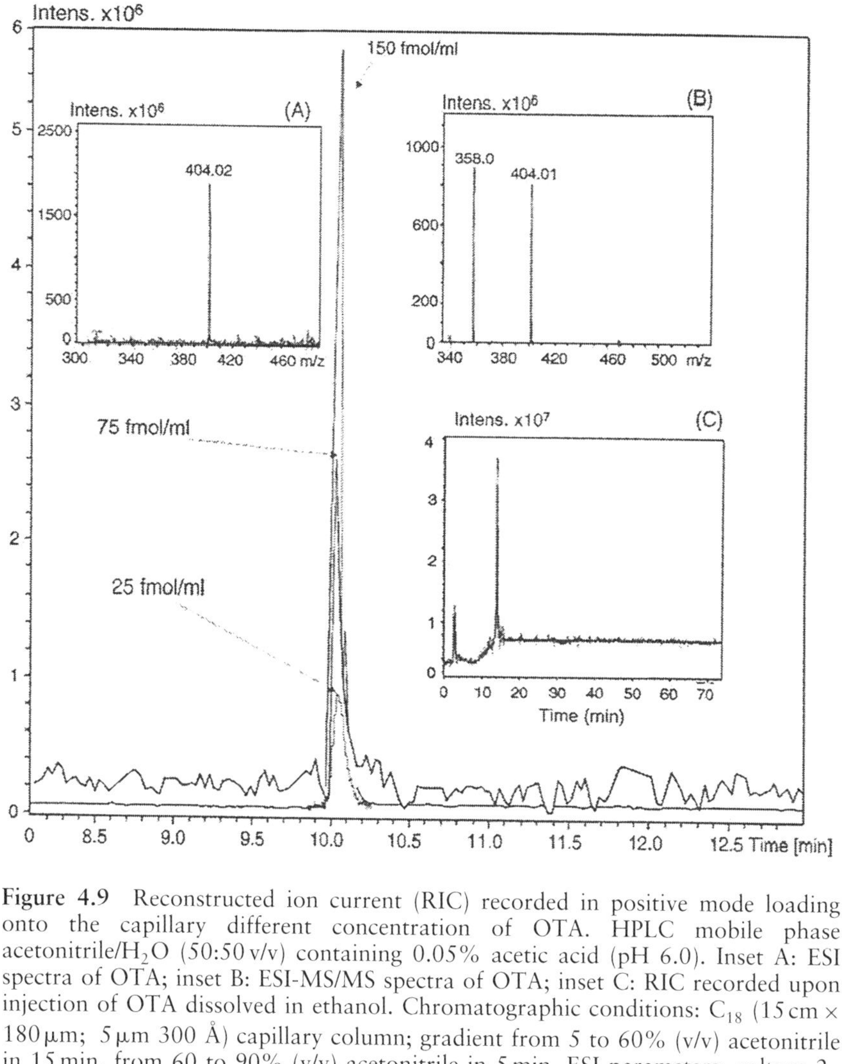

Figure 4.9 Reconstructed ion current (RIC) recorded in positive mode loading onto the capillary different concentration of OTA. HPLC mobile phase acetonitrile/H_2O (50:50 v/v) containing 0.05% acetic acid (pH 6.0). Inset A: ESI spectra of OTA; inset B: ESI-MS/MS spectra of OTA; inset C: RIC recorded upon injection of OTA dissolved in ethanol. Chromatographic conditions: C_{18} (15 cm × 180 μm; 5 μm 300 Å) capillary column; gradient from 5 to 60% (v/v) acetonitrile in 15 min, from 60 to 90% (v/v) acetonitrile in 5 min. ESI parameters: voltage 2–3 kV; dry gas 5 L/min at 300 °C; nebulizer gas 20 psi; collision gas He. (Reprinted from *Journal of Chromatography* B **832**, Timperio *et al.*, Assay of ochratoxin A in grape by high- pressure liquid chromatography coupled on line with an ESI-mass spectrometry, p. 130, Copyright 2006, with permission from Elsevier)

aminoacids, addictive drugs, pharmaceutical drugs, and steroids, can be employed in the wide eluent flow rate range 100–2000 μL/min in conjunction with high flow gradient chromatography. This technique is based on the presence of metallic catalyst set at a low potential in an ionization chamber. The solvent from the chromatographic column is vaporized by heating and directed to the metallic surface by a nitrogen flow. At the surface, charged positive, the solvent molecules turn and transfer a proton to the analyte according to their proton affinity forming $[M+H]^+$ or $[M+nH]^{n+}$ species of analyte. An analogous mechanism occurs when the surface is charged negatively: as a consequence of increased proton affinity of the solvent a proton is transferred from the analyte molecule to the solvent forming $[M-H]^-$ or $[M-nH]^{n-}$ species. Ionization by traditional ESI and atmospheric-pressure-chemical-ionization (APCI) techniques employs high potentials (2–5 kV) and the solvent is extra-charged. With SACI, a lower chemical noise due to the solvent charged clusters is observed. Moreover, the high ionization yields provided from SACI increase the signal/noise ratio in the analysis of many compounds (Cristoni *et al.*, 2003).

For analysis of OTA in wine, the sample is directly injected into the column without performing sample concentration or purification steps.

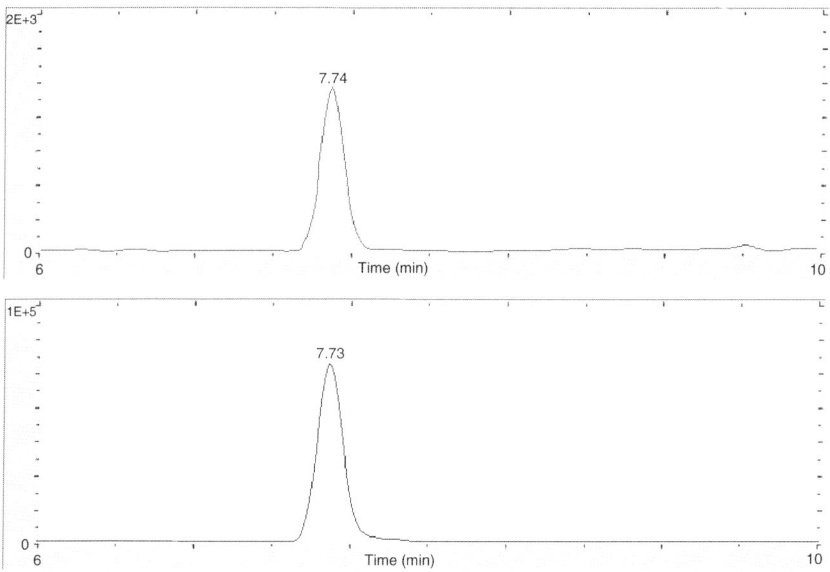

Figure 4.10 OTA and ZAN LC-MS^3 extracted ion chromatograms of a wine spiked with OTA at a concentration of 0.1 ng/mL and internal standard ZAN (10 ng/mL). Above: signal m/z 239 + 341 of OTA; below: signal m/z 207 + 189 + 163 of ZAN. Analytical conditions: C_{18} (3 × 250 mm; 5 μm) column, binary solvent A) H_2O/0.1% formic acid/sodium acetate 0.6 mM and B) methanol/0.1% formic acid. Gradient program: 50% A for 1 min, from 50% to 20% of A in 7 min, isocratic for 4 min, 20–50% A in 3 min, isocratic for 3 min (flow rate 0.5 mL/min). SACI vaporizer temperature 400 °C; entrance capillary temperature 150 °C; SACI surface voltage 50 V; surface temperature 110 °C; nebulizing sheath gas N_2 at flow rate 9 L/min; curtain gas 2 L/min; spray needle voltage set to 0 V. (Adapted from Flamini *et al.*, 2007, RCM) **21**, (22), 3737–3742

The use of medium high chromatographic flows strongly reduces the matrix effect allowing a limit of quantification at least 20-fold lower than the maximum CE Regulation limit (2 ppm). Figure 4.10 shows the LC-MS^3 extracted ion chromatograms recorded in the analysis of a 0.1 ng/mL OTA-spiked sample. Above, the signal resulting from the sum of m/z 239 + 341 OTA signals is reported; below is reported the signal resulting from the sum m/z 207 + 189 + 163 of ZAN at concentration 10 ng/mL. The figure shows that in the chromatographic conditions used the two compounds eluted from the column with the same retention time.

The most abundant OTA daughter ion at m/z 358, produced by MS^2 experiments using $[M+H]^+$ as precursor ion (for ^{35}Cl-containing molecules), is formed by HCOOH loss. Figure 4.11 shows the MS^3 spectra of OTA and ZAN in the analysis of a spiked wine where both compounds are in concentration 10 ng/mL. Additive confirmation is provided by MS^3 experiments performed on the isotopic species containing ^{37}Cl.

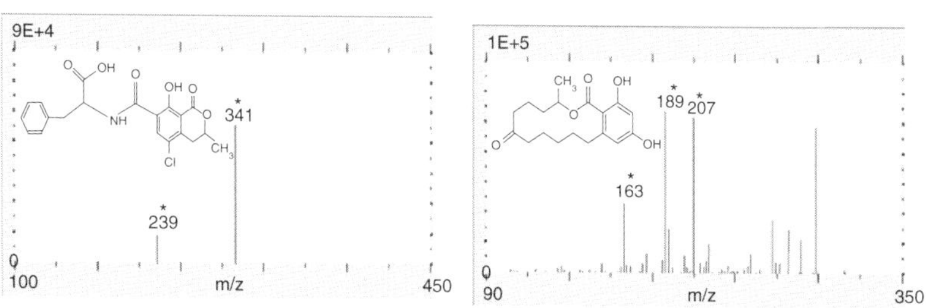

Figure 4.11 MS^3 spectra produced from collision induced of the OTA daughter ion at m/z 358 (left) and of the ZAN daughter ion at m/z 303 (right). The marked signals of OTA (m/z 239 and 341) and ZAN (m/z 163, 189 and 207) are used for quantitative analysis. (Spectra acquired in positive ion mode; collision energy applied to the parent ion 80% of maximum value; MS^3 experiments of daughter ions at 100% of maximum collision energy). (Adapted from Flamini *et al.*, 2007, RCM), **21** (22), 3737–3742

4.3. Determination of Biogenic Amines in Grape and Wine

4.3.1. Preparation of Samples

Preparation of grapes can be performed by adding an adequate aliquot of 10% $HClO_4$ to the destemmed sample, homogenizing using Ultra-turrax and centrifuging at about 10000 rpm for 10–15 minutes. Before analysis the supernatant is filtered through a 0.45 or 0.22 μm membrane cartridge (Kiss *et al.*, 2006). Alternatively, BAs can be extracted by directly crushing and pressing grape at 2.5 bar (Bertoldi *et al.*, 2004; Nicolini *et al.*, 2003). Concentration of BAs in grape is often reported on the dry matter basis.

Wine samples are generally degassed and filtrated. The BAs level in wine and juice is analytically relatively low, and various interferences are present, due to complexity of matrix. Polyphenolic compounds can interfere in the analysis of red wines, amino acids in the analysis of grape juices. Extraction steps, albeit essential, are generally time-consuming and can reduce analytical recoveries. Several methods for isolating BAs from wine and juice samples have been proposed, including liquid-liquid (LL) and SPE. Almy and co-workers (1983) performed LL extraction with butanol on samples preliminarily concentrated and adjusted at pH 1.5. If SPE is chosen, strong cation exchange (SCX) in acid conditions is generally preferred to strong anion exchange (SAX) or C_{18} cartridge clean-up (Yamamoto *et al.*, 1982; Busto *et al.*, 1995). Some authors have suggested the removal of phenolic compounds using polyvinylpyrrolidone (PVP) before performing the SPE and derivatization (Busto *et al.*, 1994).

4.3.2. Methods of Analysis

Gas Chromatography (GC) and HPLC are the traditional methods used for BAs analysis in grape and wine. Capillary Electrophoresis is more rarely used in the oenological field (Ma *et al.*, 1992; Waterval *et al.*, 2000). Most of these methods include the sample derivatization to improve the chromatographic separation and/or to enhance the method's performance.

High performance liquid chromatographic methods

o-phthalaldehyde

Most BAs do not show fluorescence or UV adsorption and a labelling procedure is indispensable to allow their detection. One of the most frequently studied fluorescent derivatizing agents in both the pre- and post-column methods is *o*-phthalaldehyde (OPA). Despite the limited stability of derivatives, the main advantages of OPA are a very fast reaction and great sensitivity in fluorescence detection at excitation wavelength 335–350 nm and emission wavelength 420–450 nm. Derivatization of BAs with OPA is obtained using different alkylthiols as a nucleophile in the cycling reaction, such as 2-mercaptoethanol (Lehtonen *et al.*, 1992; Busto *et al.*, 1995), 3-mercaptopropionic acid (Graser *et al.*, 1985; Herbert *et al.*, 2001), ethanethiol and *tert*-butylthiol (Allison *et al.*, 1984). The properties of isoindole products (see Figure 4.12) in terms of fluorescence and stability are markedly dependent on the nature both of amines and thiols.

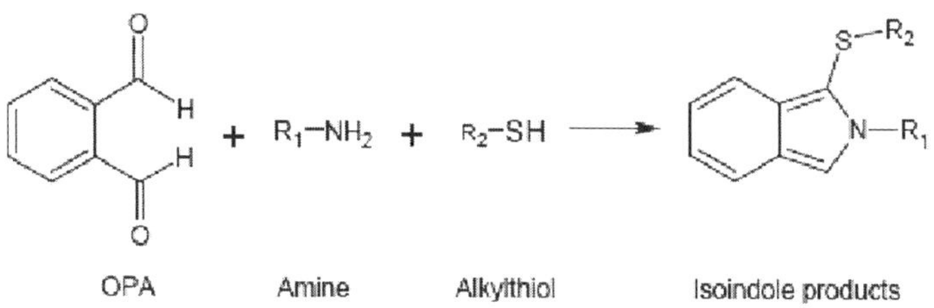

Figure 4.12 Synthesis of fluorescent isoindole complex from *o*-phthalaldehyde (OPA), alkylthiol and primary amine

Because of the instability of these products, instrumental automation of the derivatization and injection procedures is required to ensure the highest reproducibility in pre-column derivatization methods. A volume of 1 μL wine, 1 μL of OPA-alkylthiol derivatizing solution (45 mg OPA dissolved in 0.5 mL methanol containing 0.1 mL mercaptoethanol) and 5 μL of buffer solution (0.5 g H_3BO_3 dissolved in 19 mL water and the solution is adjusted to pH 10.4 with KOH) are consecutively sampled and automatically mixed for 2 minutes in the needle of the HPLC system before being injected (Lehtonen *et al.*, 1992).

Bertoldi and colleagues (2004) performed analysis of BAs in grape juice by automatic pre-column derivatization with OPA solution (45 mg OPA, 200 μL 2-mercaptoethanol, 1 mL methanol, sodium tetraborate decahydrate 0.1 M pH 10.5 buffer solution added up to final volume of 10 mL). Chromatographic separation of derivatives (see Figure 4.13) was achieved on a C_{18} (4.6 × 150 mm, 5 μm) column kept at 40 °C at flow rate 1 mL/min. Eluent A was 0.05 M sodium acetate buffer/tetrahydrofuran 96:4 (v/v) and eluent B was methanol. The gradient program is reported in Table 4.1.

In Table 4.2 the results of a survey of the natural content of some BAs in juices from fully ripe grapes of 10 different varieties, are

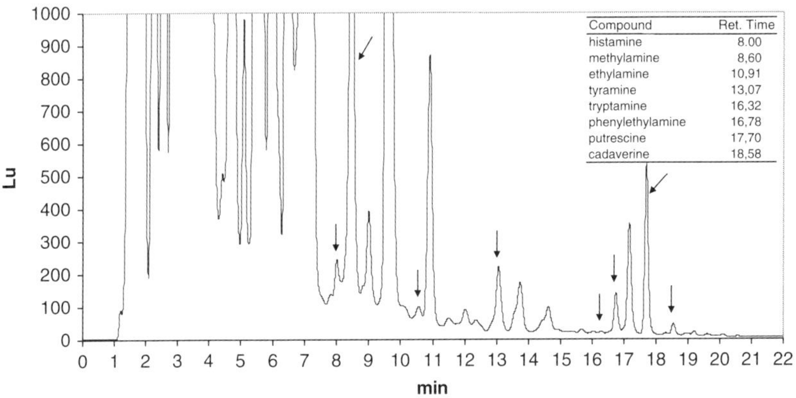

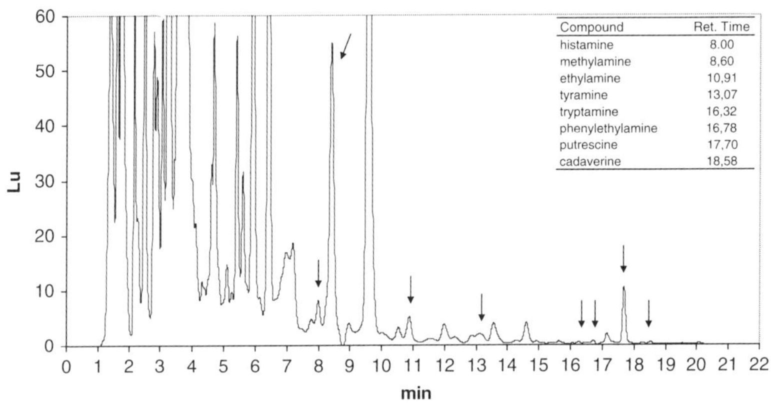

Figure 4.13 HPLC analysis of OPA-derivatives of BAs in grape juice with fluorimetric detection at excitation wavelength of 336 nm and emission wavelength 445 nm (above), and by diode array at 338 nm (below) (Bertoldi *et al.*, 2004)

Table 4.1 Eluent gradient program for HPLC analysis of biogenic amine-OPA derivatives. Solvent A) 0.05 M sodium acetate buffer/tetrahydrofuran 96:4 (v/v); solvent B) methanol; flow rate 1 mL/min (Busto *et al.*, 1995; Bertoldi *et al.*, 2004).

Time (min)	Eluent A (%)	Eluent B (%)
0	80	20
1	65	35
8.5	50	50
11	45	55
18.5	20	80
19.5	0	100
22.8	0	100
24	80	20
24.5	80	20

Table 4.2 Mean contents (µg/L) of principal biogenic amines in juices from different grape varieties.

		Nosiola	Chardonnay	Pinot gris	Lagrein	Marzemino	Rebo	Shiava	Teroldego	Cabernet Sauvignon	Merlot
°Brix	25th percentile	15,5	18,8	18,9	18,7	19,0	20,7	17,0	20,0	19,6	20,2
	Median	17,0	19,5	19,7	19,2	19,4	21,7	17,7	20,2	20,6	21,7
	75th percentile	18,2	20,2	20,5	20,4	20,3	22,1	18,2	21,9	21,0	22,1
Putrescine	25th percentile	460	411	694	1354	608	610	1250	725	852	859
	Median	582	765	893	2161	876	758	1682	1113	1297	1104
	75th percentile	771	952	1186	2444	2034	1303	1842	2239	1581	1371
Methylamine	25th percentile	109	124	96	113	108	92	55	135	111	100
	Median	120	142	127	153	140	94	104	376	142	160
	75th percentile	157	202	201	163	167	136	109	545	224	218
Ethylamine	25th percentile	53	35	143	164	230	98	158	139	52	69
	Median	137	80	172	199	376	101	385	476	83	82
	75th percentile	353	110	224	272	457	168	651	1097	123	133
Histamine	25th percentile	30	48	79	61	168	35	37	57	96	49
	Median	41	63	101	80	217	44	42	154	128	59
	75th percentile	124	76	124	104	350	59	53	218	236	80

reported (Bertoldi *et al.*, 2004). Putrescine, methylamine, ethylamine and histamine were found in quantifiable levels in all samples, while cadaverine, 2-phenylethylamine, tyramine and tryptamine were in lower concentrations, being quantifiable only in 75%, 43%, 35% and 6% of the analyzed samples, respectively. Positive correlations were observed between putrescine content and the analytical parameters indicating increasing ripeness.

A survey of the BAs content in red and white Italian wines measured with the method proposed by Soleas and co-workers (1999) is reported in Table 4.3. Significant differences for histamine, ethylamine, thyramine, phenylethylamine, putrescine and cadaverine, were observed.

Table 4.3 Distribution of amine contents (μg/L) in red and white wines. (# = significance at HSD Tukey's test for unequal samples, *** = $p < 0.001$, **= $p < 0.01$; n.s. = not significant).

	sign. (#)	Wine colour	No. Obs.	Min	25° percentile	Median	75° percentile	Max
Methylamine	n.s.	White	82	< 0.1	0.12	0.19	0.34	2.11
		Red	67	< 0.1	0.05	0.09	0.15	3.41
Histamine	***	White	147	< 0.1	0.20	0.34	0.56	7.70
		Red	142	< 0.1	1.00	2.11	3.50	12.30
Ethylamine	***	White	82	0.2	0.42	0.58	0.74	2.15
		Red	67	0.1	0.63	0.84	1.07	2.02
Thyramine	***	White	82	< 0.1	0.09	0.27	1.12	11.66
		Red	68	0.1	1.16	2.05	4.07	7.38
Tryptamine	n.s.	White	82	< 0.1	0.01	0.02	0.03	0.92
		Red	67	< 0.1	0.01	0.03	0.05	0.43
Phenylethylamine	**	White	82	< 0.1	0.09	0.16	0.24	0.47
		Red	68	< 0.1	0.14	0.23	0.35	1.44
Putrescine	***	White	82	0.5	1.32	2.08	3.47	10.32
		Red	67	1.3	6.55	8.99	12.45	17.21
Cadaverine	***	White	78	< 0.1	0.03	0.04	0.08	0.42
		Red	65	< 0.1	0.12	0.17	0.22	3.46

Dabsyl chloride

Another labelling agent suitable for determination by HPLC of histamine, tyramine, phenylethylamine, tryptamine, cadaverine, putrescine, spermidine and spermine in wines is 4-(4-dimethylaminophenylazo)benzenesulfonyl (dabsyl) chloride (see Figure 4.14) (Bockhardt *et al.*, 1996; Romero *et al.*, 2000).

A volume of 1.5 mL of wine opportunely diluted is adjusted to pH 8.2 with a buffer solution (1.06 g of Na_2CO_3 dissolved in 50 mL of water) and 3.8 mL water are added. 1.6 mL of dabsyl chloride solution (40 mg of dabsyl chloride in 10 mL of acetone) is added to the mixture and

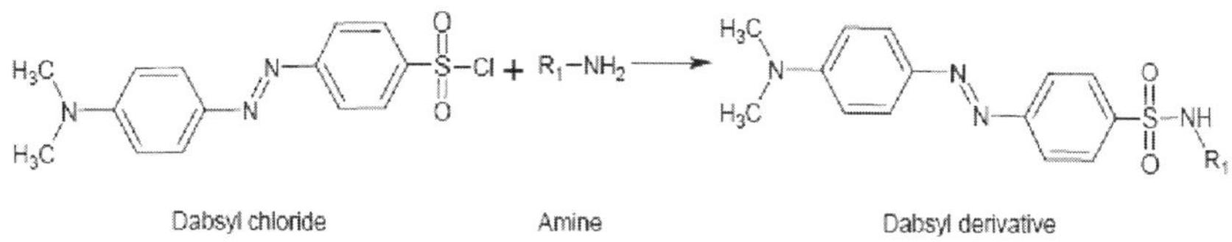

Figure 4.14 Synthesis of amine dabsyl derivatives

the reaction is carried out at 70 °C for 20 minutes. A volume of 4.6 mL of a dilution solution (50 mL acetonitrile, 25 mL of ethanol, and 25 mL of the eluent A, see below) is added to the mixture. After cooling for 20 min, HPLC analysis of dabsyl derivatives is performed by a C_{18} (244 × 4.4 mm, 5 μm) column at 40 °C, using a binary solvent mixture and the analytical conditions reported in Table 4.4 at the detection wavelength of 446 nm (Romero *et al.*, 2002).

Table 4.4 HPLC elution gradient program used for analysis of biogenic amine-dabsyl derivatives. Solvent A) 40 mM sodium acetate, 10% dimethylformamide and 0.23% (v/v/v) triethylamine adjusted to pH 5.0 with acetic acid; solvent B) 87.5% acetonitrile, 10% tert-butyl methyl ether and 2.5% water (v/v/v); flow rate 1.0 mL/min.

Time (min)	Eluent A (%)	Eluent B (%)
0	45	55
3	45	55
16	25	75
26	0	100
36	0	100
41	45	55
56	45	55

Dansyl chloride

The main advantages of dansyl (5-dimethylamino-1-naphthalenesulfonyl) chloride are the high reactivity and stability of derivatives formed with both primary and secondary amines (see Figure 4.15). On the other hand, due to its high reactivity dansyl chloride also forms derivatives with aliphatic alcohols, sugars and phenols, making it very difficult to achieve effective chromatographic separation. Dansyl chloride derivatives can be quantified using fluorescence with excitation at 340–360 nm and emission at 460–550 nm wavelengths, or UV detection at 250 nm (Lehtonen, 1986; Ibe *et al.*, 1991). For analysis of BAs in

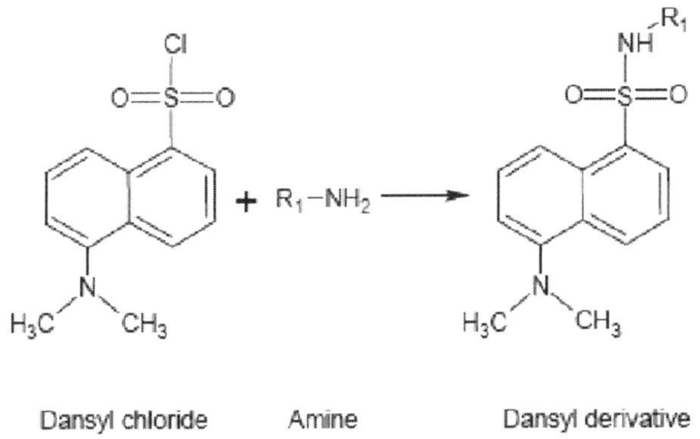

Figure 4.15 Synthesis of amine dansyl derivatives

wines, amines can be isolated by liquid-liquid extraction using Freon or butanol/dichloromethane (70:30 v/v) and concentrated to solid salts under vacuum. Volumes of 200 μL water, 50 μL of 20% sodium hydrogen carbonate solution, and 0.5 mL of 0.2% dansyl chloride in acetone are added to the concentrated residue and the reaction is carried out at 65 °C for 20 min. After acetone evaporation under nitrogen stream, 2 mL of water is added and the derivatives are extracted with 1 mL of ethyl acetate. HPLC analysis is performed with a C_{18} (200 × 4.6 mm, 5 μm) column at 40 °C, using a ternary solvent mixture at flow rate 1 mL/min and the following program of elution: from 30% acetonitrile and 70% water (v/v) to 60% acetonitrile, 15% ethanol and 25% water (v/v/v) in 40 min, finally isocratic for 10 min (Lehtonen, 1986).

1,2-naphthoquinone-4-sulfonate

García-Villar and co-workers (2005) proposed a method for determining histamine based on pre-column derivatization with sodium 1,2-naphthoquinone-4-sulfonate (NQS) (see Figure 4.16).

The labelling of the sample is carried out on-line in a three channel flow manifold, where the mixture of wine sample and derivatizing reagent is heated at 80 °C in a PTFE coil before HPLC injection. Separation of derivatives is performed on a C_{18} (150 × 4.6 mm, 4 μm) column using a binary mobile phase composed of A) 2% acetic acid in water (v/v) and B) methanol. A linear gradient is carried on in 12 min increasing eluent B from 10% to 25% at flow rate 1 mL/min. NQS-derivatives are detected at wavelength 305 nm. This approach allows automation of the analytical procedure and represents a real alternative to the conventional off-line labelling techniques.

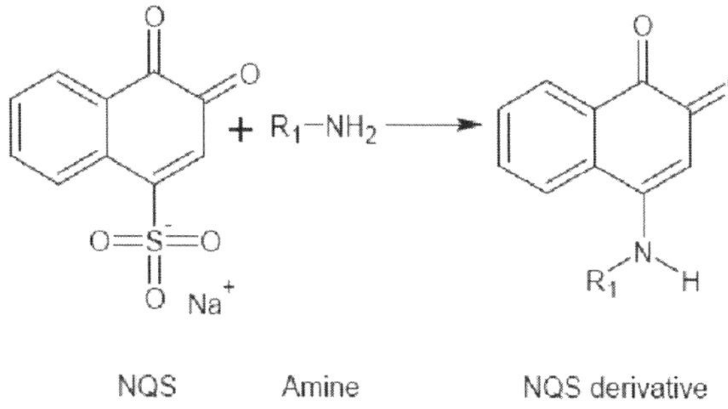

Figure 4.16 Synthesis of amine NQS derivatives

Diethyl ethoxymethylenemalonate

Diethyl ethoxymethylenemalonate (DEEMM) is usually used for amino acid analysis (Alaiz *et al.*, 1992). Recently, a HPLC-photodiode array method using DEEMM as labelling reagent for analysis of BAs in wine, has been developed (see Figure 4.17).

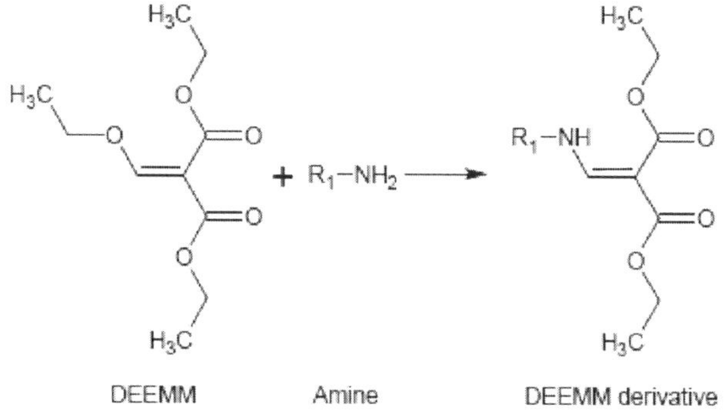

Figure 4.17 Synthesis of amine DEEMM derivatives

Derivatives are synthesized by mixing 1.75 mL of 1 M borate buffer (pH 9), 750 μL methanol, 1 mL of wine sample and 30 μL of DEEMM in a screw-cap tube. The solution is kept in an ultrasound bath for 30 min, then is heated at 70 °C for 2 hours in order to allow complete degradation of the DEEMM excess. Derivatives are stable for about one week. Analysis of DEEMM derivatives is performed with a C_{18} (250 × 4.6 mm, 5 μm) column at 16 °C using a binary solvent mixture and gradient program and conditions reported in Table 4.5. Detection is performed at wavelengths 269, 280 and 300 nm (Gómez-Alonso *et al.*, 2007).

Table 4.5 HPLC eluent program used for determination of biogenic amine-diethyl ethoxymethylenemalonate (DEEMM) derivatives. Eluent A) 25 mM acetate buffer (pH 5.8) with 0.02% sodium azide; eluent B) 80% acetonitrile and 20% methanol (v/v); flow rate 0.9 mL/min.

Time (min)	Eluent A (%)	Eluent B (%)
0.0	90	10
20.0	90	10
30.5	83	17
33.5	83	17
65.0	60	40
73.0	28	72
78.0	18	82
82.0	0	100
85.0	0	100

Labelling under phase-transfer conditions

An interesting alternative to conventional techniques of liquid derivatization is to operate under phase-transfer conditions, performing the BAs derivatization directly on the solid phase of a SPE column used for purification and pre-concentration of the sample (Molins Legua *et al.*, 2003). An efficient derivatization method using reagent benzoyl chloride for analysis of BAs in wine, was proposed (Paleologos and Kontominas, 2004). The labelling reagent is previously injected and retained on a C_{18} (10 × 4.6 mm i.d.) guard column, and 10 mL of the wine sample are passed through the column where the reaction occurs. Then the HPLC run starts and the mobile phase carries benzoyl derivatives from the cartridge into the HPLC column C_{18} (244 × 4.4 mm, 5 μm) where analytes are separated by a binary mobile phase using the gradient program and conditions reported in Table 4.6. Benzoyl derivatives are detected at wavelength 254 nm. Using this approach, detection limits of putrescine, cadaverine, agmatine, tyramine, tryptamine, phenylethylamine, spermine, spermidine and histamine were lower than those obtained using classic methods.

Post-column derivatization

Kiss and Sass-Kiss (2005) developed an on-line post-column derivatization method using OPA and 2-mercaptoethanol as reagents. HPLC

Table 4.6 HPLC elution gradient program used for analysis of biogenic amine benzoyl chloride derivatives. Eluent A) 0.40 M sodium dodecyl sulfate aqueous solution buffered to pH 3.0 with 0.02 M phosphate buffer, eluent B) acetonitrile.

Time (min)	Eluent A (%)	Eluent B (%)
1	70	30
2	58	42
4	58	42
6	50	50
8	50	50

separation of BAs was performed by a C_{18} (300 × 3.9 mm, 10 μm) column forming ion pairs with octanesulfonic acid. The gradient program used is reported in Table 4.7; mobile phase flow rate 1 mL/min and OPA on-line reagent flow rate 0.8 mL/min, were used. BAs were detected by fluorescence at excitation wavelength 345 nm and emission wavelength 455 nm.

Table 4.7 HPLC elution gradient program used for analysis of biogenic amines with the method of post-column OPA derivatization. Eluent A) 0.165 M sodium acetate (pH 5.25) and 10 mM octane sulfonate; eluent B) 0.2 M sodium acetate (pH 4.5) and 10 mM octane sulfonate in acetonitrile/H_2O (66:34); eluent C) 0.01 M sodium acetate (pH 5.25) and 10 mM octane sulfonate; flow rate 1 mL/min.

Time (min)	Eluent A (%)	Eluent B (%)	Eluent C (%)
4	100	0	0
5	100	0	0
13	85	15	0
14	85	15	0
25	75	25	0
48	0	40	60
49	0	52	48
63	0	52	48
75	0	70	30
80	0	100	0
90	0	100	0
92	100	0	0

A post-column derivatization method was recently proposed for analysis of BAs–NQS derivatives in wine. A C8 (150 × 4.6mm, 4 μm) column is used with a binary gradient composed of A) 15% acetonitrile and 85% aqueous solution (15 mM sodium heptanesulfonate and 10 mM phosphoric acid) (v/v), and B) 70% acetonitrile and 30% aqueous solution (8 mM sodium heptanesulfonate and 10 mM phosphoric acid) (v/v). Eluent A is programmed at 100% for 8 min, then linearly decreases to 55% in 24 min at a flow-rate of 0.8 mL/min. The column outlet is coupled to a labelling manifold where BAs are mixed with NQS (4 mM NQS and 0.1M HCl) in a PTFE coil at 75 °C where the reaction occurs (Hlabangana *et al.*, 2006).

Gas chromatographic methods

Sample preparation and derivatization methods for GC analysis of BAs have been also proposed. In a method developed by Daudt and Ough (1980), amines are distilled from the alkalized grape juice or wine sample and trapped in an acidified solution. After concentration under vacuum, methylamine, dimethylamine, ethylamine, diethylamine, n-propylamine, isobutylamine, α-amylamine, isoamylamine, pyrrolidine, and 2-phenethylamine in their salt form are derivatized with trifluoroacetic (TFA) anhydride. TFA derivatives are extracted with ethyl ether and analyzed by GC-MS with a capillary fused silica poly(ethylene) glycol (PEG) column and the following oven temperature program: 8 min at 70 °C, 1 °C/ min to 160 °C, isotherm for 90 min.

4.4. Determination of 2,4,6-Trichloroanisole in Wine

4.4.1. Sample Preparation

Analysis of TCA and 2,4,6-trichlorophenol (TCP) in wines can be performed by SPE sample preparation using a C_{18} cartridge followed by GC/MS. The cartridge (e.g. 500 mg) is conditioned by three consecutive passages of ethyl acetate, ethanol, and 10% ethanol aqueous solution. A volume of 50 mL of wine is passed through the cartridge, then the stationary phase is dried. TCA and TCP are recovered with 0.5 mL of dichloromethane. The first 200 μL of eluate containing TCA and TCP is mixed with 200 μL of acetonitrile in order to achieve a final sample 125-fold concentrated. Recoveries between 86–102% for TCA, and 82–103% for TCP, limits of detection (LOD) of 0.1 ng/L and of quantification (LOQ) 2 ng/L for TCA, and 0.7 and 4 ng/L for TCP, are achieved (Soleas *et al.*, 2002).

A fast and sensitive method for determination of TCA in wine is to perform solid phase-microextraction (SPME) and GC/MS analysis. Extraction is performed on 10-mL sample headspace added of 2 g NaCl in a 20-mL vial. A 100-μm polydimethylsiloxane (PDMS)

1-cm length fiber, is used. After complete dissolution of the salt, the sample extraction is performed for 20 min at 30–35 °C. The fiber is then desorbed in the injector port of the GC/MS system at 250 °C for 3 min.

4.4.2. Gas Chromatography-Mass Spectrometry Analysis

Analysis of chloroanisoles in wines and cork stoppers extract (2,4-dichloroanisole, TCA, 2,3,4,6-tetrachloroanisole, pentachloroanisole, TCP, 2,3,4,6-tetrachlorophenol, pentachlorophenol) is usually performed by GC/MS using 5% diphenyl-95% dimethyl polysiloxane (e.g. 30 m × 0.25 mm i.d., 0.25 μm film thickness) columns. When a singular quadrupole mass spectrometer is used, detection is performed working in selected ion monitoring (SIM) mode and TCA is quantified on the sum of the *m/z* 195, 197, 199, 210, 212 and 214 signals, the last two from molecules containing one or two ^{37}Cl atoms, respectively.

For analysis of TCP the signals at *m/z* 196, 198 and 200, and signals at *m/z* 229, 231, 244 and 246 for tetrachlorophenol, are recorded. By performing SPME and GC/MS-SIM analysis with a single quadrupole system, LOD and LOQ are 0.2 and 0.4 ng/L, respectively (Lizarraga *et al.*, 2004). The GC/MS-Electron Impact (EI 70 eV) fragmentation spectra of TCA is showed in Figure 4.18.

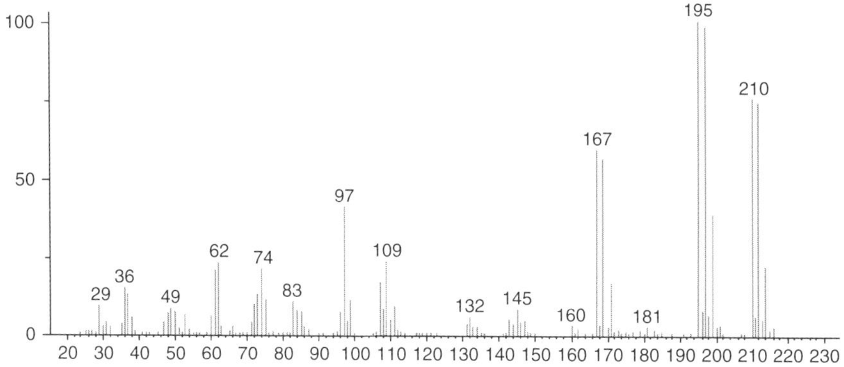

Figure 4.18 GC/MS-EI fragmentation spectra (70 eV) of 2,4,6-trichloroanisole

A chromatogram relative to GC/MS-EI analysis of TCA and TCP in wine is reported in Figure 4.19. The chromatographic conditions used are reported in the caption of the figure.

The identification, confirmation and quantification of TCA can be achieved by collision induced dissociation (CID) analysis using an ion

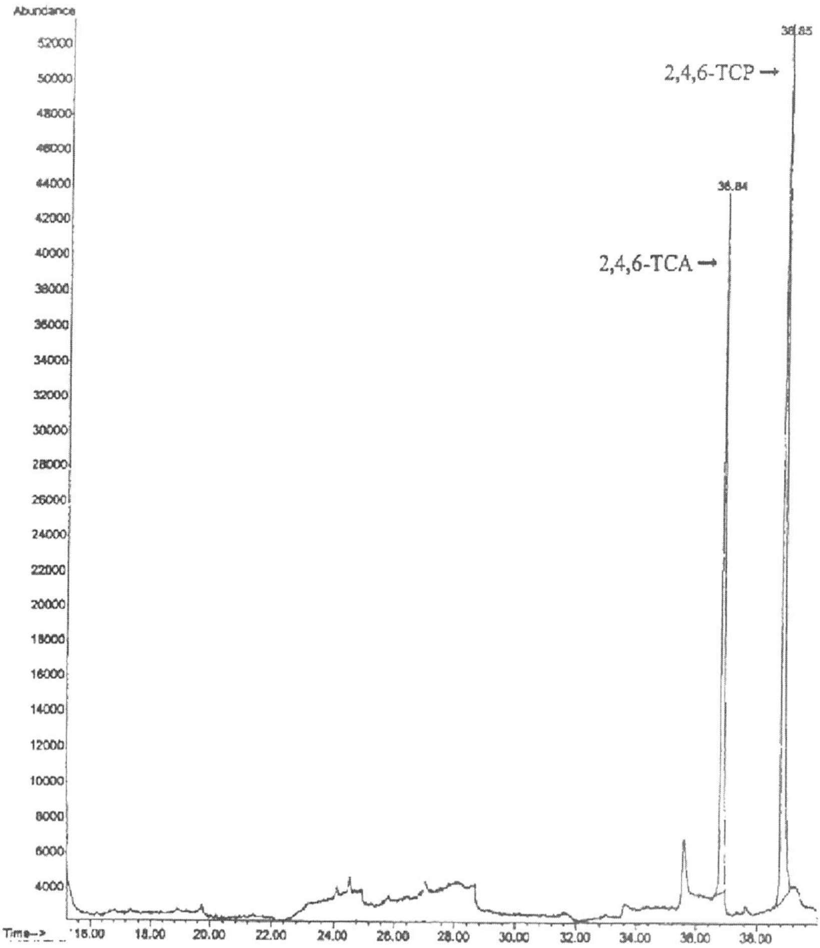

Figure 4.19 GC/MS chromatogram of total ions recorded in SIM mode in the TCA and TCP analysis of a wine extract. Chromatographic conditions: Injector and detector temperatures 200 and 240 °C, respectively; oven temperature program: 50 °C for 5 min, 1.5 °C/min until 100 °C, isotherm for 3 min, 30 °C/min until 250 °C, isotherm for 5 min. Carrier gas He; column head pressure 8 psi. (Reproduced from *J. Agric. Food Chem.*, 2002, 50, 1032–1039, Soleas *et al.*, with permission of the American Chemical Society)

trap system. Quantification is done on the signals of the daughter ions of the $M^{+\cdot}$ precursor ions at m/z 210 and 212. Depending on the type of mass spectrometer used, CID experiments can be performed in either resonant or non-resonant mode. In the first case, m/z 195 and m/z 197 are the most intense daughter ions produced, in non-resonant mode the most intense signals are at m/z 167 and m/z 169. CID of a 1 ppt TCA spiked wine shows the sum of signals m/z 167 + 169 (non-resonant) and 195 + 197 (resonant) with a signal/noise ratio of about 20. The chromatograms relative to analysis of a 1 ppt spiked red wine performed in both resonant and non-resonant mode, are shown in Figure 4.20.

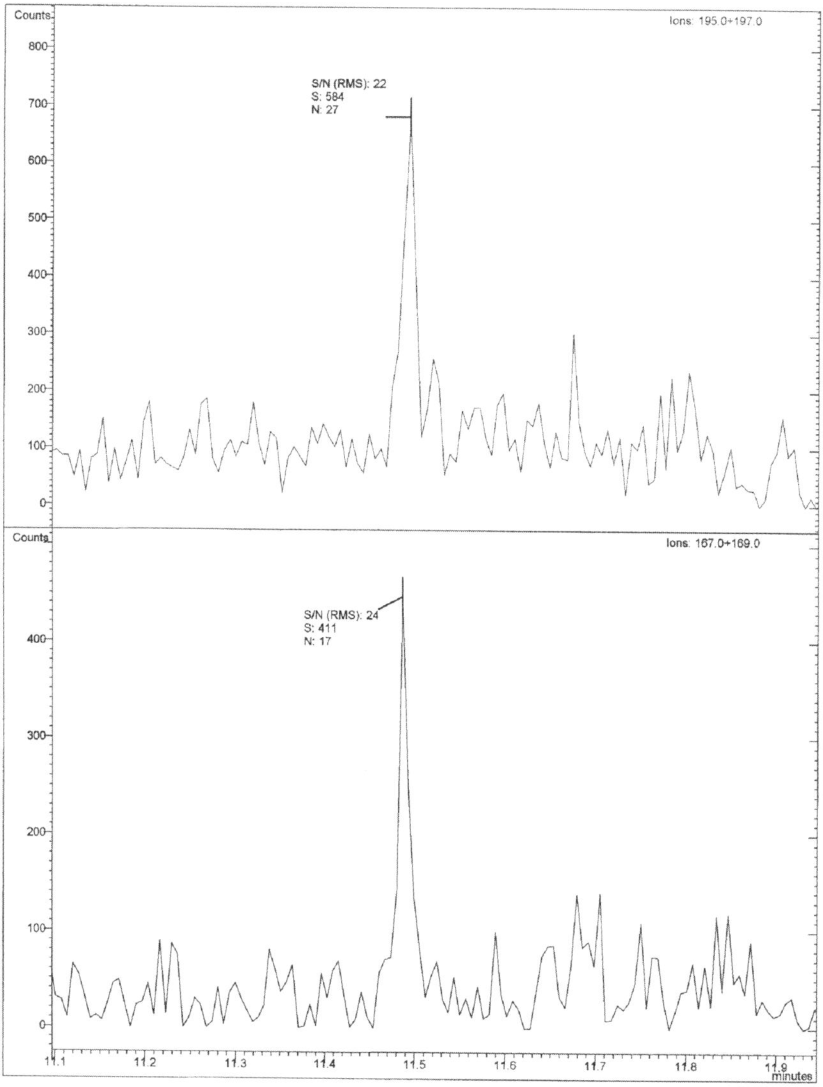

Figure 4.20 Chromatograms recorded in resonant (signal m/z 195 + 197) and not-resonant (signal m/z 167 + 169) mode in the analysis of a 1 ppt spiked red wine. Precursor ion m/z 211.9, isolation window 5 u.m.a.; excitation amplitude 80 V

4.5. Analysis of Ethylphenols in Wine

Analytically, GC equipped with a flame ionization (FID) or mass spectrometric (MS) detector is the most commonly used system for the quantification of volatile phenols, giving detection limits of a few μg/L and wide ranges of linearity of three magnitude orders (Pollnitz *et al.*, 2000; Martorell *et al.*, 2002). For the sample preparation several extraction methods have been proposed, i.e. liquid/liquid (Chatonnet *et al.*, 1992; Versini, 1985; Chatonnet *et al.*, 1995; Pollnitz *et al.*, 2000; Chatonnet and Boidron, 1988; Rocha *et al.*, 2004) and SPE techniques (Aznar *et al.*, 2001; Dominguez *et al.*, 2002), SPME and head space-SPME (Martorell *et al.*, 2002; Ferreira *et al.*, 1996; 1998; Monje *et al.*, 2002; Castro Mejías *et al.*, 2003), Stir Bar Sorptive Extraction (SBSE) (Díez *et al.*, 2004). HPLC with fluorimetric detection has also been used although less extensively than GC (Larroque *et al.*, 1987; Madigan *et al.*, 1994; Bettin *et al.*, 2002; Meyer *et al.*, 2003). Recently, two methods for the rapid measurement of volatile phenols by HPLC-fluorescence (HPLC-FLD) (Nicolini *et al.*, 2007) and HPLC-electrochemical detector (HPLC-ED) (Larcher *et al.*, 2007), without performing any sample preparation, were proposed.

4.5.1. GC-FID Analysis

Boido and co-workers (2003) proposed a method for analysis of 4-EP, 4-EG, 4-VP and 4-VG in wine. Sample preparation is performed by SPE using a cartridge packed with 1 g of highly cross-linked styrene-divinyl benzene polymer. Fifty millilitres of wine are added to 0.1 mL n-heptanol (0.2 g/L) as internal standard. The pH value of the solution is corrected to 7.0 by the addition of a diluted sodium hydroxide solution and the final volume is adjusted to 100 mL with water (Grando *et al.*, 1993). The 100-mL sample is passed through the SPE cartridge, previously rinsed with 15 mL of methanol and 20 mL of water, at a flow rate of 5 mL/min. Afterwards the cartridge is rinsed with 15 mL of water in order to remove possible matrix interferences, then analytes are eluted from the cartridge with 30 mL of methylene dichloride (5 mL/min). The eluted fraction is added to 60 mL of n-pentane, dried with anhydrous sodium sulphate, and reduced to the final volume of 2 mL at 40 °C. GC-FID analysis is performed with a PEG (30 m × 0.32 mm i.d.; 0.25 μm coating thickness) capillary fused column. This method provided good chromatographic separation in roughly 90 minutes (an example chromatogram is shown in Figure 4.21) and proved to be linear in the range between 50 and 2000 μg/L for all the 4 volatile phenols ($R^2 > 0.999$). The precision (expressed as RSD) ranged between 3% and 5%, while recoveries were higher than 90%.

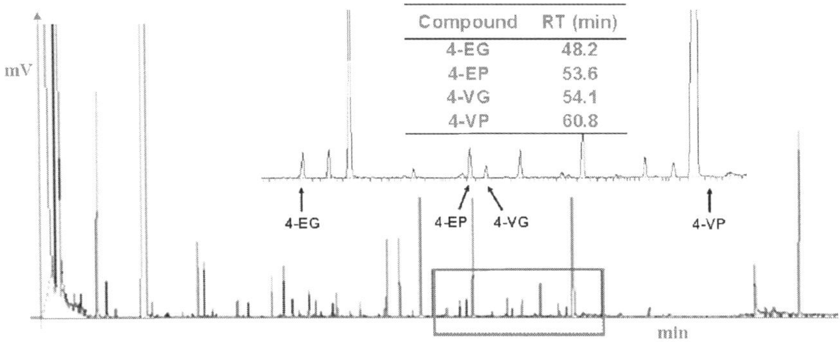

Figure 4.21 GC-FID analysis of the four VPs in a wine (analytes in exploded window). Analytical conditions: DBWax (PEG, 30 m × 0.32 mm i.d.; 0.25 μm coating thickness) capillary fused column (J&W); split injection; oven temperature program: 4 min at 40 °C, 2.5 °C/ min until 185 °C, isotherm for 15 min, 10 °C/ min until 220 °C, isotherm for 10 min; injector and detector temperature 250 °C; carrier gas He at flow rate 1.93 mL/min

4.5.2. HPLC-ED Analysis

In this method the wine sample is filtered through a 0.45 μm PTFE cartridge then analysis is performed by direct injection into the HPLC column and using an 8-electrode coulometric array detector (Larcher *et al.*, 2007). Separation is carried out under isocratic conditions with 50 mM NaH_2PO_4 (pH 3.40)/acetonitrile/methanol (65:30:5; v/v/v) eluent and a RP-C18e (125 × 3 mm, 5 μm) column kept at 25 °C. The eluent is set at a flow rate of 0.6 mL/min. Only 4 electrodes are routinely used, specifically at 340, 380, 550 and 800 mV versus Pd/H2 reference electrode, the dominant channels being at 380 mV for 4-VG and 4-EG, and at 550 mV for 4-VP and 4-EP. As a voltammetric pattern is characteristic of a specific analyte, the use of a cell with 8 electrodes (at 280, 340, 380, 430, 490, 550, 650, 800 mV) allows measurement of the specific response ratios between dominant/leading and dominant/following peaks of analytes, making possible coelution evident and electrochemically resolving it, specifically in the few cases where 4-VG partially co-elutes (see Figure 4.22).

Figure 4.23 shows hydrodynamic voltammograms of the four VPs. The oxidation of 4-VG and 4-EG starts at potentials below 300 mV,

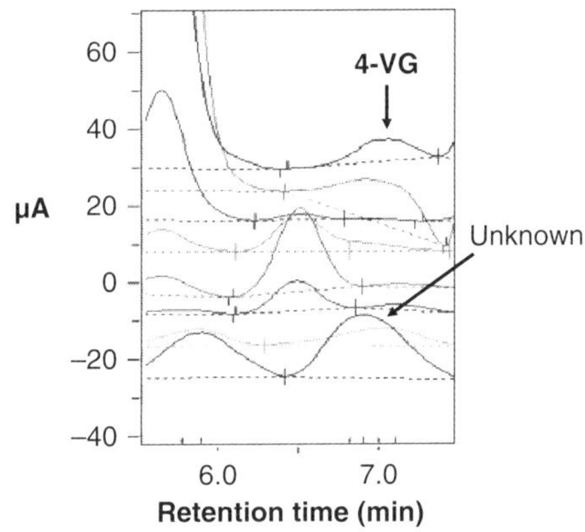

Figure 4.22 Chromatogram (cell potentials set at 280, 340, 380, 430, 490, 550, 650, 800 mV) of a wine. The co-elution of 4-vinylguaiacol with an unknown compound could be electrochemically resolved by the different response ratios between dominant/leading and dominant/following peaks

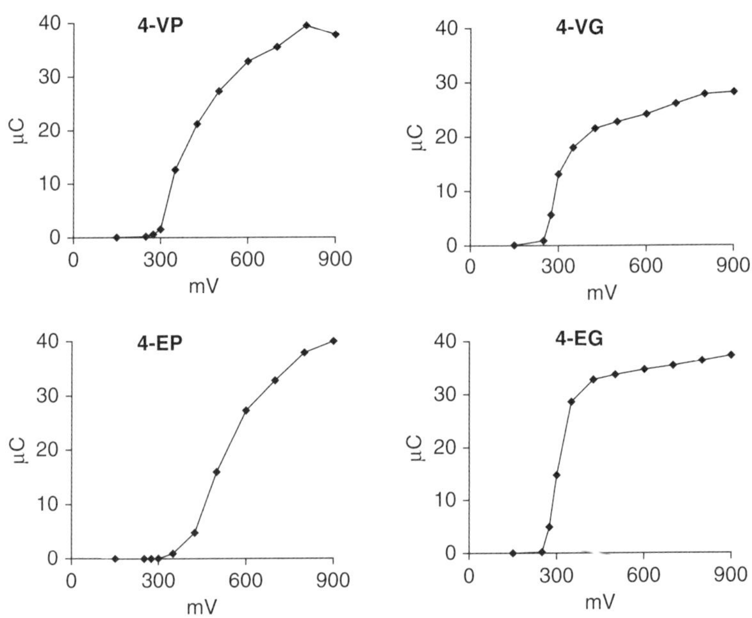

Figure 4.23 Hydrodynamic voltammograms of 4-ethylphenol (4-EP), 4-ethylguaiacol (4-EG), 4-vinylphenol (4-VP), and 4-vinylguaiacol (4-VG)

rapidly reaching plateau conditions and showing reversible or quasi-reversible wave shapes. 4-VP and 4-EF are oxidized at higher potentials, and do not reach plateau conditions within 900 mV, which is the maximum potential allowed for the graphite electrode used. Their wave shapes appear to be quasi-irreversible.

The chromatogram in Figure 4.24 shows that VPs elute in sequence, in a short time and in a zone of the chromatogram free of interfering compounds, the other electroactive compounds having far shorter retention times (RT), with the exception of eugenol. For all VPs the method is linear up to concentrations of 2000 μg/L. The RSD values are below 3% for concentrations higher than 4–7 μg/L, and below 10% for concentrations higher than 1–2 μg/L. The detection limits range between 1 and 3 μg/L and fit analytical requirements in the oenological field, being far below the sensory threshold for the compounds in wine.

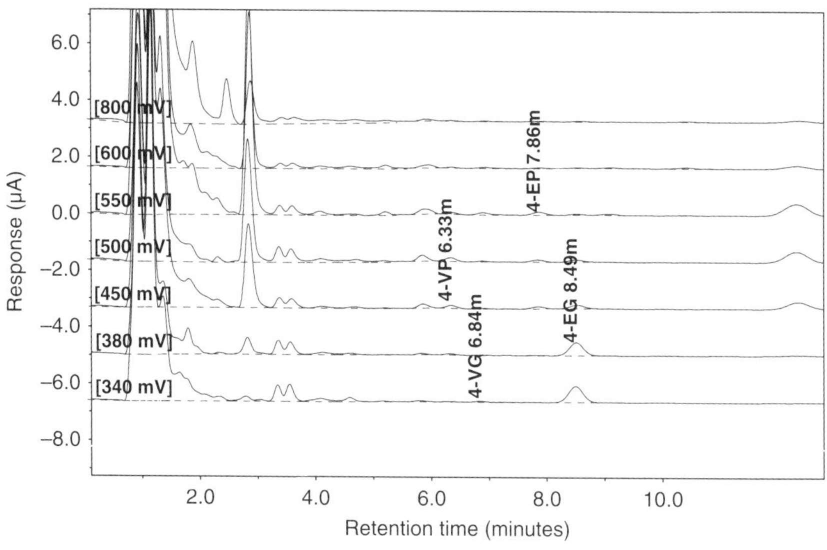

Figure 4.24 Analysis of the four volatile phenols (VPs) in a wine sample by HPLC-Coulometric array detector using 8 electrodes

4.5.3. HPLC-FLD Analysis

In this method the wine sample is filtered through a 0.45 μm PTFE cartridge and analysis is performed by direct injection into the HPLC column using a system equipped with a fluorimetric detector operating at excitation and emission wavelengths 225 and 320 nm, respectively

(Nicolini *et al.*, 2007). In Figure 4.25, the excitation and emission fluorescence spectra of VPs are shown.

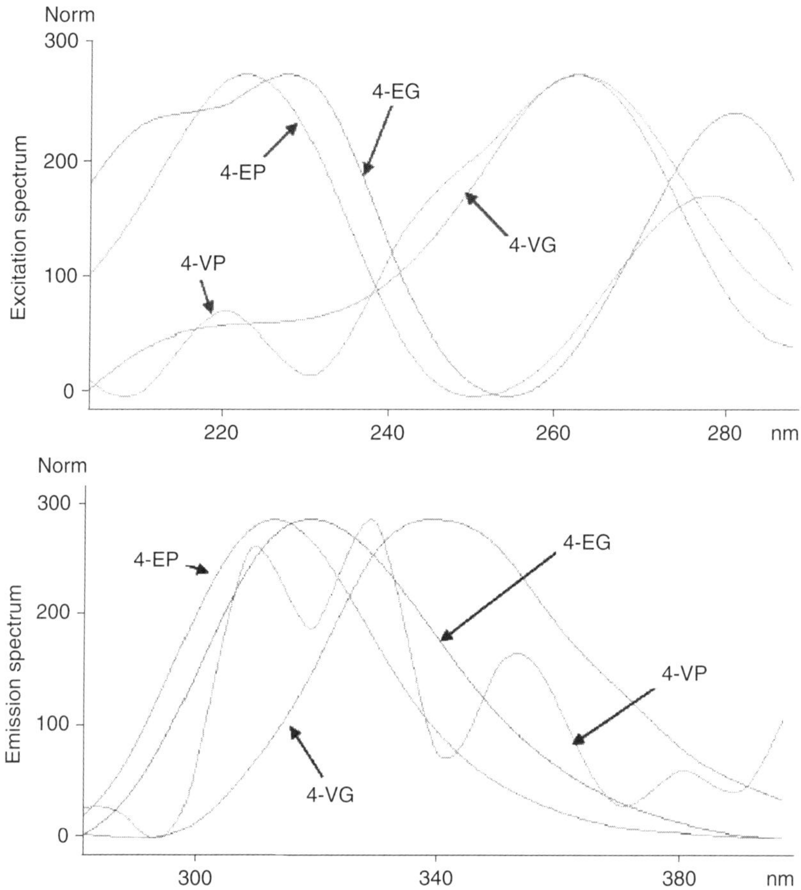

Figure 4.25 Fluorescence spectra of excitation with emission at wavelength 320 nm (above) and of emission with excitation at wavelength 220 nm (below) of the four volatile phenols (VPs)

The chromatogram in Figure 4.26 shows that VPs eluted in a very short time and in a zone apparently with no interferences. No coelution interferences are usually observed for 4-EP. Linearity of method was proved up to 2 mg/L, the critical (L_C) and detection limits are 2.0 and 4.0 μg/L, respectively. The precision (RSD %) of the HPLC-FLD method is below 3% for concentration levels higher than 20 μg/L, and below 10% for concentrations higher than 8 μg/L.

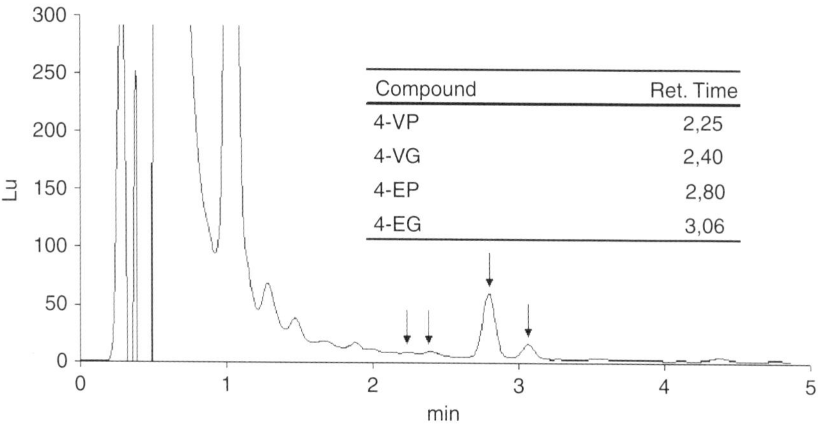

Figure 4.26 Analysis of volatile phenols (VPs) in a red wine sample by HPLC-fluorimetric detector (excitation wavelength 225 nm; emission wavelength 320 nm). Chromatographic conditions: RP-18 column (50 × 4.6 mm, 1.8 μm) at 25 °C; isocratic elution with mobile phase composed of 50 mM NaH_2PO_4 (pH 3.40)/acetonitrile/methanol 65:30:5 (v/v/v); flow rate 1.5 mL/min; injection volume 10 μL

References

Adams, D.O., Franke, K.E and Christensen, L.P. (1990) Elevated putrescine levels in grapevine leaves that display symptoms of potassium deficiency, *Am. J. Enol. Vitic.*, **41**, 121–125.

Alaiz, M., Navarro, J.L., Girón, J. and Vioque, E. (1992) Amino acid analysis by HPLC after derivatization with diethyl ethoxymethylenemalonate, *J. Chromatogr.*, **591**, 181–186.

Albagnac, G. (1975) La décarboxylation des acides cinnamiques substitués par les levures, *Ann. Technol. Agric.*, **24**, 133–141.

Alessandria, F., Borlotto, P., Cerreti, M., Echafte, M. and Roba, B. (2005) Indagine sul contenuto in etilfenoli e sulla presenza di Brettanomyces in vini albesi, *L'Enologo*, **41**(3), 101–109.

Allison, L.A., Mayer, G.S. and Shoup, R.E. (1984) *o*-Phthalaldehyde derivatives of amines for High-speed Liquid Chromatography/Electrochemistry, *Anal. Chem.*, **56**, 1089–1096.

Almy, J., Ough, C.S. and Crowell, E.A. (1983) Identification of two new volatile amines in wine, *J. Agric. Food Chem.*, **31**, 911–913.

Aung, L.H., Smilanick, J.L., Vail, P.V., Hartsell, P.L. and Gomez, E. (1996) Investigations into the origin of chloroanisoles causing musty off-flavor of raisins, *J. Agric. Food Chem.*, **44**(10), 3294–3296.

Aznar, M., Lopez, R., Cacho, J.F. and Ferreira, V. (2001) Identification and quantification of impact odorants of aged red wines from Rioja. GC-olfactometry, quantitative GC-MS, and odor evaluation of HPLC fractions, *J. Agric. Food Chem.*, **49**, 2924–2929.

Barata, A., Nobre, A., Correia, P., Malfeito-Ferreira, M. and Loureiro, V. (2006) Growth and 4-ethylphenol production by the yeast *Pichia guilliermondii* in grape juices, *Am. J. Enol. Vitic.*, **57**, 133–138.

Bauza, T., Blaise, A., Mestres, P., Teissedre, P.L., Daumas, F. and Cabanis, J.C. (1995) Teneurs en amines biogènes et facteurs de leurs variations dans les vins des côtes du Rhône, de la vallèe du Rhône et de Provence, *Sci. Aliments*, **15**, 367–380.

Bertoldi, D., Larcher, R. and Nicolini, G. (2004) Content of some free amines in grapes from Trentino. *Ind. delle Bevande*, **33**, 437–441.

Bettin, S.M., Deodato Isique, W., Franco, D.W., Andersen, M.L., Knudsen, S. and Skibsted, L.H. (2002) Phenols and metals in sugar-cane spirits. Quantitative analysis and effect on radical formation and radical scavenging, *Eur. Food Res. Technol.*, **215**, 169–175.

Bockhardt, A., Krause, I. and Klostermeyer, H. (1996) Determination of biogenic amines by RP-HPLC of the dabsyl derivates, *Z. Lebensm. Unters. Forsch*, **203**, 65–70.

Boido, E., Lloret, A., Medina, K., Fariña, L., Carrau, F., Versini, G. and Dellacassa, E. (2003) Aroma composition of *Vitis vinifera* cv. Tannat: the typical red wine from Uruguay, *J. Agric. Food Chem.*, **51**, 5408–5413.

Burdaspal, P.A. and Legarda, T.M. (1999) Ochratoxin A in wine and grape musts and juices produced in Spain and other European countries, *Alimentaria* (Madrid), (**299**), 107–113.

Burttshel, R.H., Rosen, A.A., Middleton, F.M. and Ettinger, M.B. (1951) Chlorine derivatives of phenol causing taste and odor, *J. Am. Water Works Ass.*, **51**, 205–214.

Busto, O., Guasch, J. and Borrull, F. (1995) Improvement of a solid-phase extraction method for determining biogenic amines in wines, *J. Chromatogr. A*, **718**, 309–317.

Busto, O., Valero, Y., Gulasch, J. and Borrull, F. (1994) Solid phase extraction applied to the determination of biogenic amines in wines by HPLC, *Chomatographia*, **38**, 571–578.

Careri, M., Mazzoleni, V., Musci, M. and Molteni, R. (2001) Study of electron beam irradiation effects on 2,4,6-trichloroanisole as a contaminant of cork by gas chromatography-mass spectrometry, *Chromatographia*, **53**(9–10), 553–557.

Castegnaro, M., Mohr, U., Pfohl-Leszkowicz, A., Esteve, J., Steinmann, J., Tillmann, T., Michelon, J. and Bartsch, H. (1998) Sex-and strain-specific induction of renal tumors by ochratoxin A in rats correlates with DNA adduction, *International Journal of Cancer*, **77**(1), 70–75.

Castellari, M., Fabbri, S., Fabiani, A., Amati, A. and Galassi, S. (2000) Comparison of different immunoaffinity clean-up procedures for high-performance liquid chromatographic analysis of ochratoxin A in wines, *J. Chromatogr. A*, **888**(1–2), 129–136.

Castro Mejías, R., Natera Marín, R. and García Moreno, M.d.V. (2003) Optimisation of headspace solid-phase microextraction for the analysis of volatile phenols in wine, *J. Chromatogr. A*, **995**, 11–20.

Chatonnet, P., (1993) Fenoli volatili: influenze organolettiche e metodi di prevenzione, *Vignevini*, **20**(7–8), 26–34.

Chatonnet, P. and Boidron, J.N. (1988) Dosages de phénols volatils dans les vins par chromatographie en phase gazeuse, *Sci. Aliments*, **8**, 479–488.

Chatonnet, P., Boidron, J.N. and Pons, M. (1990) Elevage des vins rouges en fûts de chêne: evolution de certains composés volatils et de leur impact aromatique, *Sci. Alim.*, **10**, 565–587.

Chatonnet, P., Dubourdieu, D. and Boidron, J.N. (1995) The influence of Brettanomyces/Dekkera sp. yeasts and lactic acid bacteria on the ethylphenol content of red wines, *Am. J. Enol. Vitic.*, **46**, 463–468.

Chatonnet, P., Dubourdieu, D., Boidron, J.N. and Lavigne, V. (1993) Synthesis of volatile phenols by Saccharomyces cerevisiae in Wines, *J. Sci. Food Agric.*, **62**, 191–202.

Chatonnet, P., Dubourdieu, D., Boidron, J.N. and Pons, M. (1992) The origin of ethylphenols in wines, *J. Sci. Food Agric.*, **60**, 165–178.

Chiodini, A.M., Scherpenisse, P. and Bergwerff, A.A. (2006) Ochratoxin A contents in wine: comparison of organically and conventionally produced products, *J. Agric. Food Chem.*, **54**(19), 7399–7404.

Coghe, S., Benoot, K., Delvaux, F., Vanderhaegen, B. and Delvaux, F.R. (2004) Ferulic acid release and 4-vinylguaiacol formation during brewing and fermentation: indications for feruloyl esterase activity in Saccharomyces cerevisiae, *J. Agric. Food Chem.*, **52**(3), 602–608.

Commission of the European Communities. (1998) Directorate General XXIV, Scientific Committee on Food, Outcome of Discussions 14, expressed on 17 September.

Couto, J.A., Campos, F.M., Figueiredo, A.R. and Hogg, T.A. (2006) Ability of lactic acid bacteria to produce volatile phenols, *Am. J. Enol. Vitic.*, **57**, 166–171.

Cristoni, S., Bernardi, L.R., Biunno, I., Tubaro, M. and Guidugli, F. (2003) Surface-activated no-discharge atmospheric pressure chemical ionization, *Rapid Commun. Mass Spectrom.* **17**(17), 1973–1981.

Daudt, C.E. and Ough, C.S. (1980) A method for detecting volatile amines in grapes and wines, *Am. J. Enol. Vitic.*, **31**, 356–360.

Delfini, C. (1989) Ability of wine lactic acid bacteria to produce histamine, *Sci. Aliments*, **9**, 413–416.

Díez, J., Domínguez, C., Guillén, D.A., Veas, R. and Barroso, C.G. (2004) Optimisation of stir bar sorptive extraction for the analysis of volatile phenols in wines, *J. Chromatogr. A*, **1025**, 263–267.

Dominguez, C., Guillén, D.A. and Barroso, C.G. (2002) Determination of volatile phenols in fino sherry wines, *Anal. Chim. Acta*, **458**, 95–102.

Dubois, P. (1983) Volatile Phenols in Wines, in *Flavour of distilled beverages, origin and development*, J.R. Piggot (Ed.) Chichester: Ellis Horwood.

Dubois, P. and Brulè, G. (1970) Étude des phénols volatiles des vins rouges, *C.R. Acad. Sci.*, **Série D.**, 1797–1798.

Eder, S., Benesic, A., Freudinger, R., Engert, J., Schwerdt, G., Drumm, K. and Gekle, M. (2000) Nephritogenic ochratoxin A interferes with mitochondrial function and pH homeostasis in immortalized human kidney ephitelial cells, *Pflug. Arch. Eur. J. Phy.*, **440**(4), 521–529.

Eder, R., Brandes, W. and Paar, E. (2002) Influence of grape rot and fining agents on the contents of biogenic amines in musts and wines. *Mitt. Klosterneuburg, Rebe Wein, Obstbau FrüchteVerwertung*, **52**, 204–217.

Etievant, P. (1991) Wine, in H. Maarse (Ed.) *Volatile compounds in foods and beverages*, New York: M. Dekker Inc.

Farias, M.E., Manca de Nadra, M.C., Rollan, G.C. and Strasser de Saad, A.M. (1993) Histidine decarboxylase activity in lactic acid bacteria from wine, *J. int. sci. vigne vin*, **27**, 191–199.

Ferreira, V., López, R., Escudero, A. and Cacho, J.F. (1998) Quantitative determination of trace and ultratrace flavour active compounds in red wines through gas chromatographic-ion trap mass spectrometric analysis of microextracts, *J. Chromatogr. A*, **806**, 349–354.

Ferreira, V., Sharman, M., Cacho, J.F. and Dennis, J. (1996) New and efficient microextraction/solid-phase extraction method for the gas chromatographic analysis of wine volatiles, *J. Chromatogr. A*, **731**, 247–259.

Flamini, R. and Panighel, A. (2006) Mass spectrometry in grape and wine chemistry. Part II: The Consumer Protection, *Mass. Spectrom. Rev.*, **25**(5), 741–774.

Flamini, R., Dalla Vedova, A., De Rosso, M. and Panighel, A. (2007) A New Sensitive and Selective Method for Analysis of Ochratoxin A in Grape and Wine by direct liquid chromatography/surface activated chemical ionization-tandem mass spectrometry, *Rapid Commun. Mass Spectrom.* **21**(22), 3737–3742.

Fuller, P. (1995) Cork taint. Closing in on an industry problem, *The Australian and New Zealand Wine Industry Journal*, **10**(1), 58–60.

Garcia Moruno, E., Sanlorenzo, C., Di Tommaso, D. and Di Stefano, R. (2004) Un metodo per la determinazione dell'ocratossina A nelle uve: applicazione allo studio dell'influenza del sistema di appassimento delle uve sul contenuto in ocratossina A, *Riv. Vitic. Enol.*, **57** (1–2), 3–11.

García-Villar, N., Saurina, J. and Hernández-Cassou, S. (2005) Determination of histamine in wines with an on-line pre-column flow derivatization system coupled to high performance liquid chromatography, *Analyst*, **130**, 1286–1290.

Geny, L., Broquedis, M., Martin-Tanguy, J. and Bouard, J. (1997) Free, conjugated, and wall-bound polyamines in various organs of fruiting cuttings of Vitis vinifera L. CV. Cabernet Sauvignon, *Am. J. Enol. Vitic.*, **48**, 80–84.

Gómez-Alonso, S., Hermosín-Gutiérrez, I. and García-Romero, E. (2007) Simultaneous HPLC analysis of biogenic amines, amino acids, and ammonium ion as aminoenone derivatives in wine and beer samples, *J. Agric. Food Chem.*, **55**, 608–613.

Grando, M.S., Versini, G., Nicolini, G. and Mattivi, F. (1993) Selective use of wine yeast strains having different volatile phenols production, *Vitis*, **32**, 43–50.

Graser, T., Godel, H., Albers, S., Földi, P. and Fürst, P. (1985) An ultra-rapid and sensitive high-performance chromatographic method for determination of tissue and plasma free amino acids, *Anal. Biochem.*, **151**, 142–152.

Herbert, P., Santos, L. and Alves, A. (2001) Simultaneous quantification of primary, secondary amino acids, and biogenic amines in musts and wines using opa/3-mpa/fmoc-cl fluorescent derivatives, *JFS: Food Chemistry and Toxicology*, **66**(9), 1319–1325.

Hlabangana, L., Hernández-Cassou, S. and Saurina, J. (2006) Determination of biogenic amines in wines by ion-pair liquid chromatography and post-column derivatization with 1,2-naphthoquinone-4-sulphonate *J. Chromatogr. A*, **1130**, 130–136.

Hohler, D. (1998) Ochratoxin A in food and feed: occurence, legislation and mode of action, *Z. Ernaehrungswiss.*, **37**(1), 2–12.

Ibe, A., Saito, K., Nakazato, M., Kikuchi, Y., Fujinuma, F. and Nishima, T. (1991) Quantitative determination of amines in wine by liquid chromatography, *J. Assoc. Off. Anal. Chem.*, **74**, 695–698.

International Organization of Vine and Wine, Measuring ochratoxin A in wine after going through an immunoaffinity column and HPLC with fluorescence detection, *Compendium of International Methods of Wine and Must Analysis* **Vol. 2** (2006).

JEFCA (Joint FAO/WHO Expert Committee on Food Additives) (2001) Ochratoxin A. In: *Safety Evaluation of Certain Mycotoxins in Food*: WHO Food Additives Series 47; FAO Food and Nutrition Paper 74; WHO: Geneva, Switzerland; p 366.

Kalač, P. and Křížek, M. (2003) A review of biogenic amines and polyamines in beer, *J. Inst. Brew.* **109** (2), 123–128.

Kiss, J., Korbász, M. and Sass-Kiss, A. (2006) Study of amine composition of botrytized grape berries, *J. Agric. Food Chem.*, **54**, 8909–8918.

Kiss, J. and Sass-Kiss, A. (2005) Protection of originality of Tokaji Aszú: Amines and organic acids in botrytized wines by high performance liquid chromatography, *J. Agric. Food Chem.*, **53**, 10042–10050.

Larcher, R., Nicolini, G., Chr. Puecher, Bertoldi, D., Moser, S. and Favaro, G. (2007) Determination of volatile phenols in wine using high-performance liquid chromatography with coulometric array detector, *Anal. Chim. Acta*, **582**, 55–60.

Larroque, M., Vian, L. and Blaise, A. (1987) Dosage du phénol par chromatographie liquide haute performance avec détection fluorimétrique. Application aux vins, *J. Chromatography*, **407**, 384–388.

Lau, B.P.Y., Scott, P.M., Lewis, D.A. and Kanhere, S.R. (2000) Quantitative determination of ochratoxin A by liquid chromatography/electrospray tandem mass spectrometry, *J. Mass. Spectrom.*, **35**, 23–32.

Lee, T.H. and Simpson, R.F. (1993) *Wine microbiology and biotechnology*, GN Fleet (Ed) Harwood Academic Publishers, Philadelphia, p. 353–372.

Lehtonen, P. (1986) Isolation and HPLC determination of amines in wine, *Z. Lebensm. Unters. Forsch.*, **183**, 177–181.

Lehtonen, P., Saarinen, M., Vesanto, M. and Riekkola, M.L. (1992) Determination of wines amines by HPLC using automated precolumn derivatisation with *o*-phthalaldehyde and fluorescence detection, *Z. Lebensm. Unters. Forsch.*, **194**, 434–437.

Lehtonen, P. (1996) Determination of amines and amino acids in wine – A review, *Am. J. Enol. Vitic.*, **47**, 127–133.

Leitao, M.C., Teixeira, H.C., Barreto Crespo, M.T. and San Romao, M.V. (2000) Biogenic amines occurrence in wine. Amino acid decarboxylase and proteolytic activities expression by *Oenococcus oeni*, *J. Agric. Food Chem.*, **48**, 2780–2784.

Leitner, A., Zöllner, P., Paolillo, A., Stroka, J., Papadopoulou-Bouraoui, A., Jaborek, S., Anklam, E. and Lindner, W. (2002) Comparison of methods for the determination of ochratoxin A in wine, *Anal. Chim. Acta*, **453** (1), 33–41.

Lindenmeier, M., Schieberle, P. and Rychlik, M. (2004) Quantification of ochratoxin A in foods by a stable isotope dilution assay using high-performance liquid chromatography-tandem mass spectrometry, *J. Chromatogr. A*, **1023** (1), 57–66.

Lizarraga, E., Irigoyen, A., Belsue, V. and González-Peñas, E. (2004) Determination of chloroanisole compounds in red wine by headspace solid-phase microextraction and gas chromatography-mass spectrometry, *J. Chromatogr. A*, **1052** (1–2), 145–149.

Lonvaud-Funel, A. (2001) Biogenic amines in wines: Role of lactic acid bacteria, *FEMS Microbiol.* Lett., **199**, 9–13.

Ma, Y., Zhang, R. and Cooper, L. (1992) Indirect photometric detection of polyamines in biological samples separated by high-performance capillary electrophoresis, *J. Chromatogr.*, **608**, 93–96.

MacDonald, S., Wilson, P., Barnes, K., Damant, A., Massey, R., Mortby, E. and Shepherd, M.J. (1999) Ochratoxin A in dried vine fruit: method development and survey, *Food Addit. Contam.*, **16** (6), 253–260.

Madigan, D., McMurrough, I. and Smyth, M.R. (1994) Rapid determination of 4-vinyl guaiacol and ferulic acid in beers and worts by High-Performance Liquid Chromatography, *J. Am. Soc. Brew. Chem.*, **52** (4), 152–155.

Martorell, N., Martì, M.P., Mestres, M., Busto, O. and Guasch, J. (2002) Determination of 4-ethylguaiacol and 4-ethylphenol in red wines using headspace-solid-phase microextraction-gas chromatography, *J. Chromatogr. A*, **975**, 349–354.

Mayer, K. and Pause, G. (1985) Senkung der Amingehalte in Wein durch Bentonitbehandlung - Laborversuche, *Schweizerische Zeitschrift fuer Obst- und Weinbau*, **121**, 203–208.

Medina, A., Valle-Algarra, F.M., Gimeno-Adelantado, J.V., Mateo, R. Mateo, F. and Jiménez, M. (2006) New method for determination of ochratoxin A in beer using zinc acetate and solid-phase extraction silica cartridges, *J. Chromatogr. A*, **1121**, 178–183.

Meilgaard, M.C. (1975) Aroma volatiles in beer: purification, flavour, threshold and interaction, in Geruch- und Geschmackstoffe, F., Drawert (Ed.), H., Carl, Nürnberg.

Meyer, J., Liesener, A., Goetz, S., Hayen, H. and Karst, U. (2003) Liquid chromatography with on-line electrochemical derivatization of phenols, *Anal. Chem.*, **75**, 922–926.

Molins Legua, C., Campíns Falcó, P. and Meseguer Lloret, S. (2003) Derivatization on solid supports: an alternative method for solution derivatization of amines in several matrices, *Chromatographia*, **58** (1–2), 15–27.

Monje, M.C., Chr. Privat, Gastine, V. and Nepveu, F. (2002) Determination of ethylphenol compound in wine by headspace solid-phase microextraction in conjunction with gas chromatography and flame ionization detection, *Anal. Chim. Acta*, **458**, 111–117.

Montedoro, G. and Bertuccioli, M. (1986) The flavour of wines, vermouth and fortified wines, in *Food flavours*, I.D., Morton and A.J., MacLeod (Eds.), Elsevier, Amsterdam.

Narziss, L., Miedaner, H. and Nitzsche, F. (1990) Ein Beitrag zur Bildung von 4-Vinyl-Guajakol bei der Herstellung von bayerischem Weizenbier, *Monatsschrift fuer Brauwissenschaft*, **43** (3), 96.

Nicolini, G., Larcher, R. and Bertoldi, D. (2003) Free amines in grape juices of Vitis vinifera L. wine varieties", *J. Comm. Sci.*, **42** (2), 67–78.

Nicolini, G., Larcher, R., Moser, S., Carlin, S. and Versini, G. (2006) Cantina e ricerca, tra caratterizzazione, sviluppo e tutela, *OICCE Times*, 7 (28), 13–17.

Nicolini, G., Larcher, R., Bertoldi, D., Chr. Puecher and Magno, F. (2007 *in press*) Rapid quantification of 4-ethylphenol in wine using high-performance liquid chromatography with a fluorimetric detector, *Vitis*.

Paleologos, E.K. and Kontominas, M.G. (2004) On-line solid-phase extraction with surfactant accelerated on-column derivatization and micellar liquid chromatographic separation as a tool for the determination of biogenic amines in various food substrates, *Anal. Chem.*, **76**, 1289–1294.

Pfohl-Leszkowicz, A., Pinelli, E., Bartsch, H., Mohr, U. and Castegnaro, M. (1998) Sex- and strain-specific expression of cytochrome P450s in ochratoxin A-induced genotoxicity and carcinogenicity in rats, *Mol. Carcinogen.*, **23** (2), 76–85.

Pollnitz, A.P., Pardon, K.H. and Sefton, M.A. (2000) Quantitative analysis of 4-ethylphenol and 4-ethylguaiacol in red wine, *J. Chromatogr. A*, **874**, 101–109.

Radler, F. and Fath, K.-P. (1991) Histamine and other biogenic amines in wines, in J. Rantz (Ed.) *Proceedings of International Symposium on Nitrogen in Grapes and Wine*, American Society for Enology and Viticulture, Davis (CA).

Rapp, A., Versini, G. (1996) Flüchtige phenolische Verbindungen in Wein, *Deutsche Lebensmittel-Rundschau*, **92** (2), 42–48.

Rocha, S.M., Rodrigues, F., Coutinho, P., Delgadillo, I. and Coimbra, M.A. (2004) Volatile composition of Baga red wine. Assessment of the identification of the would-be impact odourants, *Anal. Chim. Acta*, **513**, 257–262.

Romero, R., Gazquez, D., Bagur, M.G. and Sanchez-Vinas, M. (2000) Optimization of chromatographic parameters for the determination of biogenic amines in wines by reversed-phase high-performance liquid chromatography, *J. Chromatogr. A*, **871** (1–2), 75–83.

Romero, R., Sanchez-Vinas, M., Gazquez, D. and Bagur, M.G. (2002) Characterization of Selected Spanish Table Wine Samples According to Their Biogenic Amine Content from Liquid Chromatographic Determination, *J. Agric. Food Chem.*, **50**, 4713–4717.

Sáez, J.M., Medina, A., Gimeno-Adelantado, J.V., Mateo, R. and Jiménez, M. (2004) Comparison of different sample treatments for the analysis of ochratoxin A in must, wine and beer by liquid chromatography, *J. Chromatogr. A*, **1029** (1–2), 125–133.

Sauvant, C., Silbernagel, S. and Gekle, M. (1998) Exposure to ochratoxin A impairs organic anion transport in proximal-tubule-derived opossum kidney cells, *J. Pharmacol. Exp. Ther.*, **287**, 13–20.

Schwerdt, G., Freudinger, R., Mildenberger, S., Silbernagl, S. and Gekle, M. (1999) The nephrotoxin ochratoxin A induces apoptosis in cultured human proximal tubule cells, *Cell Biol. Toxicol.*, **15** (6), 405–415.

Serra, R., Mendonça, C., Abrunhosa, L., Pietri, A. and Venâncio, A. (2004) Determination of ochratoxin A in wine grapes: comparison of extraction procedures and method validation, *Anal. Chim. Acta*, **513** (1), 41–47.

Shalaby, A.R. (1996) Significance of biogenic amines to food safety and human health, *Food Research International*, **29**, 675–690.

Shephard, G.S., Fabiani, A., Stockenström, S., Mshicileli, N. and Sewram, V. (2003) Quantitation of Ochratoxin A in South African Wines, *J. Agric. Food Chem.*, **51** (4), 1102–1106.

Soleas, J., Carey, M. and Goldberg, D.M. (1999) Method development and cultivar-related differences of nine biogenic amines in Ontario wines, *Food Chem.*, **64**, 49–58.

Soleas, G.J., Yan, J., Seaver, T. and Goldberg, D.M. (2002) Method for the gas chromatographic assay with mass selective detection of trichloro compounds in corks and wines applied to elucidate the potential cause of cork taint, *J. Agric. Food Chem.*, **50** (5), 1032–1039.

Stratton, J.E., Hutkins, W.R. and Taylor, S.L. (1991) Biogenic amines in cheese and other fermented foods. A review, *J. Food Protect.*, **54**, 460–470.

Taylor, S.L. (1985) *Histamine Poisoning Associated with Fish, Cheese and Other Foods*, FAO/WHO Monograph VPH/F05/85.1, 1–47.

Timperio, A.M., Magro, P., Chilosi, G. and Zolla, L. (2006) Assay of ochratoxin A in grape by high- pressure liquid chromatography coupled on line with an ESI-mass spectrometry, *J. Chromatogr. B*, **832** (1), 127–133.

Tonus, T., Dalla Vedova, A., Spadetto, S. and Flamini, R. (2005) Studio dei metodi chimici per il monitoraggio dell'ocratossina A (OTA) nell'uva. *Riv. Vitic. Enol.*, **58** (1), 3–13.

van Wyk, C.J. and Rogers, I.M. (2000) A 'phenolic' off-odour in white table wines: causes and methods to diminish its occurrence, *S. Afr. J. Enol. Vitic.*, **21**, 52–57.

Versini, G. (1985) Sulĺ Aroma del vino Traminer Aromatico o Gewürztraminer, *Vignevini*, **12** (1–2), 57–65.

Versini, G., Dalla Serra, A., Falcetti, M. and Sferlazzo, G. (1992) Rôle du clone, du millésime et de l'époque de la récolte sur le potentiel aromatique du raisin de Chardonnay, *Revue des Oenologues*, **18** (65S), 19–23.

Visconti, A., Pascale, M. and Centonze, G. (1999) Determination of ochratoxin A in wine by means of immunoaffinity column clean-up and high-performance liquid chromatography, *J. Chromatogr. A*, **864** (1), 89–101.

Waterval, J.C.M., Lingeman, H., Bult, A. and Underberg, W.J.M. (2000) Derivatization trends in capillary electrophoresis, *Electrophoresis*, **21** (18), 4029–4045.

Webb, A.D. (1967) Wine flavor: volatile aroma compounds of wines, in Schultz, H.W., Day, E.A. and Libbey, L.M. (Eds.) *The Chemistry and Physiology of Flavours*, The AVI Publishing Company Inc., Westport Co.

Yamamoto, S., Itano, H., Kataoka, H. and Makita, M.J. (1982) Gas-liquid chromatographic method for analysis of di- and polyamines in foods, *J. Agric. Food Chem.*, **30**, 435–439.

Zimmerli, B. and Dick, R. (1996) Ochratoxin A in table wine and grape juice: Occurrence and risk assessment, *Food Addit. Contam.*, **13** (6), 655–668.

Zöllner, P., Leitner, A., Luboki, D., Cabrera, K., Lindner, W. (2000) Application of a Chromolith SpeedROD RP-18e HPLC column: Determination of ochratoxin A in different wines by high-performance liquid chromatography-tandem mass spectrometry, *Chromatographia*, **52** (11–12), 818–820.

5

Analysis of Aroma Compounds in Wine

Giuseppe Versini, Eduardo Dellacassa, Silvia Carlin, Bruno Fedrizzi and Franco Magno

5.1. Aroma Compounds in Wines

Monoterpenes, norisoprenoids – some C_{13} megastigmane compounds – some benzenoid compounds called shikimates because of their origin from shikimic acid, and some polyfunctional sulphured compounds, are the most important grape-derived aroma compounds. In the winemaking process, these molecules are transferred to the wine both as free and bound forms.

Monoterpenes[1], mostly present as alcohols, ethers and hydrocarbons, represent the most important category associated with basic floral aroma, beginning from linalool recognized in Muscat grapes about 50 years ago (Cordonnier, 1956). Free and bound monoterpenes are

[1] a wide list of these compounds was reported by Strauss *et al.* (1986), then updated with *cis* and *trans* 8-hydroxylinalool (Rapp *et al.*, 1986) and 6,7-dihydro form (Versini *et al.*, 1988), some menthanediols (Versini *et al.*, 1991), the monoterpenoid acid (*E*)-2,6-dimethyl-6-hydroxyocta-2,7-dienoic acid (Wintherhalter *et al.*, 1998) and some ethers like 2-*exo*-hydroxy-1,8-cineol (Bitteur *et al.*, 1990), wine-lactone (Guth, 1995) and 1,8-cineol (Fariña *et al.*, 2005), structures reported in Figure 5.1a.

Hyphenated Techniques in Grape and Wine Chemistry Edited by Riccardo Flamini

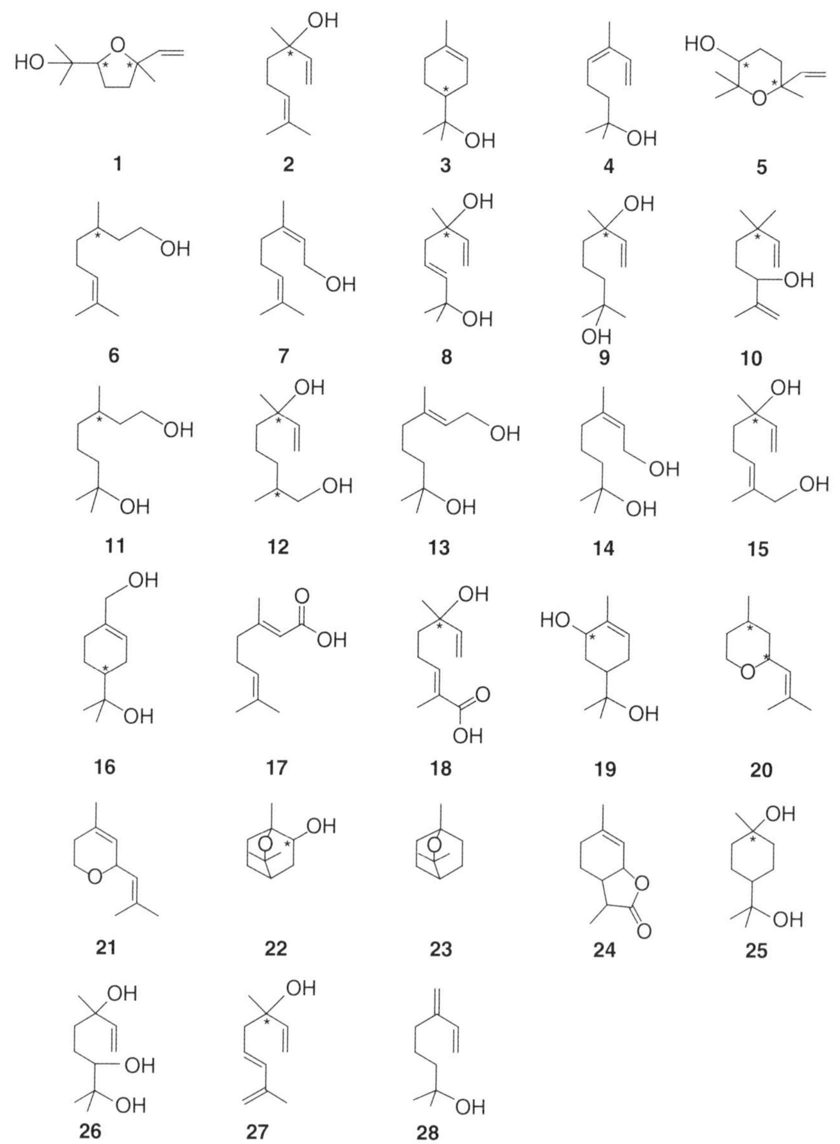

Figure 5.1a (1) *cis* and *trans* linalool oxide (5-ethenyltetrahydro-α,α,5-trimethyl-2-furanmethanol) (furanic form); (2) linalool (3,7-dimethyl-1,6-octadien-3-ol); (3) α-terpineol (α,α,4-trimethyl-3-cycloexene-1-methanol); (4) (Z,E) ocimenol (2,6-dimethyl-5,7-octen-2-ol); (5) *cis* and *trans* linalool oxide (pyranic form); (6) citronellol (3,7-dimethyl-6-octen-1-ol); (7) nerol (Z), geraniol (E) (3,7-dimethyl-2,6-octadien-1-ol); (8) Ho-diendiol I (3,7-dimethyl-1,5-octadiene-3,7-diol); (9) endiol (3,7-dimethyl-1-octene-3,7-diol); (10) Ho-diendiol II (3,7-dimethyl-1,7-

synthesized in the berry during its ripening just as diastereoisomeric and enantiomeric forms (Luan and Würst, 2002; Luan *et al.*, 2005). They are important markers of the grape varieties (Rapp *et al.*, 1978). The profile of aglycones remains a good marker even in aged wine (Versini *et al.*, 1997). In the pH of must and wine some monoterpenes are involved in chemical acid catalyzed reactions that can reduce, or change, the aroma contribution: e.g. alcohols like linalool (floral scenting) are partially transformed into α-terpineol and successively in 1,8-terpines, geraniol and nerol are transformed into linalool and α-terpineol, etc., some non-floral diols and a triol in part generating sensorially active substances like hydrocarbons cyclic ethers as neroloxide, roseoxide, anhydrofuranes and anhydropyranes (Strauss *et al.*, 1986). Recently, the formation of the highly odorant wine-lactone from (*E*)-2,6-dimethyl-6-hydroxyocta-2,7-dienoic acid was evidenced (Wintherhalter *et al.*, 1998).

Most alcoholic compounds are present, sometimes in dominant amounts, as β-glycoconjugates forms, mostly as monoglucosides like β-D-glucopyranoside, and disaccharides as 6-O-α-L-arabinofuranosyl-, 6-O-β-D-apiofuranosyl-, and 6-O-α-L-ramnopyranosyl-β-D-glucopyranosides (Ribéreau-Gayon *et al.*, 1998), which can be converted into free forms principally by hydrolysis in the case of α,β-unsaturated tertiary alcohols, or by enzymatic reaction by the yeasts or supplied exogenous enzymes in the other cases (Skouroumounis and Sefton, 2000). In these reactions, traces of potent flavourant ethyl ethers can arise (Strauss *et al.*, 1985). Other compounds are generated during the fermentation, e.g. (*R*)-(+)-citronellol from geraniol and nerol with an enantiospecific reduction operated by the yeast (Gramatica *et al.*, 1982; Di Stefano *et al.*, 1992).

Figure 5.1a (Continued) octadiene-3,6-diol); (11) hydroxycitronellol (3,7-di methyloctane-1,7-diol); (12) 8-hydroxydihydrolinalool (2,6-dimethyl-7-octene-1,6-diol); (13) 7-hydroxygeraniol (*E*-3,6-dimethyl-2-octene-1,7-diol); (14) 7-hydroxynerol (*Z*-3,6-dimethyl-2-octene-1, 7-diol); (15) *cis* and *trans* 8-hydroxy linalool (*E*- and *Z*-2,6-dimethyl-2,7-octadiene-1,6-diol); (16) p-menthenediol I (p-menth-1-ene-7,8-diol); (17) *E*-geranic acid (3,7-dimethyl-2,6-octadienoic acid); (18) *E*-2,6-dimethyl-6-hydroxyocta-2,7-dienoic acid; (19) *E*- and *Z*-sobrerol or p-menthenediol II (p-menth-1-ene-6,8-diol); (20) *cis* and *trans* rose oxide; (21) nerol oxide; (22) 2-*exo*-hydroxy-1,8-cineol; (23) 1,8-cineol; (24) wine lactone; (25) *cis* and *trans* 1,8-terpin; (26) triol (2,6-dimethyloctane-2,3,6-triol); (27) hotrienol [(5*E*)-3,5-dimethylocta-1,5,7-trien-3-ol]; (28) myrcenol (2-methyl-6-methylene-7-octen-2-ol)

Profiling and content of such compounds in grape products depends on the grape variety and how it interacts with berry ripeness and climatic and agronomic factors (Marais *et al.*, 1992).

The same factors influence the profile of another chemical group of compounds – norisoprenoids – whose contribution is important in determining some specific ageing aroma due to the evolution of free and bound polyhydroxylate C_{13} megastigmane structures (Williams *et al.*, 1989 and 1992; Winterhalter *et al.*, 1990; Wintherhalter, 1992; Knapp *et al.*, 2002; Versini *et al.*, 2002). Among them, the most important flavouring compounds are reported in Figure 5.1b, such as 1,1,6-trimethyl-1,2-dihydronaphthalene (TDN), vitispiranes, actinidols and β-damascenone with kerosene, resinous/eucalyptus-like, woody and rose-like scents, respectively. Their precursors seem to derive from carotenoid cleavage dioxygenase in relation to C_{13}-norisoprenoid accumulation (Mathieu *et al.*, 2007). Some of such compounds are considered as important contributors to the typical evolution aroma of some white (*Chardonnay*, *Semillon* and *Sauvignon blanc*) and red (*Shiraz, Grenache, Merlot*, *Cabernet Sauvignon*) variety wines (Williams *et al.*, 1989; Abbot *et al.*, 1991; Gunata *et al.*, 2002; Williams and Francis, 1996; Francis *et al.*, 1998) and grapes (Flamini *et al.*, 2006).

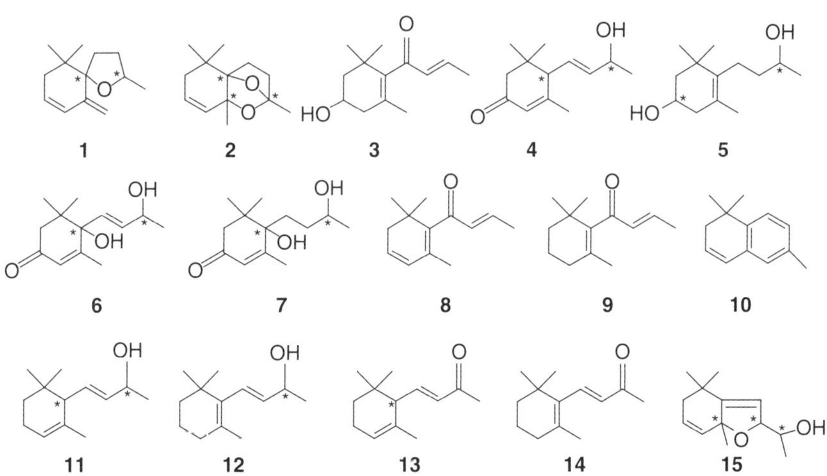

Figure 5.1b (1) vitispiranes (spiro [4.5]-2,10,10-trimethyl-6-methylene-1-oxa-7-decene); (2) riesling acetal (2,2,6-tetramethyl-7,11-dioxatricyclo[6.2.1.$0^{1,6}$]un dec-4-ene); (3) 3-hydroxy-β-damascone; (4) 3-oxo-α-ionol; (5) 3-hydroxy-7,8-di hydro-β-ionol; (6) vomifoliol; (7) dihydrovomifoliol; (8) β-damascenone; (9) β-damascone; (10) TDN (1,1,6-trimethyl-1,2-dihydronaphthalene); (11) α-ionol; (12) β-ionol; (13) α-ionone; (14) β-ionone; (15) actinidols

Other aroma contributors can be shikimates compounds, i.e. benzenoid substances, that result as free and bound forms (Williams *et al.*, 1983; Williams *et al.*, 1989; Winterhalter *et al.*, 1990). The most indicated flavouring compounds are zingerone and zingerol, vanillin, methyl vanillate and, recently, methyl salicylate, which is particularly abundant in an Italian native variety, the *Verdicchio* or *Trebbiano di Soave*, where it has been found to have a content ranging between 0.5 and 1 mg/L (expressed as aglycone). In this variety of grape, a level of benzyl alcohol ranging between 2–3 mg/L was also found (Versini *et al.*, 2005).

The majority of volatiles in wines are, however, fermentation compounds, starting with the so-called higher alcohols and going on to important flavouring compounds such acetates as of higher alcohols and the ethyl ester of C_4-C_{10} fatty acids for the fruity scents. The free fatty acids, when in abundance, contribute to the 'goaty' flavour (Meilgaard, 1975).

Other aroma compounds are the C_6 alcohols such as 1-hexanol and *cis* and *trans* 3-hexen-1-ol, sometimes at a rather high level as in the *Müller-Thurgau* wines (Nicolini *et al.*, 1995), and other ones such as 2-phenylethanol and relevant acetate, rose-like scenting. The level of most such compounds is determined by the technologies and procedures used in the winemaking (vinification for white or red wines, skin contact, clarification, fermentation temperature, yeast strain and assimilable nitrogen level). In the last case, a positive correlation has been proved between the nitrogen level in must and acetates, and a negative correlation with higher alcohols, 2-phenylethanol and methionol (Rapp and Versini, 1991). Wide literature references are reported in the review papers of Schreier (1979) and Rapp (1988).

5.2. Methods for Analysis of Non-Sulphured Aroma Compounds in Wines

In the last 15–20 years, several analytical methods have been reported for extraction of aroma compounds from wine finalized to replace the time-consuming (10–24 hours), continuous liquid-liquid methods. Liquid-liquid extraction was usually performed either with dichloromethane/pentane (2:1 v/v) (Drawert and Rapp, 1968) or Freon 11, sometimes in a mixture Freon 11/dichloromethane 9:1 (v/v) (Hardy, 1969; Rapp and Hastrich, 1978; Marais, 1986), both approaches being considered as rather extensive and repeatable for volatiles profiling of wines.

Alternative methods are the fast liquid-liquid discontinuous extraction with 1,1,2-trichloro-1,2,2-trifluoroethane (Freon 113 or 'Kaltron') (Ferreira *et al.*, 1993; Rapp *et al.*, 1994; Genovese *et al.*, 2005), and the solid-phase-extraction (SPE) with a polystyrenic resin like Amberlite XAD-2 (Gunata *et al.*, 1985; Versini *et al.*, 1988), or with a C_{18} reverse phase (Williams, 1982; Di Stefano, 1991). SPE approaches have the advantage that both the 'free' forms and the so-called glycosidically-bound compounds can be isolated and separated in the same process. More recently, a highly crosslinked hydroxylated polystyrenic phase like ISOLUTE®ENV^{+} has been marketed as a more active adsorbing material and applied to the wine aroma analysis (Boido *et al.*, 2003), as well as a highly crosslinked ethylvinylbenzene-divinylbenzene copolymer like LiChrolut EN (Genovese *et al.*, 2005). The supercritical fluid extraction (SFE) has not been popular: methanol is sometimes added to CO_2 to achieve a better recovery of the more polar compounds (Blanch *et al.*, 1995).

In the early '90s, a new technique called solid-phase-micro extraction (SPME), was developed (Arthur and Pawliszyn, 1990). The key-part component of the SPME device is a fused silica fiber coated with an adsorbent material such as polydimethylsiloxane (PDMS), polyacrylate (PA) and carbowax (CW), or mixed phases such as polydimethylsiloxane-divinylbenzene (PDMS-DVB), carboxen-polydimethylsiloxane (CAR-PDMS) and carboxen-polydimethylsiloxane-divinylbenzene (CAR-PDMS-DVB). The sampling can be made either in the headspace (Vas *et al.*, 1998) or in the liquid phase (De la Calle *et al.*, 1996) of the samples. The headspace sampling in wine analyses is mainly useful for quantifying trace compounds with a particular affinity to the fiber phase, not easily measurable with other techniques. Exhaustive overviews on materials used for the extraction-concentration of aroma compounds were published by Ferreira *et al.* (1996), Eberler (2001), Cabredo-Pinillos *et al.* (2004) and Nongonierma *et al.* (2006). Analysis of the volatile compounds is usually performed by gas chromatography (GC) coupled with either a flame ionization (FID) or mass spectrometry (MS) detector.

In the next part of this chapter some of these techniques are compared, pointing out possible limitations and advantages in free and bound forms aroma analysis, both in the research and winemakers' laboratories. In one paragraph, recent quantification methods for some sulphur compounds groups are discussed.

5.2.1. Sample Preparation

Liquid-liquid extraction with Freon 113 (Kaltron)

Liquid-liquid extraction is carried out in a 10 mL centrifuge tube, where 1 g of sodium chloride, 0.1 mL of 2-octanol solution as internal standard (230 ppm in 50% v/v hydroalcoholic solution) and 50 μL of Freon 113 are added to 5 mL of the wine sample (Rapp *et al.*, 1994). The solution is stirred for 15 min and then centrifuged. A volume of 2 μL of the Freon 113 extract is analysed by GC. An exhaustive presentation of this method, applying also a fully automated system with a large volume injection, was performed by Rapp *et al.* (1996).

Solid-phase-extraction with a polystyrenic resin

The experimental conditions used for wine analysis are reported in the literature (Gunata *et al.*, 1985; Versini *et al.*, 1988; Voirin *et al.*, 1992), where both the best particle dimension (i.e. 0.1–0.3 mm) and eluent for the free forms (i.e. pentane/dichloromethane 2:1 v/v), were established.

To eliminate impurities, the Amberlite® or XAD-2®-like resin is washed with methanol, diethyl ether and pentane/dichloromethane 2:1 (v/v) and water in sequence, taking care to continue the elution with pentane/dichloromethane until the impurities level is noticeably reduced (Versini *et al.*, 1988). A glass column (30 × 1 cm i.d.) packed with approximately 12 cm of prepurified resin (0.2–0.25 mm) and conditioned with methanol (80 mL), diethyl ether (40 mL) and distilled water (100 mL), is used. A volume of 100 mL of wine is diluted with 100 mL of distilled water; 0.2 mL of 2-octanol internal standard solution (230 ppm) is then added and percolated through the column. Possible wine residual are washed with 100 mL distilled water. Free aroma components are eluted with 100 mL of pentane/dichloromethane (2:1 v/v), the solution is dried with sodium sulphate and concentrated to 1 mL distilling in a water bath at 40 °C using a short Vigreux column. Bound forms are eluted from the resin by a second washing with 100 mL of ethyl acetate/methanol (9:1 v/v) and evaporated to dryness. The residual is dissolved in 3 mL of citrate buffer at pH 5 and treated with a β-glycosidase enzyme with a convenient activity at 40 °C overnight. Aglycones are extracted three times with 3 mL of pentane/dichloromethane (2:1 v/v) and the organic phase is dried and concentrated to 0.2 mL.

Solid-phase-extraction with pre-packed polyhydroxylated styrene-divinylbenzene polymer and silica-based C_{18} cartridges

A highly cross-linked polyhydroxylated styrene-divinylbenzene polymer ENV$^{+®}$ cartridge (1 g × 40–140 μm; Isolute, IST Ltd., Mid Glamorgan, UK) is conditioned with methanol (15 mL) and distilled water (20 mL). A volume of 0.1 mL of internal standard 2-octanol (230 ppm in 50% hydroalcoholic solution) is added to 100 mL of the sample – composed of wine diluted with distilled water to achieve an ethanol content lower than 8% v/v. This solution is then percolated through the cartridge in about 10 min working under partial vacuum with a multiple sample processing system in order to facilitate and standardize the process. After the sample has passed through the cartridge, the stationary phase is washed with 15 mL of distilled water.

The free aroma compounds are eluted with 30 mL of dichloromethane, then the bound forms with 30 mL of methanol. The dichloromethane solution is concentrated by distillation after approximately double the volume of pentane is added to obtain the azeotropic mixture with a boiling point of 31 °C. The methanol solution containing bound forms is evaporated to dryness; treatment by β-glycosidase enzyme is performed as described in the previous paragraph.

SPE extraction of wine aroma can also be performed by using a C_{18} 300-mg cartridge previously activated with 2 mL methanol and 2 mL distilled water. Ten millilitres of wine is diluted with 30 mL of water and 2-octanol (internal standard) is added to it. The solution is passed through the cartridge, stationary phase is rinsed with 3 mL water and then dried with flushing air. Free aroma compounds are recovered with 3 mL of dichloromethane and the solution is concentrated under nitrogen flow to about 100 μL. Glycoside compounds are recovered with 3 mL of methanol (Di Stefano, 1991).

Headspace solid-phase-microextraction

The headspace (HS) SPME analysis can be performed by means of a 100 μm PDMS fiber (Carlin, 1998). Sodium chloride (0.9 g) and 8 μL 2-octanol solution as internal standard (230 ppm) are added to 4 mL of wine in a 10 mL vial; this is stirred (1500 rpm) for 10 min in a water-bath at 30 °C to stabilize the conditions of the solution; then the fiber is placed in the sample headspace. After stirring for 40 min, the fiber is taken out and inserted into the injection port of the GC to perform gas chromatographic analysis.

Recently, the effectiveness of a three-phase fiber for aroma compounds analysis has been proved. In particular, the CAR-PDMS-DVB phase (50/30 μm × 1cm) appears to be effective in overcoming the lack of selectivity towards some compounds of one-phase or two-phase fiber. Furthermore, the use of an autosampling system permits an easier development of methods and a larger applicability for quality control in the laboratory (Fedrizzi and Versini, unpublished data). In this case, 1 g of NaCl in a 20 mL vial was added to 5 mL of wine and 5 μL methyl heptanoate (200 ppm in a methanolic solution) as internal standard. The sampling was performed during stirring at 40 °C for 20 min.

5.2.2. Gas Chromatographic Analysis

For unambiguous identification of compounds also present at trace level, the gas chromatography-mass spectrometry (GC-MS) approach is compulsory. By extensive optimal single compound peaks separation at fixed working conditions as proved for once by a GC-MS control, GC-flame ionization detector (FID) analysis is a suitable and reasonable routine method to be considered. Quite apolar capillary columns have been used – mostly in the past – to detect compounds with a very large boiling point range as in the case of aglycones profiling of benzenoids and norisoprenoids. Recently, bound phase polar capillary columns stable at rather high temperatures are the most suitable for wine aroma analysis, especially of the free forms in presence of short and longer-chain aliphatic acids.

Analyses can be carried out with FID using a polar poly(ethylene) glycol (PEG) (30 m × 0.32 mm i.d., 0.3 μm film thickness) fused silica capillary column operating with carrier gas H_2(2 mL/min), split ratio 30:1, injector temperature 250 °C, detector temperature 240 °C, and oven temperature programme starting at 65 °C isotherm for 4 min, from 65 to 185 °C at a rate of 2.5 °C/min, 185 °C isotherm for 20 min.

The organic samples can also be injected directly into a GC without any preconcentration using a 'large volume' on column system injection. This was tested in the case of liquid-liquid extraction with pentane/dichloromethane (2:1 v/v), by comparing the peak areas obtained with both injection methods for different compound categories present in wines in different levels, such as isoamyl acetate, acetoin, n-hexanol, *trans* 3-hexenol, *cis* 3-hexenol, ethyl octanoate, linalool, diethyl succinate, hexanoic and octanoic acids, N(3-methylbutyl)-acetamide, monoethyl succinate, 4-vinylphenol and 4-vinylguaiacol

(Versini *et al.*, 1995). Repeatability of the GC areas by large volume injection method (six repetitions) ranged from 0.81 to 1.89% for RSD.

GC-MS analysis can be performed employing the same capillary column used for GC-FID but reducing the internal diameter to 0.25 mm and working with the same oven temperature program, or by a nearly apolar phase (6% cyanopropylphenyl – 94% methylpolysiloxane, e.g. PS 264 Mega, Legnano, Italy) (25 m × 0.25 mm i.d., 0.5 μm film thickness) fused silica capillary column. In this case, a suitable oven temperature program is to start at 50 °C isotherm for 1 min, from 50 to 60 °C at a rate of 10 °C/min, 60 °C isotherm for 0.5 min, to 200 °C at a rate of 2.5 °C/min, finally, 200 °C isotherm for 30 min. Other operating conditions are carrier gas He (1.5 mL/min), injection mode splitless, injector temperature 250 °C, MS interface temperature 220 °C. To perform SPME analysis the fiber is desorbed in the GC injector for 5 min.

5.2.3. Comparison Among the Analytical Methods

Polystyrene-resin (XAD-2) extraction and HS-SPME sampling

For the precision of the HS-SPME method we can refer to the work of Vas *et al.* (1998) with a peaks evaluation based on percentage, and to the data of Carlin (1998) obtained more properly referring to internal standard method. Ulrich *et al.* (1997) also achieved precise results by using SPME analysis on fruit juices. Table 5.1, referring to the working conditions of one of the chromatograms in Figure 5.2, reports a study of repeatability for some wine compounds adsorbed in a HS-SPME 1 cm PDMS fiber and analysed by GC-MS.

GC-MS chromatograms of a *Müller-Thurgau* wine sample extracted with both polystyrenic resin (XAD-2) and a 100 μm PDMS HS-SPME fiber, are shown in Figure 5.2.

Differences in retention times in the chromatograms are due to the different oven temperature program used. Some remarkable differences in the profiles evidence the different enrichment obtained by the two methods. Data reported in Table 5.2 and Figure 5.3 show evidence of generally good linear correlations between XAD-2 extraction and HS-SPME enrichment methods. In the conditions used for SPME/GC-FID analysis, some compounds such as octanoic and decanoic acids partially overlap ethyl myristate and ethyl palmitate, respectively.

By HS-SPME using a PDMS coated fiber, a rather low adsorption for 2-phenylethanol and also, perhaps tenfold more, for 1-hexanol, is

Table 5.1 Repeatability (mean Ax/Ai.s. value of 5 data) for some compounds in a young *Gewürztraminer* wine adsorbed by HS-SPME in a 1 cm PDMS fiber and analysed by GC-MS using a PEG capillary column. * Values referred to Ax/Ai.s. (*m/z* 45) with 2-octanol as internal standard. S.D.: standard deviation.

Compounds*	Ax/[Ai.s. (*m/z* 45)]	S.D.	Cv%
Isoamyl acetate	28.5	1.1	3.7
n-Hexyl acetate	5.7	0.27	4.7
2-Phenethyl acetate	1.6	0.06	4.1
Ethyl hexanoate	25.8	0.60	2.3
Ethyl octanoate	257	11.3	4.4
Ethyl nonanoate	0.42	0.02	5.7
Ethyl decanoate	223	13.0	5.8
Ethyl 9-decenoate	14.3	1.2	8.2
Hexanol	0.33	0.02	6.8
2-Phenylethanol	0.61	0.04	5.8
Vitispiranes (*m/z* 192)	0.004	0.0003	8.3
β-Damascenone (*m/z* 69)	0.024	0.003	10.7
Linalool (*m/z* 71)	0.026	0.002	6.5
Citronellol	0.34	0.02	6.4
Nerol (*m/z* 93)	0.004	0.001	13.0
Geraniol (*m/z* 69)	0.020	0.002	8.0
Linalyl ethylether	0.57	0.05	8.6
α-Terpenyl ethylether	0.23	0.01	6.4
Geranyl ethylether	0.53	0.04	7.1
Neryl ethylether	0.34	0.03	8.8
Methyl *trans* geraniate (*m/z* 69)	0.035	0.001	3.2
Citronellyl acetate	0.50	0.03	6.6
Nerolidol	0.11	0.01	10.9

observed. Tertiary alcohols such as linalool are much more extractable, i.e. about 200, and 20 times more of 2-phenylethanol and 1-hexanol, respectively. Geraniol and, to a greater extent, citronellol, are also rather well adsorbed (Carlin, 1998). In a recent paper reporting the HS-SPME/GC-MS analysis of varietal compounds in grape homogenate (Sanchéz-Palomo *et al.*, 2005), a more reproducible adsorption efficiency has been described by working with another apolar phase such as the mixed PDMS/DVB, obtaining adsorption yields for monoterpenols and benzenoids including 2-phenylethanol in a similar relationship to the one reported in Figure 5.3. Among the compounds measurable in more favourable analytical conditions by HS-SPME are the aliphatic esters whose enrichment, related to the same internal standard, can range from 2 to 30-fold that obtained with XAD-2 method. Lower absorption on the fiber is observed for 2-phenylethyl acetate and, in particular, for diethyl succinate (Figure 5.3).

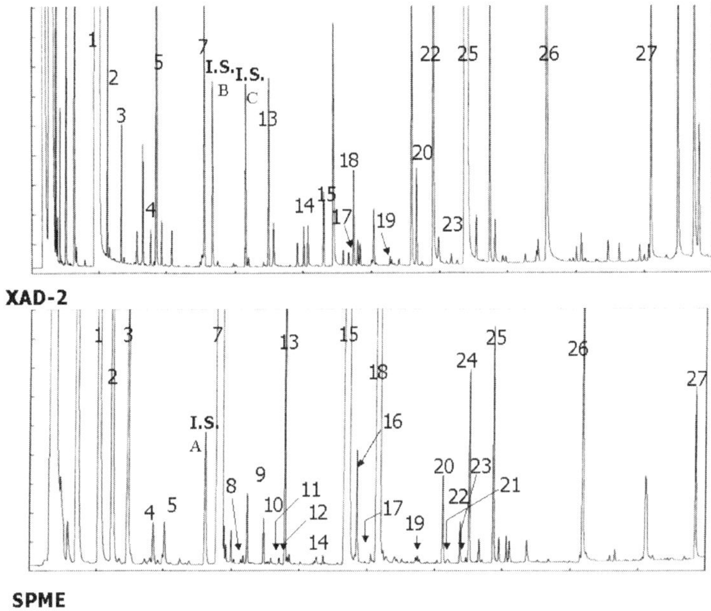

Figure 5.2 GC-MS aroma profiling of a *Müller-Thurgau* wine using SPE/XAD-2 and HS-SPME (PDMS fiber) volatiles enrichment techniques. 1. 2- and 3-methylbutanols (isoamyl alcohols); 2. ethyl hexanoate; 3. hexyl acetate; 4. linalyl ethylether; 5. 1-hexanol; 7. ethyl octanoate; 8. neryl ethylether; 9. ethyl 7-octenoate; 10. geranyl ethylether; 11. vitispiranes; 12. a diol ethylether; 13. linalool; 14. ho-trienol; 15. ethyl decanoate; 16. citronellyl acetate+isoamyl octanoate; 17. α-terpineol; 18. ethyl 9-decenoate; 19. citronellol; 20. 2-phenethyl acetate; 21. β-damascenone; 22. hexanoic acid; 23. geraniol; 24. ethyl dodecanoate; 25. 2-phenylethanol; 26. octanoic acid; 27. decanoic acid

Finally, we would like to emphasize the HS-SPME technique as a very useful tool for evaluating non-polar compounds in traces as shown in Figure 5.2, such as esters like ethyl 9-decenoate or monoterpenes, as well as some norisoprenoids and monoterpenols derivatives such as the ethoxyethers (Carlin, 1998; Versini *et al.*, 1999). HS-SPME efficiency in the higher alcohols quantification was verified by evaluating results with those obtained by direct injection of the wine distillate.

Recently, other mixed phases for SPME have been considered, like CAR-PDMS (Sánchez-Palomo *et al.*, 2005) and CAR-PDMS-DVB (Sánchez-Palomo *et al.*, 2005; Fedrizzi and Versini, unpublished data) (Figure 5.4). Calibration data calculated by performing extraction of an ethanol 12% (v/v) synthetic wine solution containing several compounds representing different functional groups, and methyl heptanoate as

Table 5.2 Linear relationships for some wine aroma components between GC-peak areas ratios (Ax/Ai.s.) for Kaltron liquid-liquid or HS-SPME aroma extraction *vs* SPE/XAD-2 (mg/L) extraction as reference. *y =[Area component]/[Area i.s.] by Kaltron or HS-SPME. X=[Area component]/[Area i.s.] x RF by XAD-2 (mg/L). a=angular coefficient. **In italics are presented the relationships involving HS-SPME.

compounds	*linear correlation* $y = ax + b$*	*determination coefficient* (R^2)
1-hexanol	0.13x + 0.002 *0.093x – 0.02***	0.965 *0.996*
linalool	1.04x + 0.03 *1.51x + 0.0002*	0.874 *0.997*
2-phenylethyl alcohol	0.02x + 0.02 *0.01x + 0.008*	0.898 *0.937*
isoamyl acetate	1.38x – 0.02 *1.73x + 0.001*	0.998 *0.999*
hexyl acetate	1.66x – 0.004 *6.26x + 0.006*	0.996 *0.997*
ethyl hexanoate	1.55x – 0.02 *7.06x – 0.17*	0.989 *0.997*
ethyl octanoate	1.54x – 0.09 *28.70x – 0.79*	0.991 *0.992*
ethyl decanoate	1.62x – 0.03 *29.47x + 0.37*	0.988 *0.986*
ethyl 9-decenoate	*18.62x + 0.19*	*0.973*
2-phenylethyl acetate	1.28x – 0.04 *0.80x + 001*	0.981 *0.981*
diethyl succinate	0.50x + 0.09 *0.09 + 0.02*	0.996 *0.995*
hexanoic acid	0.17x – 0.17	0.864
octanoic acid	1.09x – 0.55	0.975
decanoic acid	1.54x – 0.19	0.974

internal standard, using a 1 cm CAR-PDMS-DVB fiber, are reported in Table 5.3.

As a consequence of the sensitivity of the method, higher values of the slope are reported for the hexanoate, octanoate and decanoate ethyl esters, and secondly for the isoamyl and 2-phenethyl acetates. On the contrary, 2-phenylethanol and ethyl acetate showed the smallest enrich-

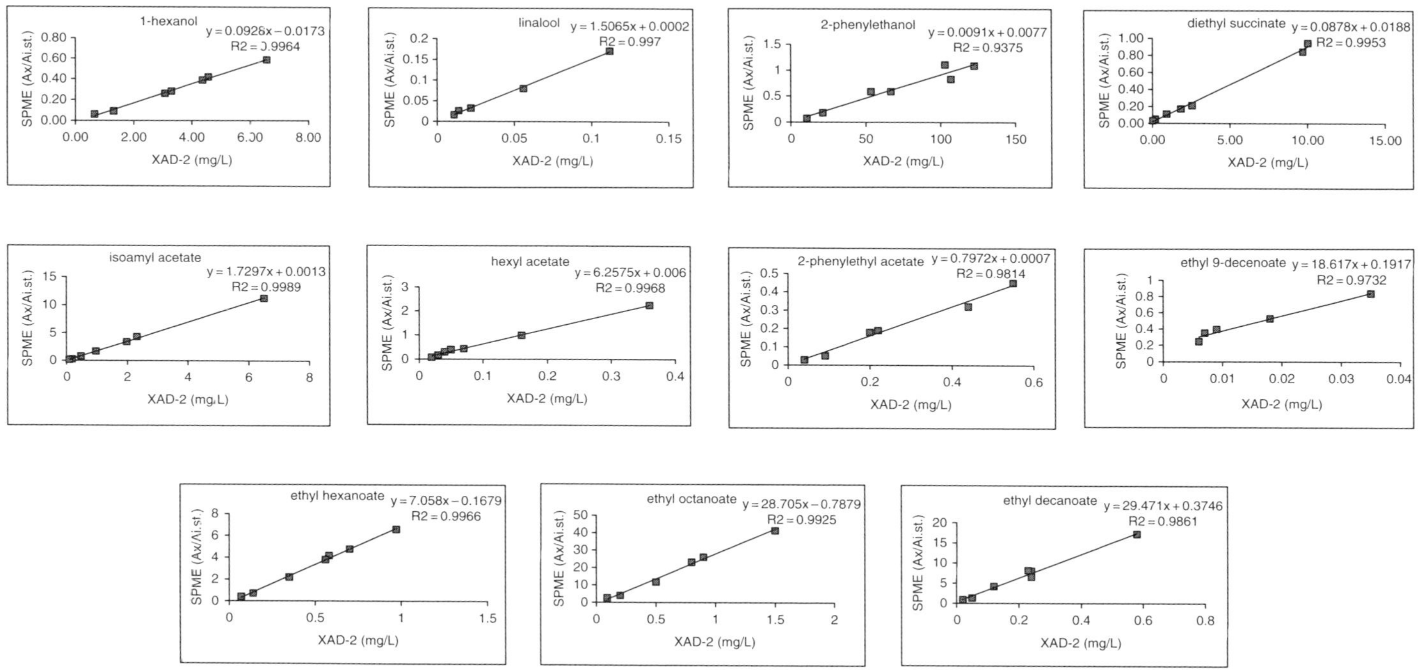

Figure 5.3 Linear relationships between concentrations (mg/L) of some aroma compounds evaluated by GC-FID after XAD-2 enrichment technique in different wines and the relevant ratios of the compounds areas to that of internal standard (2-octanol evaluated on the fragment at m/z 45) determined by GC-MS after HS-SPME (PDMS fiber) extraction (*some scores could be hidden*)

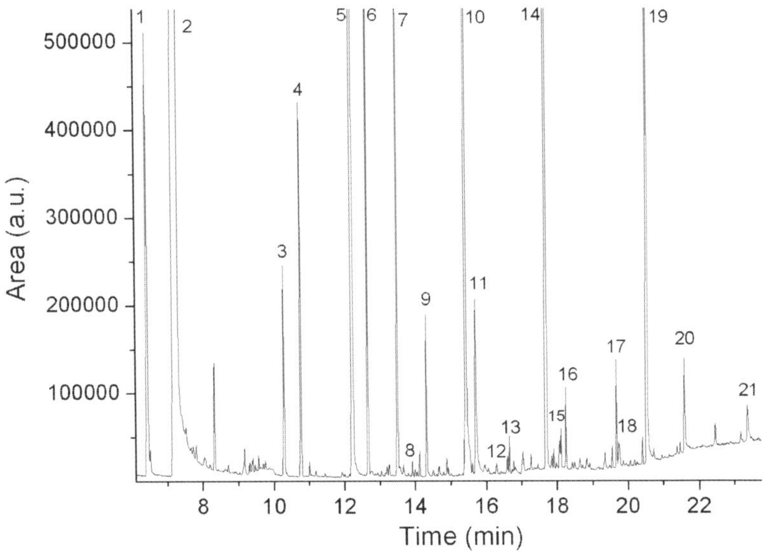

Figure 5.4 GC-MS total ion current (TIC) chromatogram of a red Italian wine (*Corvina*) after HS- SPME automated aroma enrichment using a 1 cm CAR-PDMS-DVB fiber. 1. ethyl acetate, 2. ethanol, 3. isobutyl alcohol, 4. isoamyl acetate, 5. isoamyl alcohols, 6. ethyl hexanoate, 7. methyl heptanoate (I.S.), 8. linalyl ethylether, 9. 1-hexanol, 10. ethyl octanoate, 11. acetic acid, 12. geranyl ethylether, 13. linalool, 14. ethyl decanoate, 15. diethylsuccinate, 16. ethyl 9-decenoate, 17. 2-phenethyl acetate, 18. hexanoic acid, 19. 2-phenylethanol, 20. octanoic acid, 21. decanoic acid

Table 5.3 Calibration parameters of some aroma compounds in a synthetic wine solution enriched by automated HS-SPME with a 1 cm CAR-PDMS-DVB fiber and analysed by GC-MS. S.D.: standard deviation.

Compound	Slope	S.D. Slope	Intercept	S.D. Intercept	S.D.	R^2
Ethyl acetate	0.75	0.09	0.08	0.01	0.30	0.996
Isoamyl acetate	43.07	1.02	1.01	0.11	2.72	0.997
Ethyl hexanoate	160.91	2.56	1.12	0.12	1.40	0.998
Ethyl octanoate	731.64	8.64	1.38	0.31	4.01	0.997
Ethyl decanoate	703.03	8.95	1.76	0.16	6.09	0.998
2-Phenylethyl alcohol	4.87	0.12	0.28	0.03	0.55	0.995
2-Phenethyl acetate	24.33	0.41	0.13	0.42	1.11	0.998

ment but the R^2 factor was not affected and the results were still good. The automated approach seems to improve remarkably the precision of the method, as well as the optimized volume of liquid extracted with respect to the fiber adsorption capability, thus avoiding a possible overloading of the fiber due to a rather high concentration of compounds in the sample volume. To perform analysis by the triphase fiber using an autosampling system, a PEG (60 m × 0.32 mm; 0.25 μm film thickness) fused silica capillary column with oven program temperature 50 °C isotherm for 4 min, from 50 °C to 250 °C at a rate of 10 °C/min, 250 °C for 5 min and desorption time of 5 min, can be used.

Kaltron and polystyrene-resin (XAD-2) extraction

Applications of the Kaltron method and GC-MS in wine aroma analysis were reported by Rapp *et al.* (1996). The study of the data repeatability in the analysis performed using both Kaltron (Rapp *et al.*, 1994) and polystyrene XAD-2 resin (Gunata *et al.*, 1985; Versini *et al.*, 1988; Voirin *et al.*, 1992) extraction, has already been performed in the cited papers. Besides this, a test with eight repetitions of the whole analytical process using XAD-2 resin has been performed both on a non-floral (*Pinot blanc*) and a floral (*Morio-Muskat*) varietal wine for the quantification of 25 varietal and fermentation compounds (Carlin, 1998): for each wine sample the mean CV % values ranged from about 7.0 to 7.5 with a standard deviation from about 3 to 4.7 %, respectively, depending on the different level of some compounds, mostly monoterpenes, in each wine type.

In Table 5.2 and Figure 5.5, data for a quantification comparison between the two techniques referred to some compounds important for the wine quality such as esters, alcohols and fatty acids relative to analysis of different wines, are reported. Figure 5.5 shows the straight line regressions between the concentrations measured after XAD-2 extraction and the ratios of GC-FID area of each compound to that of the internal standard applying the Kaltron method.

Among alcohols (Figure 5.5) rather good sensitivity and correlations are evidenced for 1-hexanol and linalool, compounds representing indexes of the grape pressing and floral scenting, respectively. A lower extraction of 2-phenylethyl alcohol, an interesting contributor to the rose-like scent of wines at concentrations higher than 50 mg/L, is achieved by Kaltron. The level of this compound is negatively related with the whole amino acid level – except proline – in the must (Rapp and Versini, 1991), or linked to the metabolism of particular yeast strains such as *Saccharomyces cerevisiae r.f. uvarum* (Ciolfi and Di Stefano, 1983). Enrichment by Kaltron can be extended to the isoamyl alcohols; for these compounds, results obtained by the use of XAD-2 resin are

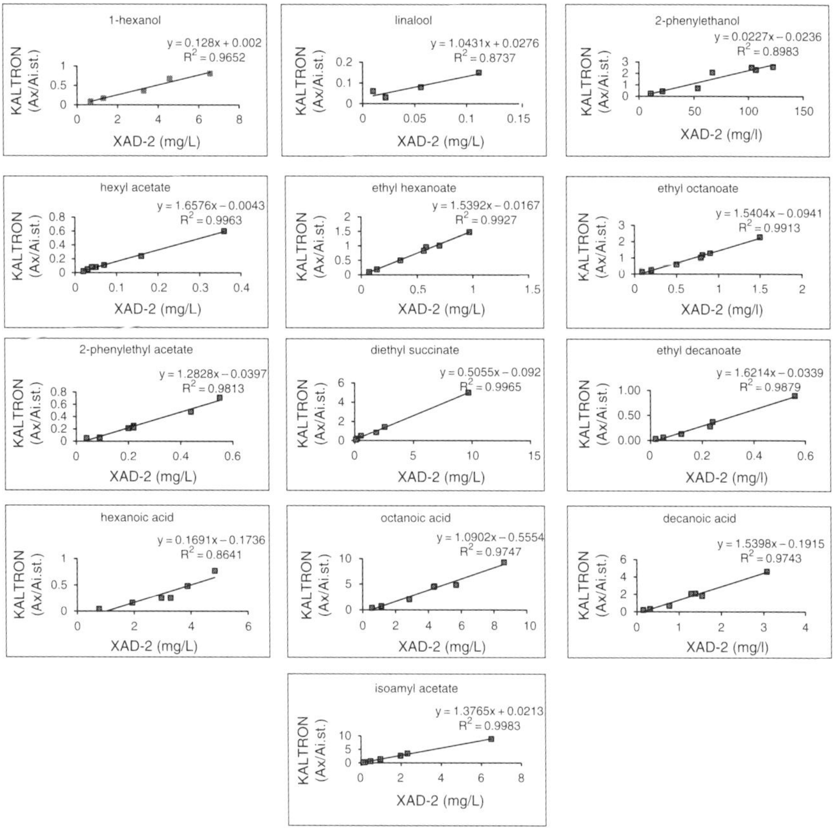

Figure 5.5 Linear relationships between concentrations (mg/L) of aroma compounds evaluated in 7 different wines by GC-FID analysis after XAD-2 enrichment and the Ax/Ai.s. ratios (i.s. 2-octanol) after Kaltron liquid-liquid extraction (*some scores could be hidden*)

not usually reproducible and the same behaviour is observed for ethyl lactate.

Among esters, compounds relevant in the characterization of the fruity aroma of wine are isoamyl acetate, hexyl acetate, 2-phenylethyl acetate and the ethyl esters of the hexanoic, octanoic and decanoic acids. Diethyl succinate is considered as an index of wine ageing and bacterial activity. For all these compounds and the corresponding fatty acids, there is a good linear correlation between XAD-2 enrichment and Kaltron liquid-liquid extraction. A little higher extraction, related to the internal standard, of Kaltron with respect to XAD-2 resin is obtained for most esters and acids, but not for the diethyl succinate and the short-chain acids such as butanoic and isovaleric acid. The choice of 2-octanol as internal standard seems to be a good compromise among the different classes of compounds.

A recent comparison between Kaltron extraction and SPE with LiChrolut EN resin has been performed by Genovese *et al.* (2005). LiChrolut EN seems to be, due to its highly crosslinked ethylvinylbenzene – divinylbenzene polymer, more effective than XAD-2 in the adsorption of free forms. For all analyzed compounds with largely different polarity, both methods showed a good linearity for the calibration curves in different model wine solutions. The exhaustive adsorption of free forms in wines by LiChrolut EN has been confirmed by Gomez-Minguez *et al.* (2007). For Ibarz *et al.* (2006), the extraction of bound (glycoside) compounds in wine by this resin resulted between two and six-fold higher than that which was achieved by XAD-2 and C_{18} sorbents (see later).

SPE by polyhydroxylated styrene-divinylbenzene (ENV^+) or C_{18} cartridge and polystyrene-resin (XAD-2) extraction

As may be seen from the data in Table 5.4, good repeatability of the quantitative data by ENV^+ cartridge extraction, comparable with that by polystyrene resin, is achieved, except of some higher boiling ethyl esters and some more polar esters as ethyl lactate, succinate and 4-hydroxybutyrate for which a better fixation on the ENV^+ cartridge, was observed (Carlin, 1998).

All the less polar compounds are almost completely fixed on the ENV^+ cartridge (>99%) as was proved by analyzing the extract achieved by elution of analytes from the cartridge using dichloromethane. The solution not retained from the cartridge during the sample passage, was added with 1-heptanol as internal standard, and passed through the cartridge a second time. In the case of more polar compounds, recoveries higher than 90% for isobutyric and butyric acids, ethyl 3-hydroxybutyrate, ethyl 4-hydroxybutyrate, 3-(methylthio)-1-propanol (methionol), 70% for 1-butanol, 2-methyl-1-butanol, ethyl lactate, and lower than 50% for 3-ethoxy-1-propanol, 1,3-propandiol monoacetate and γ-butyrolactone, were observed. Other authors reported a higher efficiency in the recovery of some glycols and acids (Lindstrom *et al.*, 2004). We observed that some impurities are released from the cartridge and can interfere, in a few cases, with the qualitative and quantitative GC-FID analysis. This occurs mostly by using a polyethylene glycol (PEG) capillary column and with particular oven temperature programs.

Table 5.4 Repeatability of four extractions of free aroma compounds from different wines using a polyhydroxylated styrene-divinylbenzene (ISOLUTE®ENV$^+$) cartridge, and comparison with mean values of data of two repeated extractions by polystyrene resin XAD-2. *Value Ax/Ai.s. after GC analysis; i.s. 2-octanol.

Compounds*	mean	S.D.	Cv %	polystyrene resin
Isoamyl acetate	6.49	*0.25*	3.8	6.71
Hexyl acetate	0.22	*0.01*	2.4	0.29
2-Phenethyl acetate	0.44	*0.04*	9.2	0.46
Ethyl hexanoate	1.52	*0.09*	5.9	1.41
Ethyl octanoate	3.19	*0.10*	3.2	2.30
Ethyl decanoate	1.14	*0.15*	12.9	0.67
Ethyl lactate	8.35	*0.67*	8.0	1.83
Diethyl succinate	0.51	*0.02*	4.3	0.47
Ethyl 4-hydroxybutyrate + nerol	3.50	*0.14*	4.1	1.17
Ethyl monosuccinate	14.7	*0.67*	4.5	3.83
Hexanoic acid	7.33	*0.40*	5.5	7.22
Octanoic acid	12.7	*0.53*	4.2	12.5
Decanoic acid	3.96	*0.18*	4.7	3.51
Isovaleric acid	0.68	*0.04*	6.6	0.61
Butanol	1.22	*0.05*	4.1	0.27
Isoamyl alcohols	291	*8.09*	2.8	232
Hexanol	2.42	*0.11*	4.7	2.46
trans 3-Hexenol	0.34	*0.02*	6.8	0.30
cis 3-Hexenol	0.14	*0.01*	5.9	0.11
Benzyl alcohol	0.11	*0.02*	16.5	0.09
2-Phenylethanol	27.1	*1.02*	3.8	25.9
4-Vinylguaiacol	0.56	*0.03*	6.2	0.55
4-Vinylphenol	0.30	*0.013*	4.3	0.35
Terpenes				
Linalool	0.09	*0.00*	4.12	0.08
α-Terpineol	0.05	*0.00*	7.86	0.04
Geraniol	0.11	*0.01*	11.29	0.11
Ho-diendiol (I)	0.28	*0.02*	6.8	0.23
trans Geranic acid	0.47	*0.02*	5.2	0.31

On the other hand, no impurity problem exists in the GC aglycones analysis on the solution from β-glycosidase treatment of bound forms and extracted by hand with pentane/dichloromethane (2:1 v/v). Also for the heterosides, as observed for the free forms, the recoveries from ENV$^+$ cartridge are quantitative, contrary to what was observed by working with XAD-2 resin (Voirin *et al.*, 1992). Lower percentage recoveries are obtained by polystyrene resin XAD-2: 44% for benzyl alcohol, 56% for isoamyl alcohol, from 71% to 86% for 2-phenylethanol, and for some dihydroxylated monoterpenols like Ho-diendiol I, *trans* and *cis* 8-hydroxylinalool and 7-hydroxygeraniol, and *trans* and *cis* pyranic linalooloxides. The repeatability for the aglycones quantification by repeating analysis four times (Table 5.5), is not so good as is found by working with XAD-2 resin (Carlin, 1998).

The mean CV % value referred to the 20 aglycones reported in Table 5.5 – with concentrations included from 30–1900 μg/L – ranges from 4.8% (isoamyl alcohol) to 29.3% (*trans* pyran linalooloxide). A possible variation source could be linked to a badly reproduced enzymatic reaction usually carried out 'overnight' and not for a fixed number of hours. The GC aroma profile achieved by ENV^+ cartridge is more similar to that achieved after a lengthy liquid-liquid extraction. Moreover, the advantage of being able to continue sample preparation for the glycoside forms analysis is evident. Figure 5.6 shows

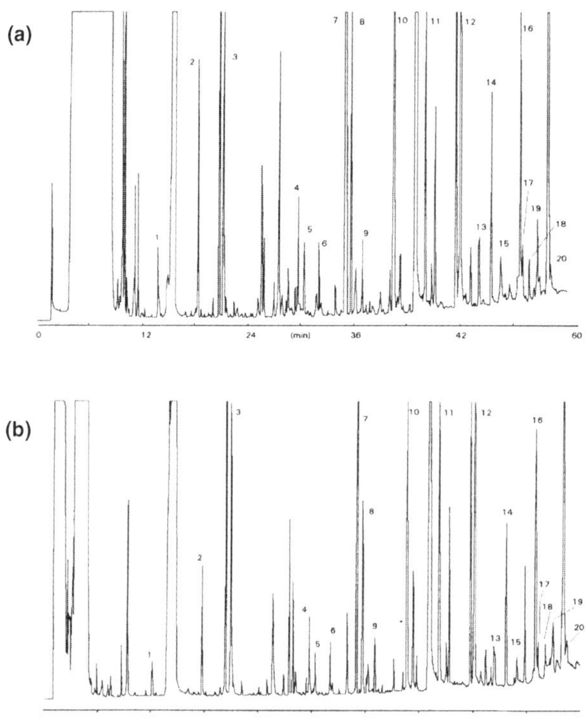

Figure 5.6 GC-MS analyses of a *White Muscat* wine by SPE wine aroma enrichment using a C_{18} pre-packed cartridge (**a**) and a XAD-2 hand-made cartridge (**b**) performed in the same analytical conditions (Versini *et al.*, 1995). 1. isoamyl acetate, 2. ethyl hexanoate, 3. 1-hexanol, 4. linalool, 5. ho-trienol, 6. ethyl decanoate, 7. diethyl succinate, 8. α-terpineol, 9. *trans* pyran linalooloxide, 10. hexanoic acid, 11. Ho-diendiol (I), 12. octanoic acid, 13. Ho-diendiol (II), 14. diethyl 2-hydroxyglutarate, 15. hydroxycitronellol, 16. decanoic acid, 17. *trans* 8-hydroxylinalool, 18. *cis* 8-hydroxylinalool, 19. *trans* geranic acid, 20. 4-vinylphenol

the GC-chromatograms relative to a *White Muscat* wine analysis after sample preparation by solid phase extraction using C_{18} or XAD-2. In general, two analyses are quite similar for all the compounds.

Quantitative data of compounds reported in the chromatograms in Figure 5.6 shows comparable results for the two methods. As may be inferred from Table 5.5 reporting the comparison among the sample enrichment by ENV^+ and XAD-2 resin, some higher boiling compounds resulted underestimated of about 20% by XAD-2 resin.

Table 5.5 Repeatability in the analysis, four times repeated, of aglycons from heterosides adsorbed on polyhydroxylated styrene-divinylbenzene cartridge (ISOLUTE®ENV^+) compared with mean data of two repeated extractions performed by polystyrene resin (XAD-2). ** Value Ax/Ai.s., i.s. 1-heptanol.

Compounds**	mean	S.D.	Cv %	polystyrene resin
Isoamyl alcohols	0.43	*0.02*	*4.8*	0.27
Hexanol	0.17	*0.02*	*13.6*	0.15
Benzyl alcohol	1.90	*0.33*	*17.6*	0.88
2-Phenylethyl alcohol	0.49	*0.07*	*15.4*	0.52
3-Oxo-α-ionol	0.14	*0.02*	*13.4*	0.15
Terpenes				
trans Furan linalooloxide	0.08	*0.01*	*16.7*	0.07
cis Furan linalooloxide	0.09	*0.02*	*17.1*	0.08
trans Pyran linalooloxide	0.04	*0.01*	*29.3*	0.03
cis Pyran linalooloxide	0.05	*0.01*	*26.7*	0.04
Linalool	0.06	*0.01*	*19.6*	0.06
α-Terpineol	0.06	*0.01*	*16.3*	0.05
Citronellol	0.04	*0.01*	*22.9*	0.04
Nerol	0.31	*0.06*	*18.0*	0.33
Geraniol	1.01	*0.18*	*17.9*	0.99
Ho-diendiol (I)	0.03	*0.00*	*14.4*	0.04
OH-citronellol	0.03	*0.01*	*26.7*	0.02
trans 8-OH linalool	0.08	*0.02*	*18.5*	0.07
cis 8-OH linalool	0.40	*0.06*	*14.8*	0.28
7-OH-geraniol	0.15	*0.02*	*11.8*	0.12
trans Geranic acid	0.13	*0.02*	*15.3*	0.11

Concluding remarks

In general, aroma enrichment by using both polyhydroxylated styrene-divinylbenzene/ENV^+ and C_{18} cartridge or polystyrene/XAD-2 resin are less time-consuming, require less solvent vapour protected working space, are generally more healthy and consume less organic solvent than traditional methods of continuous liquid-liquid extraction. Analytical precision, referred to the most representative wine compounds included in different chemical categories for functional group and polarity, is similar for the three solid-phase methods, and results comparable to those obtained using the rapid Kaltron liquid–liquid extraction method. However, this method does not permit the recovery of the glycoside forms. These methods are also suitable for routine applications in the control quality laboratories.

The use of either ENV$^+$ – and highly sorbing phases as reported for LiChrolut EN – or reverse phase cartridge allows a more exhaustive free-forms profile, including the most polar compounds. In both these approaches, by performing a second step, an exhaustive and rapid separation of the glycoside forms of aroma in wines can be accomplished, which after performing an enzymatic hydrolysis, are determined in the aglycone form.

For analysis of free compounds, the headspace analysis with a multiphase fiber is even more interesting and less time-consuming. This approach can replace the majority of the quantification of free compounds measured by SPE techniques, considering also the possibility of the sampling automation using a GC-MS system which can be coupled to a statistical treatment of fragments abundance (Kinton *et al.*, 2003; Cozzolino *et al.*, 2006). Moreover, HS-SPME/GC-MS is a very effective and efficient method to analyze specific compounds present in trace levels at about ppt level, because they can be better and selectively enriched in the headspace. This method is employed nowadays to quantify some important and peculiar sensory compounds such as ethyl and vinylphenols, pyrazines, cork off-flavour substances (TCA, etc.) and other contaminants such as geosmine (Riu *et al.*, 2002; Chatonnet *et al.*, 2006) and, as shown below, sulphur volatiles.

5.3. Volatile Sulphur Compounds in Wines

5.3.1. Introduction

Organic sulphur compounds usually play a considerable role in the sensory characteristics of food and beverages, as they are frequently contributors to the character of impact-scents. Examples of these products include truffle (Diaz *et al.*, 2003; Aprea *et al.*, 2007), garlic and onion (Yan *et al.*, 1993; Bocchini *et al.*, 2001), cooked meat (Hinrichsen and Pedersen, 1995; Andres *et al.*, 2002; Carrapiso *et al.*, 2002), coffee (Shimoda and Shibamoto, 1990; Semmelroch and Grosch, 1996), fruit juices (Winter *et al.*, 1976; Boelens and van Gement, 1993; Hinterholzer and Schieberle, 1998) and some distillates (Leppanen *et al.*, 1979; Cardoso *et al.*, 2004; Ledauphin *et al.*, 2006). In fermented drinks, several papers looked at beer (Burmeister *et al.*, 1992; Dercksen *et al.*, 1992) and wine (Schreier and Drawert, 1975; Würdig *et al.*, 1989; Ribereau-Gayon *et al.*, 1998; Flanzy, 1998) mainly to investigate and justify off-flavours resembling onion, garlic, cooked cabbage, rubber and putrefaction. Around the last decade, the important positive contribution to aroma of some compounds with scents of box-tree and tropical fruits has been emphasized in wines obtained from different grape varieties, in particular by peculiar winemaking techniques (Darriet, 1993; Murat *et al.*, 2001; Tominaga *et al.*, 1998a; Fedrizzi *et al.*, 2007b).

Sulphur compounds are formed through several pathways implying both enzymatic and non-enzymatic processes. The first ones mainly concern yeast fermentation biochemistry and involve inorganic sulphates and sulphites, and mostly sulphur containing amino acids and oligopeptides (e.g. glutathione), as well as organic sulphur compounds such as cysteinyl or glutathionyl derivatives (Tominaga *et al.*, 1998b, Peyrot des Gachons *et al.*, 2002a; 2002b), while the non-enzymatic mechanisms regard chemical, photochemical and thermal reactions during winemaking and storage (Mestres *et al.*, 2000).

Usually considered off-flavouring, sulphur compounds in wine are thiols, sulphides, thioesters, and heterocyclic compounds, frequently divided into 'light' (b.p. $< 90\,^{\circ}$C) and 'heavy' (b.p. $> 90\,^{\circ}$C) compounds (Ribereau-Gayon *et al.*, 1998; Mestres *et al.*, 2000). In the relevant quantification concurrent HS-SPME/GC-MS method, discussed here, we include compounds whose b.p. range from 35 °C of ethylmercaptan to 231 °C of 5-(2-hydroxyethyl)-4-methylthiazole. With other methods including SPE/GC-MS, HS-SPME/GC-MS and Purge and Trap/GC-MS techniques, we quantify the two poly-functional mercaptans 3-mercaptohexan-1-ol and 3-mercaptohexyl acetate, with a tropical fruit-like scent typifying some wines such as *Sauvignon blanc*, and several others with a level depending on the variety and the technological procedure adopted (Fedrizzi *et al.*, 2007b; Fedrizzi *et al.*, 2007c), in particular by the so-called reductive vinification. 3-Mercaptohexan-1-ol and 4-methyl-4-mercaptopentan-2-one (this latter at ppt levels or fractions has a very sensitive box-tree like aroma typical of *Sauvignon blanc* wines), are mostly present in grapes as S-cysteine conjugates (Ribereau-Gayon *et al.*, 1998; Flanzy, 1998) and are made free to a small extent (less than 5%) together with the production of 3-mercaptohexyl acetate by yeast metabolism (Swiegers and Pretorius, 2007; Swiegers *et al.*, 2007).

In the following paragraphs, some of the most recent original results concerning the two different quoted groups of compounds, as well as relevant quantification methods, are presented and discussed.

'Common Fermentative Sulphur Volatiles' (CFSV) in wine: recent results in qualification and discrimination of different varietal wines at different storage time

The first group of sulphur compounds mentioned above can be defined as 'common fermentative sulphur volatiles' (CFSV) and includes in our presentation: ethylmercaptan (EtSH), dimethyl sulphide (DMS), diethyl sulphide (DES), dimethyl disulphide (DMDS), diethyl disulphide (DEDS), methyl thioacetate (MTA), ethyl thioacetate (ETA), 2-mercaptoethanol (ME), 2-(methylthio)-1-ethanol (MTE), 3-(methylthio)-1-propanol (MTP), 4-(methylthio)-1-butanol (MTB), benzothiazole (BT) and 5-(2-hydroxyethyl)-4-methylthiazole (HMT) (see Table 5.6).

Table 5.6 Sensory thresholds and levels of common fermentative sulphur volatile compounds in wines from Trentino (Northern Italy) in the international grape varieties *Chardonnay* and *Merlot* and the red berries native grape varieties *Marzemino* and *Teroldego*. Samples were from the five vintages 1998 and 2001–2004 (Fedrizzi *et al.*, 2007a).

Analyte	Mean ($\mu g\,L^{-1}$)	Min ($\mu g\,L^{-1}$)	Max ($\mu g\,L^{-1}$)	S.D.	Sensory Threshold	Literature values ($\mu g\,L^{-1}$)	Chemical structure
DMS	26.9	2.4	78.2	18.8	1.1 $\mu g\,L^{-1}$	0–480	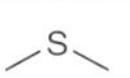
EtSH	1.6	0.1	10.5	1.9	0.13–10 $\mu g\,L^{-1}$	0–12	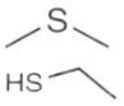
DES	6.1	0.9	17.2	3.4	10–160 $\mu g\,L^{-1}$	1–2	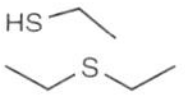
MTA	9.8	1.4	29.0	5.4	1.01 $\mu g\,L^{-1}$	0–20	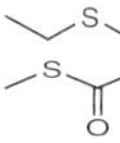
DMDS	5.2	0.2	31.0	6.9	10–160 $\mu g\,L^{-1}$	0–22	
ETA	2.6	0.6	7.1	1.4	0.93–18 $\mu g\,L^{-1}$	0–56	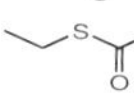
DEDS	3.9	2.0	13.7	2.9	20–45 $\mu g\,L^{-1}$	0–80	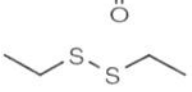
ME	15.6	0.8	47.5	11.6	4.3–40 $\mu g\,L^{-1}$	0–180	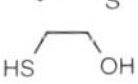
MTE	23.0	3.8	61.9	15.6	–	0–70	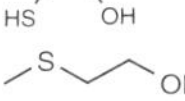
MTP	2551	862	4914	865.9	1.2–4.5 $mg\,L^{-1}$	0–4500	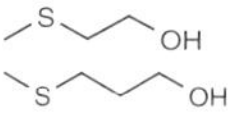
MTB	38.7	5.3	107.6	24.9	0.1 $mg\,L^{-1}$	0–180	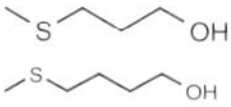
BT	5.2	1.0	14.1	3.5	50–350 $\mu g\,L^{-1}$	0–14	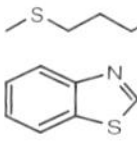
HMT	2.6	0.9	6.1	1.1	0.1–1 $mg\,L^{-1}$	5–50	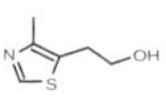

Dimethyl sulphide-d_6 (d_6-DMS), dipropyl disulphide (DPDS), 3-(methylthio)-1-hexanol (MTH) and 4-methylthiazole (MT) have been used as internal standards for different compounds or groups of them.

Recent research showed that all these compounds can be present in a variety of typical wines at levels lower than, or around, the commonly accepted sensory thresholds, contributing to the variety discrimination of wines themselves and likely also to their sensory profiling (Fedrizzi *et al.*, 2007d). Otherwise, it is well known that occasional off-flavouring higher levels of some of such sulphur volatiles (Rauhut *et al.*, 1998) are basically induced either by particular nutritional situations of the yeasts or by a wrong technological conduction of the vinification (Ribereau-Gayon *et al.*, 1998; Karagiannis and Lanardis, 1999; Kotseridis and Baumes, 2000), this fact easily transforming a typical wine into an anomalous product. Relevant sensory thresholds and examples of the CFSVs levels in wines are reported in Table 5.6.

Levels of CFSV in wine can change during storage and this fact is important in the definition and in the understanding of the wine aroma evolution. It is well known that the DMS level increases with ageing (Simpson, 1979; Segurel *et al.*, 2005) and that methyl and ethyl thioacetates hydrolyse in the first months of storage with the concomitant increase of the relevant thiols and disulphides (Rauhut, 1996).

In a recent research, some results of the literature have been confirmed, as well as other new ones put in evidence (Fedrizzi *et al.*, 2007d). Table 5.7 reports the evolution of CFSV in wines of four different varietal types considered altogether and referred to five vintages during a period of seven ageing years. As may be seen from the table, besides DMS, also MTP, ME, EtSH, DES and DEDS contents change in the time decreasing or increasing.

Figure 5.7 shows the evolution of DMS and ME using a straight line model in different wines. Two compounds – DMS and ME – show similar trends during the ageing of *Merlot* and *Chardonnay* wines; this trend is different to that shown in the slope observed in *Marzemino* and *Teroldego* wines. The level of DMS increases due to S-methyl methionine degradation (Segurel *et al.*, 2005); the decrease of ME is due to its oxidation by dimerization (Anocibar Beloqui *et al.*, 1995a; Silva Ferreira *et al.*, 2003).

Also MTP increases likely for a degradation of methionine via Strecker mechanism (Ribereau-Gayon *et al.*, 1998). Furthermore, EtSH and DES show opposite changing in the time resulting for DEDS a similar increasing tendency as for DES; thus a shift towards the oxidized form

Table 5.7 Evolution of CFSV in 80 wines from four different grape varieties considered altogether and referred to five vintages during a period of ageing of seven years. Different letters as index indicate a statistically significant ($P<0.05$) difference; ns: not significant.

Analyte	1998	2001	2002	2003	2004
($\mu g\ L^{-1}$)	mean	mean	mean	mean	mean
DMS	53.41^{a}	35.61^{b}	22.47^{c}	15.11^{d}	7.98^{e}
EtSH	0.70^{b}	1.45^{ab}	1.48^{ab}	1.77^{ab}	2.49^{a}
DES	8.92^{a}	7.41^{ab}	4.86^{bc}	5.67^{bc}	3.47^{c}
MTA	8.21$^{n.s.}$	8.32$^{n.s.}$	10.39$^{n.s.}$	10.30$^{n.s.}$	11.58$^{n.s.}$
DMDS	8.73^{a}	3.96^{ab}	3.46^{b}	5.26^{ab}	4.47^{ab}
ETA	1.83^{b}	2.32^{ab}	2.75^{ab}	2.70^{ab}	3.23^{a}
DEDS	5.99^{a}	3.36^{b}	3.48^{b}	3.71^{b}	2.76^{b}
ME	4.77^{d}	10.92^{c}	13.77^{c}	20.96^{b}	27.72^{a}
MTE	26.43^{a}	25.47^{ab}	20.76^{ab}	23.21^{ab}	18.93^{b}
MTP	3386^{a}	2807^{b}	2543^{bc}	2170^{cd}	1850^{d}
MTB	38.71$^{n.s.}$	42.60$^{n.s.}$	41.04$^{n.s.}$	41.14$^{n.s.}$	30.18$^{n.s.}$
BT	6.60^{a}	4.40^{ab}	3.81^{b}	5.89^{ab}	5.05^{ab}
HMT	2.12$^{n.s.}$	3.01$^{n.s.}$	2.46$^{n.s.}$	2.59$^{n.s.}$	2.64$^{n.s.}$

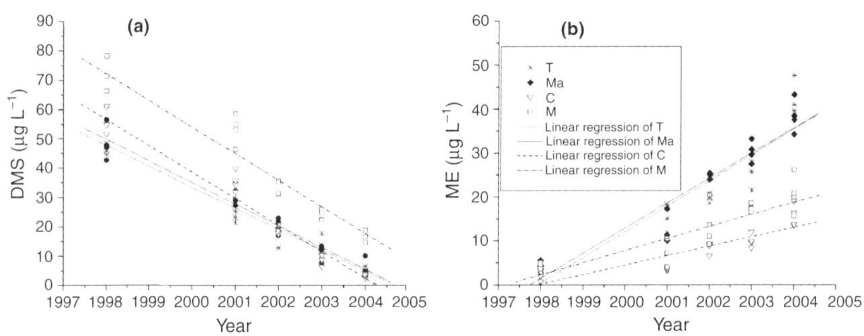

Figure 5.7 Evolution of DMS (**a**) and ME (**b**) during the period 1998–2005 in a straight line model for varietal wines from grape *Chardonnay* (C), *Merlot* (M), *Marzemino* (Ma) and *Teroldego* (T): four wines per vintage were considered

in the redox equilibrium between thiols and disulphide (Bobet *et al.*, 1990) is confirmed.

There is also dependence of sulphur compounds on the grape variety (Anocibar Beloqui *et al.*, 1995b). Data in Table 5.8 show that some sulphur compounds are more abundant in some wine varieties than in others. In particular, the concentration of DMS and MTB in *Merlot* wines is significantly higher than in all the other

Table 5.8 Mean values of CFSV compounds in four *Merlot*, *Teroldego*, *Marzemino* and *Chardonnay* wines with different aging times. Indexes letters show the variety discrimination significance by Tukey-test.

Analyte	*Merlot*	*Teroldego*	*Marzemino*	*Chardonnay*
(μg L^{-1})	mean	mean	mean	mean
DMS	39.9^{a}	22.1^{b}	21.9^{b}	23.8^{b}
EtSH	0.7^{b}	3.2^{a}	1.5^{b}	0.8^{b}
DES	7.2^{ab}	5.2^{bc}	7.5^{a}	4.3^{c}
MTA	8.5^{b}	13.7^{a}	7.6^{b}	9.2^{b}
DMDS	9.0^{a}	0.7^{b}	1.1^{b}	10.0^{a}
ETA	$2.1^{n.s.}$	$2.7^{n.s.}$	$2.7^{n.s.}$	$2.7^{n.s.}$
DEDS	4.9^{a}	2.2^{b}	2.3^{b}	6.0^{a}
ME	11.5^{b}	20.7^{a}	22.3^{a}	8.0^{b}
MTE	44.7^{a}	11.5^{c}	24.2^{b}	11.4^{c}
MTP	3024^{a}	2569^{b}	2861^{ab}	1749^{c}
MTB	75.1^{a}	28.9^{b}	26.0^{b}	24.9^{b}
BT	9.0^{a}	2.9^{c}	3.2^{c}	5.5^{b}
HMT	2.3^{b}	2.0^{b}	2.4^{b}	3.6^{a}

varieties, thus justifying a possible remarkable contribution to the important grassy/truffle-like scent for DMS and the earthy-like one for MTB, both descriptors commonly used for the *Merlot* aroma. Also, a higher level of MTE and BT are characteristic of *Merlot* variety wines. EtSH and MTA, likely causing rotten eggs/garlic/onion-like scents when over the threshold levels, are higher in *Teroldego* wine and positively evaluated. For boiled potatoes-scenting MTP, *Merlot* is again the leading variety, while *Chardonnay* has the lowest content, as expected for a white wine (Seeber *et al.*, 1991).

Principal Component Analysis (PCA) can be a useful support to demonstrate the possible differentiation of the varietal wines on the basis of sulphur compounds content. The scattering of the scores in the plan determined by the first two eigenvectors indicates prevalent clustering according to three variety groupings, i.e. *Chardonnay* (C), *Merlot* (M) and *Teroldego* (T) with *Marzemino* (Ma) (Figures 5.8) (Fedrizzi *et al.*, 2007a).

The first component (PC1) is mainly positively related in Figure 5.8b with EtSH, ME, MTA and ETA (group α) and negatively with MPT, DES, MTE, MTB and DMS (group β). Furthermore, the loadings evaluation shows an anticorrelation between EtSH and DEDS (group γ), thus emphasizing a redox correlation between two compounds.

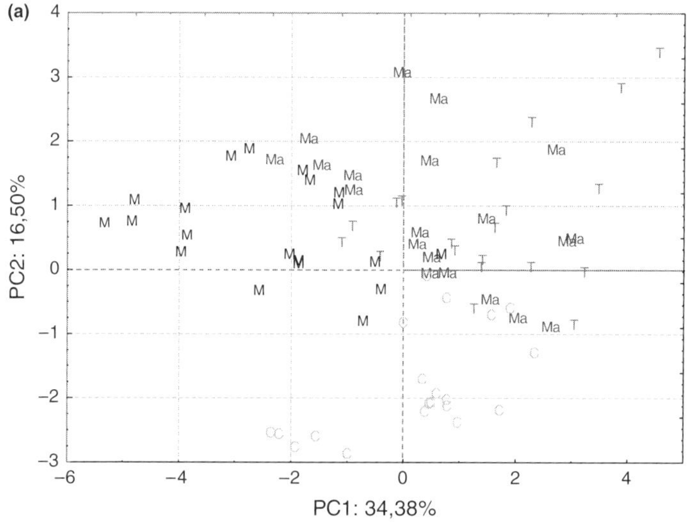

Figure 5.8a Principal Component Analysis on the four varietal wines. PC1 *vs* PC2 by-plot of the scores indicates prevalent clustering according to three variety groupings: *Chardonnay* (C), *Merlot* (M), and *Teroldego* (T) together with *Marzemino* (Ma)

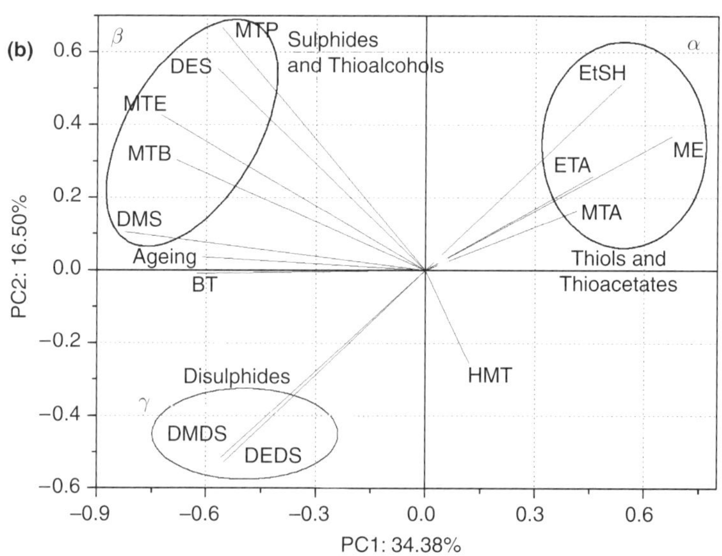

Figure 5.8b Principal Component Analysis on the four varietal wines. Representation of the loadings of the variables in PC1 and PC2 in the by-plot schema of Figure 5.8a

The first component results mostly connected with the ageing, therefore a younger wine is more charged by the variables of the group α, while an aged one by those of the group β. It also contributes to distinguishing *Merlot* from *Marzemino* and *Teroldego* wines. The second function is positively related with almost all the variables, and distinguishes the *Chardonnay* wines from the red ones. By PCA it is confirmed that the group α of variables, above all represented by the classes of thiols and thioacetates, mostly model the wines of the variety *Teroldego* (T) and in part also those of *Marzemino* (Ma), while group β including sulphides, disulphides and most thioalcohols fits the group of *Merlot* wines (M) better. As discussed above, these different groups of compounds could determine different scents in the relevant wines. The difficulty in separating the scores of *Teroldego* from those of *Marzemino* represents rather well the practical sensory situation by which it is often quite difficult to distinguish such wine-types, both well distinguishable from *Merlot* wines. Besides, the particular dominance of thiols and thioacetates in *Teroldego* and *Marzemino* wines confirms the frequent off-flavouring possibility for them in the direction of the corresponding sensory descriptors of such compounds.

'Tropical Fruit Scenting Sulphur Volatiles' (TFSV) in wine: recent additional results in quantification and ranging in different varietal wines of Italy

The most important Tropical Fruit Scenting Sulphur Volatiles (TFSV) in wine are 3-mercaptohexan-1-ol (3-MH), 3-mercaptohexanol acetate (3-MHA) and 4-methyl-4-mercaptopentan-2-one (4-MPP). The main problem for quantification of TFSV is their possible very low concentration which may be only a few ng/L. Therefore, only the use of selective isolation and pre-concentration steps allows this difficulty to be overcome. Another problem occurring in thiols analysis is their possible reactivity towards oxidation induced by the working conditions (Rapp *et al.*, 1985) and/or by some components of the matrix in analysis (Murat *et al.*, 2003; Blanchard *et al.*, 2004).

Three methods are reported to quantify these TFSV-thiols analytes with separation and pre-concentration steps (Tominaga *et al.*, 1998c; Schneider *et al.*, 2003; Mateo-Vivaracho *et al.*, 2007). For all such approaches, the first step of the analysis is their recovery from the wine either by liquid-liquid extraction with methylene chloride (with a rather low yield) or by a more efficient SPE enrichment using a highly cross-linked polystyrene resin followed by an elution with dichloromethane/methanol 99:1 (v/v) (Mateo-Vivaracho *et al.*, 2007). Thereafter, there is a selective trapping of TFSV either with a *p*-hydroxymercuribenzoate water solution (Tominaga *et al.*, 1998c; Mateo-Vivaracho *et al.*, 2007) or on a cross-linked agarose gel containing phenylmercuric chloride, called Affi-Gel 501 (Schneider *et al.*, 2003). An on-fiber derivatization method after HS-SPME sampling was also proposed (Mateo-Vivaracho *et al.*, 2006) using negative chemical ionization to improve MS-sensitivity: very sensitive results have been obtained only for the quantification of 2-furanmethanethiol – an additional sulphured odorant of wine coming from the 'elevage' in barrique – at concentrations below 0.1 ng/L, and for 3-mercaptohexyl acetate (3-MHA), but not for 3-mercaptohexan-1-ol (3-MH).

We focused our attention on performing a simple, less time-consuming method with minimum risk to human health. A HS-SPME/GC-MS approach without on-fiber derivatisation was developed, which allows quantitative analysis for quantification of about a dozen ng/L of each analyte (values close to the sensory thresholds of both 3-MH and 3-MHA). This method can be applied in routine analysis in the winemaking process, in both monitoring and quality control. (Fedrizzi *et al.*, 2007a; 2007b). Due to the very low sensory threshold in wines of 4-methyl-4-mercaptopentan-2-one (about 0.4 ng/L) and the consequent very low detection limit required in the method to obtain useful data, this compound was not considered in this approach. Both such compounds can be determined at a similar detection level also by a SPE sampling procedure using ENV$^+$-type resin (Boido *et al.*, 2003; Fedrizzi *et al.*, 2007a), resulting in a very good correlation between 3-MH and 3-MHA data as shown in Figure 5.9.

Finally, an improvement in the sensitivity of the detection limit for such compounds can be reached by using a Purge and Trap well-optimized enrichment approach instead of the HS-SPME technique (Fedrizzi *et al.*, 2007b). This method provides enough sensitivity, and analysis times are suitable to perform routine analysis. The measured amounts are, in fact, within the variability ranges reported in the literature for both compounds in wine, and close to the sensory thresholds. Using both such methods, it has been possible for the first time to measure the contents and relevant ratios of such compounds in Italian wines of different varieties as represented in Figure 5.10.

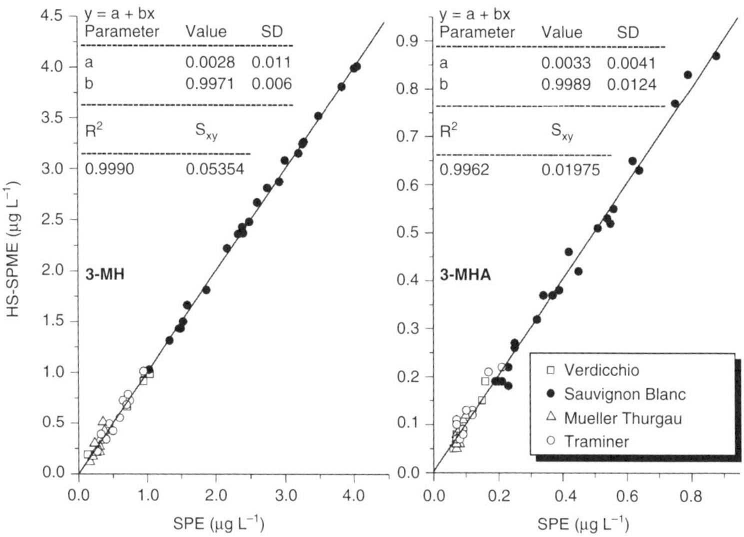

Figure 5.9 Comparison between GC-MS analysis of 3-mercaptohexan-1-ol (3-MH) and 3-mercaptohexanol acetate (3-MHA) in wine by HS-SPME without on-fiber derivatisation and by SPE using ENV$^+$-type resin enrichment. (From: Gas Chromatography-Mass Spectrometry determination of 3-mercaptohexan-1-ol and 3-mercaptohexyl acetate in wine. A comparison between Solid Phase Extraction and Headspace Solid Phase Microextraction methods, Fedrizzi *et al.*, 2007b, *Anal. Chim. Acta*, **596**(2), 291–297)

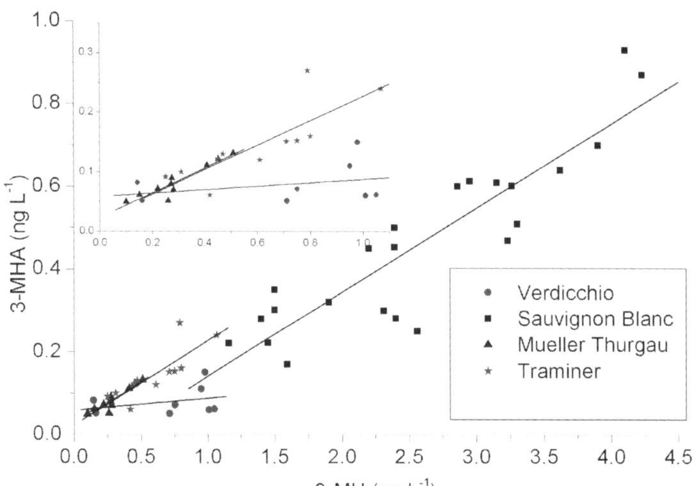

Figure 5.10 Contents and relevant correlations of 3-mercaptohexan-1-ol (3-MH) and 3-mercaptohexanol acetate (3-MHA) in 52 Italian wines from four different grape varieties determined by the Purge and Trap enrichment using optimized conditions

5.3.2. Method of Concurrent Analysis of Common Fermentative Sulphur Volatile (CFSV) Compounds in Wine

Several analytical approaches are employed to quantify sulphur volatiles in wine; among them, the headspace procedure, such as the Purge and Trap, and solid-phase microextraction methods, combined with GC coupled to different detectors, was shown to be quite effective (Mestres *et al.*, 2000; Rauhut *et al.*, 1998). Recent papers suggest that the HS-SPME technique, with an improved choice of fiber coating phases, would be one of the more promising approaches for the concurrent measurement of compounds with different boiling points (Mestres *et al.*, 2002; Fang and Qian, 2005).

Optimisation of Headspace Solid Phase Microextraction (HS-SPME) sampling for GC-MS analysis of CFSV compounds

The fiber carboxen-polydimethylsiloxane-divinylbenzene (CAR-PDMS-DVB; $50/30\,\mu m \times 2\,cm$) proved to be optimal in the GC-MS analysis in terms of the maximum signal recorded for each compound and the simpler sampling procedure and operating conditions (ionic strength, sample temperature, adsorption time) (Fedrizzi *et al.*, 2007a).

Table 5.9 Analytical GC-MS parameters used for detection and quantification of CSFV compounds in wines (chromatograms in Figure 5.11). The last four compounds in the table are used as internal standards.

Analytes	RT (min)	Quantifier ion (m/z)	Qualifier ions (m/z)
DMS	4.76	62	45,47
EtSH	4.61	62	47,61
DES	8.52	75	61,90
MTA	14.42	90	43,47
DMDS	15.01	94	64,79
ETA	15.40	104	43,60
DEDS	18.00	122	66,94
ME	22.19	78	47,60
MTE	22.61	92	47,61
MTP	24.72	106	58,61
MTB	26.04	120	61,102
BT	27.61	135	69,108
HMT	30.58	112	85,143
d_6-DMS	4.73	68	50,66
DPDS	18.68	108	66,150
MT	19.21	71	39,99
MTH	26.57	148	61,75

Before sampling, the fiber is conditioned according to the producer's instructions. In agreement with most literature, the volumes ratio used (headspace/solution) is 1:2 (Mestres *et al.*, 2000).

MS fragments recorded in selected ion monitoring (SIM) used for the evaluation of the compounds, including the several internal standards, are listed in Table 5.9. The chromatograms relative to HS-SPME/GG-MS analysis of CFSV are reported in Figure 5.11. The experimental conditions finally established for the analysis are summarized in Table 5.10.

As shown in the chromatograms in Figure 5.11, the use of a 2-cm long DVB/CAR/PDMS fiber in the quoted conditions allows the highest peaks for all CFSV compounds, completely resolved and adequately shaped also in the presence of ethanol at the usual wine proof. The best effect of the modified ionic strength of solution to induce salting-out of

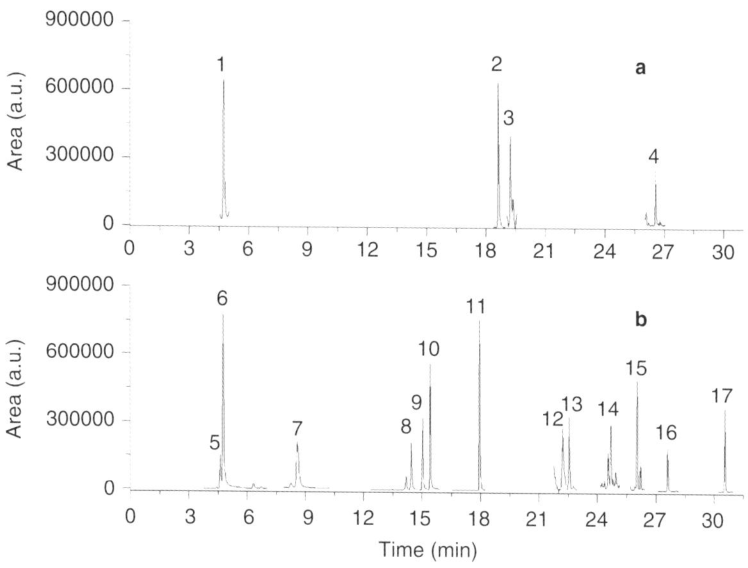

Figure 5.11 Chromatograms recorded in SIM mode in the HS-SPME/GC-MS analysis of CFSV. The experimental conditions are reported in Table 5.10. (**a**) internal standards: 1. d_6-DMS (*m/z* 68), 2. DPDS (*m/z* 108), 3. MT (*m/z* 71), 4. MTH *m/z* 148); (**b**) compounds quantified: 5: EtSH (*m/z* 62), 6. DMS (*m/z* 62), 7. DES (*m/z* 75), 8. MTA (*m/z* 90), 9. DMDS (*m/z* 94), 10. ETA (*m/z* 104), 11. DEDS (*m/z* 122), 12. ME (*m/z* 78), 13. MTE (*m/z* 92), 14. MTP (*m/z* 106), 15. MTB (*m/z* 120), 16. BT (*m/z* 135), 17. HMT (*m/z* 122). (Reprinted from *Rapid Communications in Mass Spectrometry* **21**, Fedrizzi *et al.*, Concurrent quantification of light and heavy sulphur volatiles in wine by headspace solid-phase microextraction coupled with gas chromatography/mass spectrometry, p. 710, Copyright 2007, with permission from John Wiley & Sons Ltd)

Table 5.10 Experimental conditions used for the analysis of CFSV compounds.

Fiber coating	CAR-PDMS-DVB; 2 cm
Sample temperature	35 °C
Extraction	30 min with stirring (500 rpm)
Salt addition	5 g $MgSO_4 \cdot 7H_2O$
Sample volume	20 mL
Vial volume	30 mL

most compounds, is achieved by addition of $MgSO_4 \cdot 7H_2O$ 1.0 M, with peak areas increasing by 10%–25% with respect to other salts tested. Figures 5.12(a–d) show the adsorption temperature effect on CFSV in the range 25–65 °C: a temperature of 35 °C is the best compromise between larger sampling of less volatile compounds (Figure 5.12d) and a reduced desorption of the higher volatile compounds (Figure 5.12a).

Figures 5.13(a–d) show the adsorption time effect on CFSV in the range 5–90 min. As may be seen, a sampling time of 30 min is also a good compromise among all compounds to reach enough sampling (Fedrizzi *et al.*, 2007a). It is evident that the more volatile compounds (e.g. DMS, ETA and MTA) reach an equilibrium more quickly than others, but then they show decreasing signals as a consequence of competitive desorption mechanisms (Murray, 2001; Lestremau *et al.*, 2004); the contrary is true of the behaviour of the heavier compounds (Figure 5.12c).

In the first column of Table 5.11 concentration ranges for the calibration of the CFSV compounds considered according to those naturally occurring in wines, are reported; in other columns, the relative linear regression parameters and evaluation limits of the HS-SPME/GC-MS method in wine-similar matrixes, are listed. Recoveries by HS-SPME of wines using a particular fiber can vary depending on matrix effects, in addition to the parameters influencing the volatility and adsorbtion of compounds already optimized in a wine-similar matrix. Different matrixes of a white and a red wine deprived of sulphur volatiles by treatment with $AgNO_3$ that mostly differ for volatiles, polyphenols and levels of alcohol and sugars, have to be considered. For these reasons, quantification requires methodologies involving suitable spiked internal standards, as reported in Table 5.11. Table 5.12 shows CFSV recoveries are close to 100% for all compounds when related to the relevant internal standards, as achieved by this method in different matrix conditions.

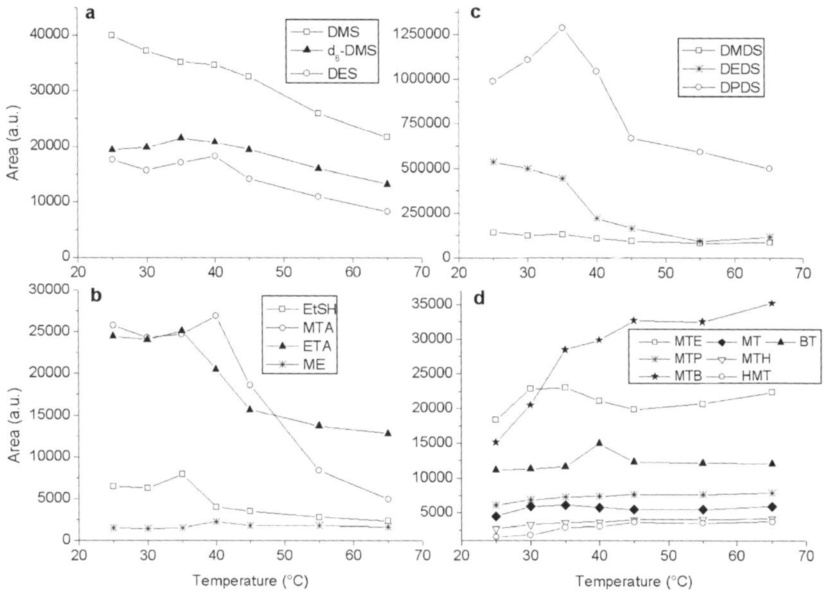

Figure 5.12 Temperature effect in the HS-SPME adsorption of the wine CFSV on DVB/CAR/PDMS fiber. (Reprinted from *Rapid Communications in Mass Spectrometry* **21**, Fedrizzi *et al.*, Concurrent quantification of light and heavy sulphur volatiles in wine by headspace solid-phase microextraction coupled with gas chromatography/mass spectrometry, p. 711, Copyright 2007, with permission from John Wiley & Sons, Ltd)

GC-MS analysis of CFSV

Calibration curves for each CFSV can be prepared according to the internal standard method using d_6-DMS 25 μg/L, DPDS 25 μg/L, MT 10 μg/L and MTH 50 μg/L as internal standards. The matrix used is a white wine (e.g. alcohol strength 10% v/v; sugar content < 4 g/L; polyphenols content 115 mg/L as (+)-catechin) treated twice with charcoal (3 g/L) in order to remove sulphur compounds and other main volatile compounds, except the most polar ones such as the higher alcohols. The total SO_2 content is reset to 100 g/L. The aim of this pretreatment is to obtain a matrix similar to the wine, but free of volatile compounds potentially affecting adsorption and widely varying among the wines.

From individual standard solutions of each sulphur compound in pure ethanol, a working ethanolic solution containing all the analytes is prepared and is stored at −16 °C. The same procedure is used to prepare the internal standard solutions at similar concentrations. The

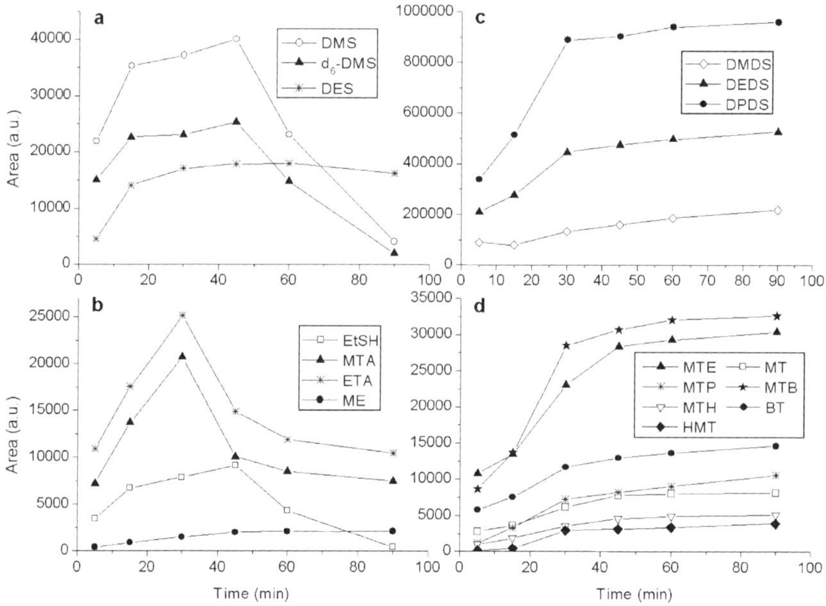

Figure 5.13 Time effect in the HS-SPME adsorption of CFSV in wine on DVB/CAR/PDMS fiber. (Reprinted from *Rapid Communications in Mass Spectrometry* **21**, Fedrizzi *et al.*, Concurrent quantification of light and heavy sulphur volatiles in wine by headspace solid-phase microextraction coupled with gas chromatography/mass spectrometry, p. 712, Copyright 2007, with permission from John Wiley & Sons, Ltd)

Table 5.11 Calibration ranges, linear regression parameters and evaluation limits for the quantification of CFSV in wine-similar matrixes (Fedrizzi *et al.*, 2007a).

Analytes	Calibration Range (μg L^{-1})	Slope	SD Slope	Intercept	SD Intercept	SD	R^2	L_C (μg L^{-1})	L_D (μg L^{-1})	L_Q (μg L^{-1})
DMS[a]	3.2 – 240.0	1.03	0.01	0.045	0.047	0.07	0.997	0.078	0.156	0.358
EtSH[a]	0.4 – 30.0	1.19	0.02	0.042	0.012	0.05	0.994	0.067	0.134	0.254
DES[a]	0.2 – 15.0	7.28	0.06	0.053	0.010	0.12	0.994	0.057	0.114	0.336
MTA[a]	0.8 – 60.0	0.85	0.01	0.006	0.009	0.04	0.997	0.153	0.306	0.643
DMDS[b]	0.2 – 15.0	0.80	0.01	0.002	0.002	0.01	0.997	0.031	0.062	0.136
ETA[a]	0.8 – 60.0	3.27	0.37	0.000	0.004	0.11	0.996	0.102	0.204	0.429
DEDS[b]	0.8 – 60.0	1.02	0.01	0.044	0.016	0.07	0.994	0.037	0.074	0.156
ME[c]	3.2 – 240.0	0.82	0.01	0.011	0.029	0.09	0.993	0.383	0.766	1.563
MTE[c]	0.8 – 60.0	8.63	0.07	0.010	0.040	0.10	0.998	0.103	0.206	0.453
MTP[c]	250 – 3250	0.54	0.00	0.102	0.020	0.07	0.996	0.826	1.652	5.862
MTB[c]	1.6 – 120.0	2.75	0.04	0.086	0.047	0.15	0.993	0.27	0.54	1.893
BT[d]	0.4 – 30.0	2.78	0.04	0.052	0.059	0.17	0.994	0.484	0.968	3.388
HMT[d]	0.4 – 30.0	0.83	0.01	0.119	0.018	0.08	0.993	0.436	0.872	3.806

L_C: critical limit; L_D: detection limit; L_Q: quantification limit; SD: standard deviation

[a] d_6-DMS as I.S.; [b] DPS as I.S.; [c] MTH as I.S.; [d] MT as I.S.

Table 5.12 Percentage recoveries of CFSV from different matrices related to each correspondent internal standard and determined by the analytical method proposed by Fedrizzi *et al.*, 2007a.

Matrix	Wine Composition	Level	DMS	EtSH	DES	MTA	DMDS	ETA	DEDS	ME	MTE	MTP	MTB	BT	HMT
White Wine	Sugars 3 g L^{-1} Alcohol 10 % v/v	High	100.4	103.9	108.8	98.9	100.9	102.0	99.5	103.1	102.8	98.6	99.0	102.3	103.7
			(5.7)	(6.9)	(6.4)	(8.1)	(7.3)	(7.0)	(5.7)	(5.9)	(4.9)	(5.6)	(5.5)	(5.3)	(6.1)
		Middle	99.4	99.3	106.1	104.1	106.3	104.1	103.5	98.4	106.1	103.6	102.5	97.9	101.0
			(4.9)	(7.6)	(7.1)	(6.7)	(8.6)	(6.1)	(8.3)	(6.3)	(8.3)	(6.1)	(6.3)	(4.0)	(6.6)
		Low	103.0	100.3	98.1	100.1	95.9	101.9	103.4	106.1	106.6	98.6	97.1	101.3	99.6
			(2.9)	(5.9)	(11.1)	(9.1)	(5.0)	(7.6)	(6.6)	(7.2)	(7.0)	(7.1)	(5.7)	(6.9)	(5.0)
White Wine	Sugars 3 g L^{-1} Alcohol 12 % v/v	High	99.4	103.2	100.1	98.6	101.6	100.4	105.5	97.6	100.9	101.8	101.1	102.4	100.4
			(3.1)	(5.5)	(8.1)	(7.6)	(10.7)	(5.3)	(7.3)	(7.9)	(4.9)	(5.6)	(9.1)	(8.3)	(7.9)
		Middle	103.8	98.3	98.8	104.5	97.9	103.8	99.9	98.1	99.5	100.6	102.6	101.8	101.9
			(5.5)	(4.9)	(9.1)	(9.7)	(7.0)	(6.9)	(10.1)	(9.8)	(10.1)	(8.7)	(7.3)	(6.1)	(11.3)
		Low	98.4	99.8	101.1	99.7	103.5	101.5	98.4	100.8	103.1	101.9	98.6	99.7	99.5
			(5.5)	(4.7)	(6.7)	(9.1)	(8.3)	(4.3)	(6.8)	(10.5)	(5.9)	(6.1)	(6.6)	(5.7)	(5.7)
White Wine	Sugars 3 g L^{-1} Alcohol 15 % v/v	High	101.6	103.7	102.3	98.4	100.8	101.2	99.8	98.7	99.7	99.9	97.9	102.9	103.6
			(2.7)	(7.8)	(7.7)	(7.3)	(5.6)	(5.8)	(6.1)	(10.0)	(6.6)	(5.9)	(8.3)	(6.7)	(7.4)
		Middle	100.5	104.4	107.8	97.9	101.2	103.6	96.6	99.4	99.5	103.1	106.7	98.3	103.4
			(3.3)	(6.1)	(6.9)	(8.5)	(4.9)	(8.0)	(4.9)	(9.1)	(8.0)	(7.3)	(6.9)	(8.3)	(8.3)
		Low	101.8	98.3	98.9	99.2	98.3	99.2	105.8	100.1	97.1	97.2	101.6	96.9	101.7
			(1.9)	(9.7)	(9.6)	(6.0)	(5.9)	(8.1)	(4.3)	(6.9)	(6.1)	(8.1)	(6.9)	(6.9)	(6.1)
		High	104.8	104.3	103.8	96.7	99.6	98.8	97.1	101.5	102.3	96.8	101.1	106.4	101.4
			(3.7)	(6.6)	(9.9)	(10.3)	(10.2)	(10.2)	(7.1)	(6.7)	(5.7)	(8.9)	(7.0)	(9.1)	(8.1)

Table 5.12 (Continued).

Matrix	Wine Composition	Level	DMS	EtSH	DES	MTA	DMDS	ETA	DEDS	ME	MTE	MTP	MTB	BT	HMT
White Wine	Sugars 50 g L^{-1} Alcohol 10 % v/v	Middle	99.5	98.1	109.2	101.4	106.3	105.8	98.6	98.4	98.5	105.9	103.9	98.6	103.2
			(5.3)	(6.6)	(10.6)	(9.3)	(8.4)	(6.9)	(5.2)	(9.4)	(7.2)	(10.3)	(4.9)	(7.3)	(5.1)
		Low	99.7	97.6	101.4	101.3	97.2	103.7	101.2	99.9	98.4	101.8	98.9	104.4	97.3
			(4.7)	(10.1)	(7.9)	(12.3)	(9.4)	(7.3)	(2.9)	(6.4)	(6.4)	(4.6)	(8.1)	(3.9)	(5.2)
		High	100.1	100.1	105.6	96.2	98.4	102.5	99.9	98.3	101.0	98.6	99.4	101.1	99.3
			(3.9)	(4.9)	(8.3)	(10.4)	(7.8)	(6.3)	(5.9)	(5.9)	(5.5)	(9.1)	(6.3)	(5.5)	(7.4)
White Wine	Sugars 100 g L^{-1} Alcohol 10 % V/v	Middle	103.8	100.4	102.6	104.5	98.7	103.8	98.8	95.3	100.9	101.9	104.1	97.1	104.2
			(6.0)	(5.9)	(7.1)	(5.9)	(6.1)	(4.9)	(6.3)	(10.3)	(6.0)	(6.1)	(6.9)	(6.1)	(5.6)
		Low	102.3	99.6	109.2	100.7	104.8	99.1	104.5	106.5	98.5	98.0	101.6	102.2	98.5
			(7.0)	(7.0)	(5.9)	(7.6)	(9.4)	(5.5)	(4.9)	(5.1)	(8.8)	(7.4)	(9.7)	(6.4)	(8.4)
		High	102.8	106.5	106.9	99.0	98.3	101.4	102.5	102.5	100.4	99.8	99.7	100.6	104.3
			(3.4)	(7.3)	(7.8)	(6.3)	(7.7)	(7.3)	(5.1)	(7.3)	(8.3)	(5.3)	(6.7)	(7.3)	(3.9)
Red Wine	Sugars 3 g L^{-1} Alcohol 10 % v/v	Middle	99.5	107.2	101.5	102.0	100.8	105.0	101.5	95.3	102.0	104.1	104.8	97.1	106.6
			(6.1)	(5.9)	(8.3)	(6.9)	(6.8)	(5.9)	(7.0)	(6.6)	(4.3)	(5.9)	(5.9)	(8.9)	(6.7)
		Low	98.7	103.2	99.8	105.1	103.5	103.8	101.3	100.1	97.3	100.4	97.1	101.8	98.5
			(5.1)	(7.1)	(11.3)	(10.7)	(9.9)	(8.7)	(3.9)	(8.1)	(6.7)	(7.6)	(7.1)	(7.3)	(8.9)

calibration curves are built-up using seven different concentrations in the range listed in Table 5.11 and five replicate solutions per level.

For the calculation of the critical (L_C) and the detection (L_D) limits (e.g. by using the Hubaux-Vos method; Hubaux and Vos, 1970), four standard solutions at very low concentrations are prepared. The signals corresponding to these dedicated solutions, together with the signal corresponding to the blank solution, are used in regression. This procedure decreases the difference between the experimentally measured blank and the intercept of the regression line (Lavagnini and Magno, 2007). The quantification limit (L_Q) can be estimated using the Eurachem approach (1993), which states that L_Q is the analyte concentration for which the experimental relative standard deviation of the responses reaches a specific level, for example level 0.1.

GC-MS (70 eV) analysis is possibly performed with a modern sensitive instrument equipped with a PEG-type bound-phase fused-silica (30 m × 0.32 mm i.d.; 0.25 μm film thickness) capillary column, working with helium as carrier gas at flow-rate of 1.2 mL/min, transfer line temperature 220 °C and MS source temperature 150 °C. Analysis is carried out in single ion recording (SIR) mode. In the optimized chromatographic conditions proposed by Fedrizzi *et al.* (2007a), the GC injector is set at a temperature of 250 °C and operates in splitless injection mode for 1 min; the oven temperature is programmed to start at 35 °C (5 min), then increased by 1 °C/min to 40 °C, and finally increased by 10 °C/min to 250 °C.

5.3.3. Optimisation of Headspace Solid Phase Microextraction (HS-SPME) Sampling for GC-MS Analysis of Tropical Fruit Scenting Sulphur Volatiles (TFSV) Compounds

HS-SPME conditions for analysis of TFSV (temperature, sampling time, pH of solution) were reported by Fedrizzi *et al.*, (2007b). Increase of the pH matrix at a value of 7, performed in order to reduce interference in the analysis due to presence of hexanoic acid, can reduce the 3-mercaptohexyl acetate hydrolysis. Unfortunately, the antioxidant protection operated from SO_2 is also reduced (due to the shift of equilibrium towards the HSO_3^- form) as a consequence phenolic acids can be easily oxidized forming the correspondent quinones (Singleton, 1987; Murat *et al.*, 2003). They can, in their turn, oxidize thiols (Rigaud *et al.*, 1991), even if with an opportunely delayed time as evidenced by Blancard *et al.*

(2004). Hydrolysis of 3-MHA and oxidation of thiols seem not clearly detectable. A constant increase of 3-MHA signal when pH is increased is not always confirmed by concomitant decrease of the correspondent alcohol 3-MH. Such behaviour could indicate a dominant matrix effect on both 3-MH and 3-MHA with a possible influence of pH, even if a rather stable signal of internal standard 6-Mercaptohexan-1-ol (6-MH) is observed. For a solution at pH 7, a maximized signal for 3-MA and a slightly reduced one for 3-MH – but still fitting the sensitivity expected for this method – in respect to the situation working at pH of wine, were observed (Fedrizzi *et al.*, 2007b). The very comparable results reached using HS-SPME and SPE enrichment can exclude measurable side-reactions on the quantified thiols by working with conditions reported in Table 5.13.

Table 5.13 Optimized HS-SPME operating conditions for analysis of TFSV in wine (Fedrizzi *et al.*, 2007b).

Fiber coating	CAR-PDMS-DVB; 2 cm
Sample temperature	40 °C
Extraction	40 min under stirring (500 rpm)
Salt addition	5 g $MgSO_4 \cdot 7H_2O$
Solution pH	7.0
Sample volume	20 mL
Vial volume	30 mL

Recent studies performed on white and red compounds wines treated to eliminate all sulphur volatile compounds by Ag salts addition (Fedrizzi *et al.*, 2007a; 2007b), showed the matrix strongly influences the apparent partition between the liquid phase and the fiber coatings. Data of peak areas reported in Table 5.14 show an inversely proportional dependence between the ethanol concentration and the thiols signals, mostly for 3-MHA.

Several other wine components may potentially affect the fiber sampling. In the presence of a remarkable sugar content an increase of the peaks is observed. Also a different presence of polyphenols may influence the adsorption on the fiber due to possible competition with redox reactions (Murat *et al.*, 2003; Blanchard *et al.*, 2004). Finally, a competition with different volatiles, in particular the fermentation esters, in the adsorption on the fiber may also occur (Murray, 2001; Lestremau *et al.*, 2004). In any case, it has taken into account that the richer polyphenols wines are also the fruity esters poorest ones. As a

Table 5.14 Data of HS-SPME/GC-MS peak areas and recoveries of 3-MHA and 3-MH in wines in dependence of their concentration and the sugars and ethanol content (relevant standard deviations in brackets).

Matrix	Wine Composition	Level	3-MHA		3-MH	
			Area	Recovery	Area	Recovery
White Wine	Sugars 3 g L^{-1} Alcohol 10 % v/v	High	1743 (5.4)	100.4 (5.7)	563 (5.0)	103.9 (5.9)
		Middle	863 (5.6)	99.4 (4.9)	222 (7.5)	99.3 (6.6)
		Low	354 (5.5)	103.0 (5.9)	131 (7.3)	100.3 (7.9)
White Wine	Sugars 3 g L^{-1} Alcohol 12 % v/v	High	1563 (4.2)	99.4 (3.1)	521 (6.0)	103.2 (4.5)
		Middle	762 (5.2)	103.8 (5.5)	220 (6.1)	98.3 (4.9)
		Low	302 (6.1)	98.4 (5.5)	124 (6.2)	99.8 (5.7)
White Wine	Sugars 3 g L^{-1} Alcohol 15 % v/v	High	1342 (3.3)	101.6 (2.7)	482 (6.0)	103.7 (5.8)
		Middle	703 (3.8)	100.5 (3.3)	212 (6.2)	104.4 (6.1)
		Low	286 (4.1)	101.8 (1.9)	115 (7.2)	98.3 (7.2)
White Wine	Sugars 50 g L^{-1} Alcohol 10 % v/v	High	1864 (4.0)	104.8 (3.7)	612 (6.6)	104.3 (6.6)
		Middle	966 (5.1)	99.5 (5.3)	251 (7.0)	98.1 (6.6)
		Low	363 (5.0)	99.7 (4.7)	112 (7.5)	97.6 (7.1)
White Wine	Sugars 100 g L^{-1} Alcohol 10 % v/v	High	2001 (4.2)	100.1 (3.9)	653 (5.5)	100.1 (4.9)
		Middle	1002 (5.8)	103.8 (6.0)	243 (5.7)	100.4 (5.9)
		Low	352 (7.1)	102.3 (7.0)	122 (6.8)	99.6 (6.2)
Red Wine	Sugars 3 g L^{-1} Alcohol 10 % v/v	High	1982 (4.7)	102.8 (3.4)	601 (5.5)	106.5 (4.3)
		Middle	876 (6.0)	99.5 (6.1)	221 (5.6)	107.2 (4.9)
		Low	351 (5.5)	98.7 (5.1)	117 (6.0)	103.2 (6.1)

3-MHA: high level = 0.75 μg/L; middle level = 0.35 μg/L; low level = 0.15μg/L.

3-MH: high level = 2.00 μg/L; middle level = 0.75 μg/L; low level = 0.30 μg/L.

consequence, a reliable quantification of TFSC by HS-SPME requires the use of a suitable internal standard, such as 6-MH.

Headspace Solid Phase Microextraction (HS-SPME), Solid Phase Extraction (SPE) and Purge and Trap (PT) analyses

Fiber type and working conditions for the HS-SPME method are reported in Table 5.13. It is worth mentioning that it is necessary to calibrate the method every time the fiber is changed using a wine pre-treated with charcoal, spiked with different quantities of standard solutions, as established in the calibration method proposed (Fedrizzi *et al.* 2007a; 2007b).

For SPE extraction (Fedrizzi *et al.*, 2007b), a cartridge with highly cross-linked styrene-divinylbenzene polyhydroxylated polymer stationary phase (1 g, ENV^+) is used. The pH value of the sample is adjusted to 7.0 to reduce the absorption of free fatty acids on the resin thus limiting possible interferences in GC-MS analysis. A volume of 100 mL of diluted wine (1:1 v/v by water) is percolated through the cartridge; recovery of analytes is performed with 30 mL of dichloromethane, then 60 mL of *n*-pentane is added to the solution to obtain an azeotropic pentane-dichloromethane mixture (2:1 v/v) with lower boiling point. This solution is dried with Na_2SO_4 and concentrated just before the analysis to about 100–150 μL by distillation using a Vigreux column and stored frozen (Boido *et al.*, 2003).

Also PT analysis of 3-MH and 3-MHA is performed by adjusting the pH solution to 7.0 to reduce the stripping of the free fatty acids and to avoid 3-MHA hydrolysis. Linearity and sensitivity have been investigated using the same pre-treated wine employed for controlling the other two quoted methods and the same calibration solutions. The optimized experimental conditions of sample temperature (St), flow rate (Fr), purge flow time (Pft), trap temperature (Tt) and desorption time (Dt) are reported in Table 5.15 and the sensitivity results in Table 5.16 (Fedrizzi *et al.*, 2007c).

The method is more sensitive than HS-SPME, mostly in the determination of 3-MHA for which L_D is about 60% lower (0.057 μg/L of HS-SPME with respect to 0.036 μg/L for PT), in the case of 3-MH L_D it is about 43% lower (0.069 with respect to 0.048 μg/L), and close to the sensory threshold. L_D by SPE are just a little higher than by HS-SPME and the difference with the PT method is more evident in the case of 3-MHA for which L_D achieved by SPE is 0.083 μg/L. In conclusion, the best improvement in sensitivity has been reached

Table 5.15 Optimized experimental conditions used for Purge and Trap analysis of 3-MH and 3-MHA in wine.

Solution pH	7.0
Sample volume	5 mL
Sample temperature	45 °C
Purge flow time	15 min
Purge flow rate	50 mL min^{-1}
Trap temperature	38 °C
Desorption time	2 min

Table 5.16 Sensitivity results in the analysis of 3-MH and 3-MHA in wine using the experimental conditions reported in Table 5.15.

Analytes	Calibration Range (μg L^{-1})	Slope	SD Slope	Intercept	SD Intercept	SD	R^2	L_C (μg L^{-1})	L_D (μg L^{-1})
3-MHA	0.25–4.00	2.63	0.03	0.015	0.009	0.05	0.992	0.019	0.036
3-MH	0.10–1.50	1.12	0.02	0.021	0.010	0.03	0.993	0.023	0.048

for 3-MHA, with a less time-consuming routine method used to quantify a compound with sensory threshold estimated around 0.004 μg/L (about nine times less than L_D). Despite this, the results are good enough to be applied in the quantification of these compounds in most wines characterized by tropical fruit scents.

Comparison of PT, HS-SPME and SPE methods for analysis of TFSV GC-MS(SIR)

Comparison of PT, SPE and HS-SPME techniques used for sampling and concentration of the volatile compounds fraction before analysis is shown in Figure 5.14, where the data plot of 52 different wines from *Sauvignon blanc, Gewürztraminer*, *Müller-Thurgau* and *Verdicchio* white grape varieties is reported. Both SPE and HS-SPME methods show similar precision and accuracy. They satisfy the oenological requirements providing a suitable L_D for each analyte. HS-SPME provides slightly higher sensitivity and it has the advantage of being more easily applicable. (Fedrizzi *et al.*, 2007c). Each point on the graph represents the concentration of a single wine sample obtained from the

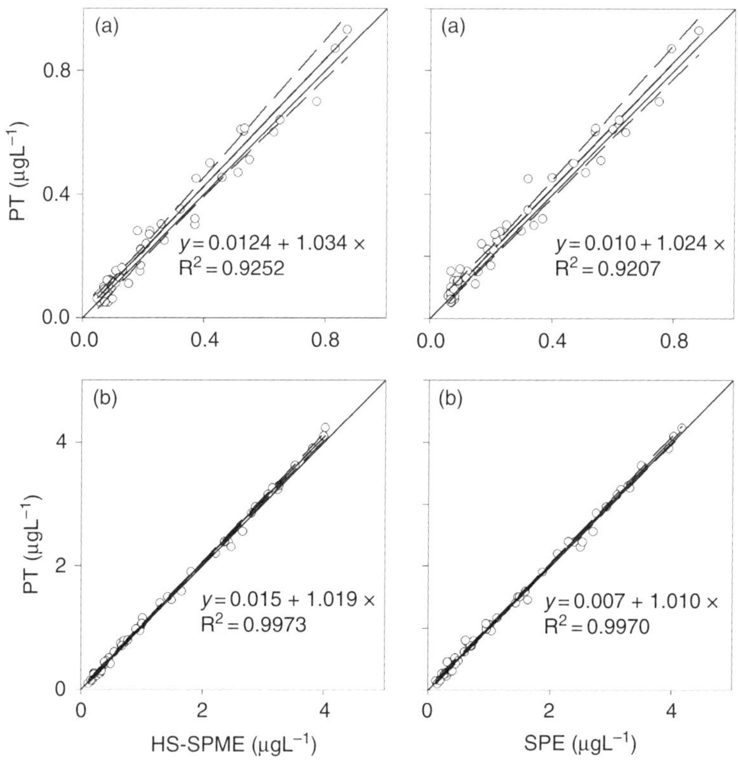

Figure 5.14 Data plot of results obtained comparing the three different methods in the TFSV analysis in 52 different wines from white grape varieties *Sauvignon blanc*, *Gewürztraminer*, *Müller-Thurgau* and *Verdicchio*. (a) data of 3-MHA; (b) data of 3-MH

relevant calibration curve of the two techniques being compared. Data of 3-MHA (Figure 5.14a) are wider spread due to their low concentration in the samples near the L_D, in respect to those of 3-MH (Figure 5.14b) well above the relevant L_D.

By performing analysis of the same sample, PT shows slightly higher data in comparison with those from the other two methods, probably due to better extraction of the analytes. PT sampling coupled with GC-MS analysis is a simple method applicable in the analysis of both 3-MH and 3-MHA; moreover, it provides lower detection limits which are close to sensory thresholds. Among the three sampling methods, PT seems to overcome the matrix effects affecting the HS-SPME and, to a lesser extent, the SPE approaches. As for SPE, we have to emphasize that, due to its exhaustive extraction for both free and bound forms of aroma compounds, it also allows quantification of a higher number of volatiles including the bound forms which can be analyzed as aglycones.

GC-MS analysis of TFSV

For analysis of 3-MH and 3-MHA in samples enriched by SPE or HS-SPME techniques a fused silica apolar column (10 m × 0.32 mm i.d.; 0.25 μm film thickness) connected to the PEG-like polar column, can be used. Possible working conditions include the following: He as carrier gas at a flow rate of 1.2 mL/min; transfer line temperature at 220 °C and MS source at 150 °C; GC injector temperature at 250 °C, injection mode splitles for 5 min, oven temperature programe: 35 °C (5 min), 1 °C/min to 40 °C, 10 °C/min to 250 °C.

PT analysis is carried out with an apolar fused-silica capillary column (30 m × 0.25 mm i.d.; 0.25 μm film thickness) with carrier gas He at flow-rate 1.0 mL/min. Other conditions are: temperature of transfer line at 220 °C, MS source at 150 °C, GC injector at 250 °C, oven temperature program: 35 °C (5 min), 1 °C/min to 40 °C, 10 °C/min to 250 °C.

Calibration curves for each TFSV are prepared according to the internal standard method using 6-MH as internal standards and the same type of pre-treated matrix, i.e. a white wine pre-treated with charcoal to eliminate most volatiles including sulphur compounds and polyphenols, restored to 100 mg/L total SO_2 and properly spiked with different quantities of standard solutions (Fedrizzi *et al.*, 2007a). The aim of this pre-treatment is to obtain a matrix similar to the wine, but free of volatile compounds potentially affecting adsorption and widely varying among the wines.

For the calculation of the critical (L_C) and the detection (L_D) limits the same criteria described for CFSV are used (see page 211). Linearity and precision are evaluated in the concentration ranges typical of 3-MH and 3-MHA in wine.

References

Abbott, N.A., Coombe, B.G. and Williams, P.J. (1991) The contribution of hydrolyzed flavor precursors to quality differences in Shiraz juice and wines. An investigation by sensory descriptive analysis, *Am. J. Enol. Vitic.*, **42**, 167–174.

Andres, A.I., Cava, R. and Ruiz, J. (2002) Monitoring volatile compounds during dry-cured ham ripening by solid-phase microextraction coupled to a new direct-extraction device, *J. Chromatogr. A*, **963**(**1-2**), 83–88.

Anocibar Beloqui, A. and Bertrand, A. (1995a) Study on sulfur compounds in wine: preliminary study, *Ital. J. Food Sci.*, **3**, 279–289.

Anocibar Beloqui, A., Guedes de Pinho, P. and Bertrand, A. (1995b) Bis-(hydroxyethyl)disulfide a new sulfur compound found in wine, *Am. J. Enol. Vitic.*, **46**(**1**), 84–87.

Aprea, E., Biasioli, F., Carlin, S., Versini, G., Märk, T. and Gasperi, F. (2007) Rapid white truffle headspace analysis by proton transfer reaction-mass spectrometry and comparison with solid-phase microextraction coupled with gas chromatography/mass spectrometry, *Rapid Commun. Mass Sp.*, **21**(**16**), 2564–2572.

Arthur, C.L. and Pawliszyn, J. (1990) Solid-phase microextraction with thermal desorption using fused silica optical fibers, *Anal. Chem.*, **62**(**19**), 2145–2148.

Bitteur, S.M., Baumes, R.L., Bayonove, C.L., Versini, G., Martin, C. and Dalla Serra, A. (1990) 2-exo-Hydroxy-1,8-cineole: a new component from grape var. Sauvignon, *J. Agric. Food Chem.*, **38**(**5**), 1210–1213.

Blanch, P., Reglero, G. and Herraiz, M. (1995) Analysis of wine aroma by off-line and online supercritical fluid extraction-gas chromatography, *J. Agric. Food Chem.*, **43**(**5**), 1251–1258.

Blanchard, L., Darriet, P. and Dubourdieu, D. (2004) Reactivity of 3-mercaptohexanol in red wine: impact of oxygen, phenolic fractions, and sulfur dioxide, *Am. J. Enol. Vitic.*, **55**(**2**), 115–120.

Bobet, R.A., Noble, A.C. and Boulton, R.B. (1990) Kinetics of the ethanethiol and diethyl disulfide interconversion in wine-like solutions, *J. Agric. Food Chem.*, **38**, 449–452.

Bocchini, P., Andalò, C., Pozzi, R., Galletti, G.C. and Antonelli, A. (2001) Determination of diallyl thiosulfinate (allicin) in garlic (*Allium sativum* L.) by high performance liquid chromatography with a post-column photochemical reactor, *Anal. Chim. Acta,* **441**(**1**), 37–43.

Boelens, M.H. and van Gemert, L. (1993) Volatile character-impact sulfur compounds and their sensory properties, *Perfumer Flavorist*, **18**(**3**), 29–39.

Boido, E., LLoret, A., Medina, K., Farina, L., Carrau, F., Versini, G. and Dellacassa, E. (2003) Aroma composition of *Vitis vinifera Cv.* Tannat: the typical red wine from Uruguay, *J. Agric. Food Chem.*, **51**(**18**), 5408–5413.

Burmeister, M.S., Drummond, C.J., Pfisterer, E.A. and Hysert, D.W. (1992) Measurement of volatile sulfur compounds in beer using gas chromatography with a sulfur chemiluminescence detector, *J. Am. Soc. Brew. Chem.*, **50**(**2**), 53–58.

Cabredo-Pinillos, S., Cedron-Fernandez, T. and Saenz-Barrio, C. (2004) Comparison of different extraction methods applied to volatile compounds in wine samples previous to the determination by gas chromatography, *Anal. Lett.*, **37**(**14**), 3063–3084.

Cardoso, D.R., Andrade Sobrinho, L.G., Lima Neto, B.S. and Franco, D.W. (2004) A rapid and sensitive method for dimethylsulphide analysis in Brazilian sugar cane sugar spirits and other distilled beverages, *J. Braz. Chem. Soc.*, **15**(**2**), 277–281.

Carlin, S. (1998) *Methods of aroma compounds enrichment for possible typifying Traminer wines (ital.)*, MSc. Thesis, University of Udine.

Carrapiso, A.I., Jurado, A., Timon, M.L. and Garcia, C. (2002) Odor-active compounds of Iberian hams with different aroma characteristics, *J. Agric. Food Chem.*, **50**(**22**), 6453–6458.

Chatonnet, P., Boutou, S. and Labadie, D. (2006) Identification et quantification rapide des principaux contaminants responsable de défauts olfactifs des vins. Check-list Excell®: dosage et identification simultanée multicontaminants par HS-SPME/GC/MS-EI-SIM, *Revue des Oenologues*, **121**, 48–50.

Ciolfi, G. and Di Stefano, R. (1983) Volatile compounds produced by the yeasts during the fermentation: comparison among strains of the same species or different species (Italian), *Vignevini*, **10**(**12**), 43–49.

Cordonnier, R. (1956) Recherches sur l'aromatisation et le parfum des vins doux naturels et des vins de liqueur, *Ann. Technol. Agric.*, **5**, 75–110.

Cozzolino, D., Smyth, H.E., Lattey, K.A., Cynkar, W., Janik, L. Dambergs, R.G., Francis, I.L. and Gishen, M. (2006) Combining mass spectrometry based electronic nose, visible-near infrared spectroscopy and chemometrics to assess the sensory properties of Australian Riesling wines, *Anal. Chim. Acta*, **563**(**1–2**), 319–324.

Darriet, P. (1993) *L'arôme et les précurseurs d'arôme du Sauvignon*, PhD Thesis, University of Bordeaux II, France.

De la Calle, G., Magnaghi, S., Reichenbächer, M. and Danzer, K. (1996) Systematic optimization of the analysis of wine bouquet components by solid phase microextraction, *J. High Res. Chromatog.*, **19**(5), 257–262.

Dercksen, A.W., Meijering, I. and Axcell, B. (1992) Rapid quantification of flavor-active sulfur compounds in beer, *J. Am. Soc. Brew. Chem.*, **50**(**3**), 93–101.

Diaz, P., Ibañez, E., Señorans, F.J. and Reglero, G. (2003) Truffle aroma characterisation by headspace solid-phase microextraction, *J. Chromatogr. A*, **1017**(**1–2**), 207–214.

Di Stefano, R. (1991) Proposition d'une methode de preparation de l'echantillon pour la determination des terpenes libres et glycosides des raisins et des vins, *Bull. O.I.V.*, **64** (**721–722**), 219–223.

Di Stefano, R., Cravero, M.C. and Gentilini, N. (1989) Metodi per lo studio dei polifenoli dei vini, *L'Enotecnico*, **5**, 83–89.

Di Stefano, R., Maggiorotto, G. and Gianotti, S. (1992) Transformation of nerol and geraniol induced by yeasts (Italian), *Rivista di Viticoltura e di Enologia*, **45**(**1**), 43–49.

Drawert, F. and Rapp, A. (1968) Gaschromatographische Untersuchung pflanzicher Aromen. I. Anreicherung, Trennung und Identifizierung von flüchtigen Aromastoffen in Traubenmosten und Weinen, *Chromatographia*, **1**, 446–458.

Ebeler, S.E. (2001) Analytical chemistry: unlocking the secrets of wine flavour, *Food Rev. Int.*, **17**(**1**), 45–64.

Eurachem (1993) *Accreditation for chemical laboratories*, Eurachem Secretariat, Teddington, UK.

Fang, Y. and Qian, M.C. (2005) Sensitive quantification of sulfur compounds in wine by headspace solid-phase microextraction technique, *J. Chromatogr. A*, **1080**(**2**), 177–185.

Fariña, L., Boido, E., Carrau, F., Versini, G. and Dellacassa, E. (2005) Terpene compounds as possible precursors of 1,8-cineole in red grapes and wines, *J. Agric. Food Chem.*, **53**(**5**), 1633–1636.

Fedrizzi, B., Magno, F., Badocco, D., Nicolini, G. and Versini, G. (2007d) Aging effects and variety dependence on the content of sulphur volatiles in wine, *J. Agric. Food Chem.*, **55**, 10880–10887.

Fedrizzi, B., Magno, F., Moser, S., Nicolini, G. and Versini, G. (2007a) Concurrent quantification of light and heavy sulphur volatiles in wine by headspace solid-phase microextraction coupled with gas chromatography/mass spectrometry, *Rapid Commun. Mass Sp.*, **21**(**5**), 707–714.

Fedrizzi, B., Versini, G., Lavagnini, I., Nicolini, G. and Magno, F. (2007b) Gas Chromatography-Mass Spectrometry determination of 3-mercaptohexan-1-ol and 3-mercaptohexyl acetate in wine. A comparison between Solid Phase Extraction and Headspace Solid Phase Microextraction methods, *Anal. Chim. Acta*, **596**(**2**), 291–297.

Fedrizzi, B., Versini, G., Lavagnini, I., Badocco, D., Nicolini, G. and Magno, F. (2007c) Hyphenated Gas Chromatography – Mass Spectrometry analysis of 3-mercaptohexan-1-ol and 3-mercaptohexyl acetate in wine. Comparison with results of other sampling procedures via a robust regression, *Anal. Chim. Acta.* DOI: 10.1016/j.aca.2007.10.040

Ferreira, V., Rapp, A., Cacho, J.F., Hastrich, H. and Yavas, I. (1993) Fast and quantitative determination of wine flavor compounds using microextraction with Freon 113, *J. Agric. Food Chem.*, **41**(9), 1413–1420.

Ferreira, V., Sharman, M., Cacho J.F. and Dennis, J. (1996) New and efficient microextraction/solid-phase extraction method for the gas chromatographic analysis of wine volatiles, *J. Chromatogr. A*, **731**(**1–2**), 247–259.

Flamini, R., Dalla Vedova, A., Panighel, A., Biscaro, S., Borgo M. and Calò, A. (2006) Caratterizzazione aromatica del Torbato (V. vinifera) e studio degli effetti dell'accartocciamento fogliare sui composti aromatici delle sue uve, *Rivista di Viticoltura e di Enologia,* **59**(**1**), 13–26.

Flanzy, C. (1998) *Œnologie – fondements scientifiques et technologiques*, TEC & DOC, Paris.

Francis, I.L., Kassara, S., Noble, A.C. and Williams, P.J. (1998) The contribution of glycoside precursors to Cabernet savignon and Merlot aroma, in A.L. Waterhouse and S.E. Ebeler (Eds), *Chemistry of wine flavor,* ACS Symp. Series 714, Am. Chem. Soc., Washington DC.

Genovese, A., Dimaggio, R., Lisanti, M.T., Piombino, P. and Moio, L. (2005) Aroma components of red wines by different extraction methods and gas chromatography-SIM/mass spectrometry analysis, *Annali di Chimica*, **95**(**6**), 383–394.

Gòmez-Míguez, M.J., Cacho, J.F., Ferreira, V., Vicario, I.M. and Heredia, F.J. (2007) Volatile components of Zalema white wines, *Food Chem.*, **100**(**4**), 1464–1473.

Gramatica, P., Manitto, P., Ranzi, B.M., Dalbianco, A. and Francavilla, M. (1982) Stereospecific reduction of geraniol to (*R*)-(+)-citronellol by Saccharomyces cerevisiae, *Experientia*, 38, 775–776.

Gunata, Y.Z., Bayonnove, C.L., Baumes, R.L. and Cordonnier, R.E. (1985) The aroma of grapes I. Extraction and determination of free and glycosidically bound fractions of some grape aroma components, *J. Chromatogr.*, **331**, 83–90.

Gunata, Z., Wirth, J.L., Guo, W. and Baumes, R.L. (2002) C_{13}-norisoprenoid aglycon composition of leaves and grape berries from Muscat of Alexandria and Shiraz cultivars, In: *Carotenoid-derived aroma compounds,* ACS Symp. Series 802, Am. Chem. Soc., Washington DC.

Guth, H. (1995) Potente Aromastoffe von Weißweinen unterschiedlicher Rebsorten - Identifizierung und Vergleich, *Lebensmittelchemie*, **49**, 107.

Hardy, P.J. (1969) Extraction and concentration of volatiles from dilute aqueous-alcoholic solution using trichlorofluoromethane, *J. Agric. Food Chem.*, **17**(3), 656–658.

Hinrichsen, L.L. and Pedersen, S.B. (1995) Relationship among flavor, volatile compounds, chemical changes, and microflora in Italian-type dry-cured ham during processing, *J. Agric. Food Chem.*, **43**(**11**), 2932–2940.

Hinterholzer, A. and Schieberle, P. (1998) Identification of the most odour-active volatiles in fresh, hand-extracted juice of Valencia late oranges by odour dilution techniques, *Flavour Frag. J.*, **13**(1), 49–55.

Hubaux, A. and Vos, G. (1970) Decision and detection limits for linear calibration curves, *Anal. Chem.*, **42**(8), 849–855.

Ibarz, M.J., Ferreira, V., Hernández, P., Loscos, N. and Cacho, J. (2006) Optimisation and evaluation of a procedure for the gas chromatographic-mass spectrometric analysis of the aromas generated by fast acid hydrolysis of flavor precursors extracted from grapes, *J. Chromatog. A*, **1116(1–2)**, 217–229.

Karagiannis, S. and Lanaridis, P. (1999) The effect of various vinification parameters on the development of several volatile sulfur compounds in Greek white wines of the cultivars Batiki and Muscat of Hamburg, *Am. J. Enol. Vitic.*, **50(3)**, 334–342.

Kinton, V.R., Collins, R.J., Kolahgar, B. and Goodner, K.L. (2003) Fast analysis of beverages using mass spectral based chemical sensor, *Gerstel AppNote,* **4**, 1–10.

Knapp, H., Straubinger, M., Stingl, C. and Winterhalter, P. (2002) Analysis of norisoprenoid aroma precursors. in *Carotenoid-derived aroma compounds,* ACS Symp. Series 802, Am. Chem. Soc., Washington DC.

Kotseridis, Y. and Baumes, R. (2000) Identification of impact odorants in Bordeaux red grape juice, in the commercial yeast used for its fermentation, and in the produced wine, *J. Agric. Food Chem.*, **48(2)**, 400–406.

Lavagnini, I. and Magno, F. (2007) A statistical overview on univariate calibration, inverse regression and detection limits. Application to gas chromatography/mass spectrometry technique, *Mass Spectrom. Rev.*, **26(1)**, 1–18.

Ledauphin, J., Basset, B., Cohen, S., Payot, T. and Barillier, D. (2006) Identification of trace volatile compounds in freshly distilled Calvados and Cognac: Carbonyl and sulphur compounds, *J. Food Compos. Anal.,* **19(1)**, 28–40.

Leppanen, O.A., Densolw, J. and Ronkainen, P.P. (1979) A gas chromatographic method for the accurate determination of low concentrations of volatile sulphur compounds in alcoholic beverages, *J. I. Brewing.*, **85**, 350–353.

Lestremau, F., Andersson, F.A.T. and Desauziers, V. (2004) Investigation of artefact formation during analysis of volatile sulphur compounds using solid phase microextraction (SPME), *Chromatographia*, **59(9–10)**, 607–613.

Lindström, A., Albertsson, A.C. and Hakkarainen, M. (2004) Development of a solid-phase extraction method for simultaneous extraction of adipic acid, succinic acid and 1,4-butanediol formed during hydrolysis of poly(butylene adipate) and poly(butylene succinate), *J. Chromatogr. A*, **1022(1–2)**, 171–177.

Luan, F., Mosandl, A., Münch, A. and Würst, M. (2005) Metabolism of geraniol in grape berry mesocarp of *Vitis vinifera* L. Cv. Scheurebe: demonstration of stereoselective reduction, E/Z-isomerisation, oxidation and glycosilation, *Phytochemistry*, **66(3)**, 295–303.

Luan, F. and Würst, M. (2002) Differential incorporation of 1-deoxy-D-xylulose into (3*S*)-linalool and geraniol in grape berry exocarp and mesocarp, *Phytochemistry*, **60(5)**, 451–459.

Marais, J.A. (1986) A reproducible capillary gas chromatographic technique for the determination of specific terpenes in grape juice and wine, *S. Afr. J. Enol. Vitic.*, **7(2)**, 21–25.

Marais, J., Versini, G., van Wyk, C.J. and Rapp, A. (1992) Effect of region on free and bound monoterpene and C13-norisoprenoids in Weisser Riesling wines, *S. Afric. J. Enol. Vitic.*, **13**, 71–77.

Mateo-Vivaracho, L., Cacho, J. and Ferreira, V. (2007) Quantitative determination of wine polyfunctional mercaptans at nanogram per litre by gas chromatography-negative chemical ionization mass spectrometric analysis of their pentafluorobenzyl derivatives, *J. Chromatogr. A*, **1146(2)**, 242–250.

Mateo-Vivaracho, L., Ferreira, V. and Cacho, J. (2006) Automated analysis of 2-methyl-3-furanethiol and 3-mercaptohexyl acetate at ng L^{-1} level by headspace solid phase microextraction with on-fibre derivatization and gas chromatography-negative chemical ionization mass spectrometric determination, *J. Chromatogr. A*, **1121(1)**, 1–9.

Mathieu, S., Terrier, N., Procureur, J., Bigey, F. and Gunata, Z. (2007) A carotenoid cleavage dioxygenase from *Vitis vinifera* L.: functional characterization and expression during grape berry development in relation to C13-norisoprenoid accumulation, *J. Exp. Bot.,* **56(420)**, 2721–2731.

Meilgaard, M.C. (1975) Aroma volatiles in beer: purification, flavour, thresholds and interaction, in F. Drawert (Ed) *Geruch- und Geschmackstoffe,* Verlag H. Carl, Nürnberg.

Mestres, M., Busto, O. and Gulasch, J. (2002) Application of headspace solid-phase microextraction to the determination of sulphur compounds with low volatility in wines, *J. Chromatogr. A*, **945(1–2)**, 211–219.

Mestres, M., Busto, O. and Guasch, J. (2000) Analysis of organic sulfur compounds in wine aroma, *J. Chromatogr. A*, **881(1–2)**, 569–581.

Murat, M.L., Tominaga, T. and Dubourdieu, D. (2001) Assessing the aromatic potential of Cabernet Sauvignon and Merlot musts to produce rose wine by assaying the cysteinylated precursor of 3-mercaptohexan-1-ol, *J. Agric. Food Chem.*, **49(11)**, 5412–5417.

Murat, M.L., Tominaga, T., Saucier, C., Glories, Y. and Dubourdieu, D. (2003) Effect of anthocyanins on stability of a key odorous compound, 3-mercaptohexan-1-ol, in Bordeaux rose wine, *Am. J. Enol. Vitic.*, **54(2)**, 135–138.

Murray, R.A. (2001) Limitations to the use of solid-phase microextraction for quantitation of mixtures of volatile organic sulfur compounds, *Anal. Chem.*, **73(7)**, 1446–1649.

Nicolini, G., Versini, G., Dalla Serra, A., Seppi, A., Amedei, E. and Falcetti, M. (1995) Aspetti compositivi di mosti e vini Mueller-Thurgau del Trentino, *Rivista di Viticoltura e di Enologia*, **48**(3), 47–61.

Nongonierma, A., Cayot, P., Le Quere, J.L., Springhett, M. and Voilley, A. (2006) Mechanisms of extraction of arma compounds from foods, using adsorbents. Effect of various parameters. *Food Rev. Int.*, **22(1)**, 51–94.

Peyrot des Gachons, C., Tominaga, T. and Dubourdieu, D. (2002a) Localization of S-Cysteine conjugates in the berry: effect of skin contact on aromatic potential of *Vitis vinifera* L. Cv. Sauvignon blanc must, *Am. J. Enol. Vitic.*, **53(2)**, 144–146.

Peyrot des Gachons, C., Tominaga, T. and Dubourdieu, D. (2002b) Sulfur aroma precursor present in *S*-glutathione conjugate form: identification of *S*-3-(hexan-1-ol) glutathione in musts from *Vitis vinifera* L. cv. Sauvignon Blanc, *J. Agric. Food Chem.*, **50(14)**, 4076–4079.

Rapp, A. (1988) Wine aroma substances from gas chromatographic analysis. In *Modern methods of plant analysis,* H.F. Linskens and J.F. Jackson (Eds), Springer Verlag, Berlin.

Rapp, A., Güntert, M. and Almy, J. (1985) Identification and significance of several sulfur-containing compounds in wine, *Am. J. Enol. Vitic.*, **36**(3), 219–221.

Rapp, A. and Hastrich, H. (1978) Gaschromatographische Untersuchungen ueber die Aromastoffe von Weinbeeren. III. Die Bedeutung des Standortes fuer die Aromastoffzusammensetzung der Rebsorte Riesling, *Vitis*, **17**, 288–298.

Rapp, A., Hastrich, H., Engel, L. and Knipser, W. (1978) Possibilities of characterising wine quality and wine varieties by means of capillar chromatography. In G. Charalambous (Ed) *Flavor of food and beverages*, Academic Press, New York.

Rapp, A., MacNamara, K. and Hoffmann, A. (1996) Determination of wine aroma compounds from simple extracts using automated Large Volume Injection with PTV solvent splitting. *Gerstel AppNote*, **1**(**1**), 1–8.

Rapp, A., Mandery, H. and Niebergall, H. (1986) Neue Monoterpendiole in Traubenmost und Wein sowie in Kulturen von *Botrytis cinerea*, *Vitis*, **23**, 79–84.

Rapp, A. and Versini, G. (1991) Influence of nitrogen compounds in grape on aroma compounds of wines, in *Proceed. Intern. Symp. on Nitrogen in Grapes and Wines*, Seattle, Washington, DC.

Rapp, A., Yavas, I. and Hastrich, H. (1994) Einfache und schnelle Anreicherung ('Kaltronmethode') von Aromastoffen des Weines und deren quantitative Bestimmung mittels Kapillargaschromatographie, *Deutsche Lebensm.-Rdsch.*, **90**, 171–174.

Rauhut, D. (1996) Qualitätsmindernde schwefelhaltige Stoffe im Wein – Vorkommen, Bildung, Beseitigung, PhD Thesis, University of Giessen, Germany.

Rauhut, D., Kuerbel, H., MacNamara, K. and Grossmann, M. (1998) Headspace GC-SCD monitoring of low volatile sulfur compounds during fermentation and in wine, *Analusis*, **26**, 142–145.

Ribéreau-Gayon, P., Glories, Y., Maujean, A. and Dubourdieu, D. (1998) *Traité d'Oenologie 2. Chimie du vin – stabilisation et traitements,* Dunod, Paris.

Rigaud, J., Cheynier, V., Souquet, J.M. and Moutounet, M. (1991) Influence of must composition on phenolic oxidation kinetics, *J. Sci. Food Agric.*, **57**(**1**), 55–63.

Riu, M., Mestres, M., Busto, O. and J. and Guasch, J. (2002) Determination of 2,4,6-trichloroanisole in wine by headspace solid-phase microextraction and gas chromatography-electron-capture detection, *J. Chromatogr. A*, **977**(**1**), 1–8.

Sánchez-Palomo, E., Díaz-Maroto, M.C. and Pérez-Coello, M.S. (2005) Rapid determination of volatile compounds in grapes by HS-SPME coupled with GC-MS, *Talanta*, **66**(**5**), 1152–1157.

Schneider, R., Kotseridis, Y., Ray, J., Augier, C. and Baumes, R. (2003) Quantitative determination of sulfur-containing wine odorants at sub parts per billion levels. Development and application of a stable isotope dilution assay, *J. Agric. Food Chem.*, **51**(**11**), 3243–3248.

Schreier, P. (1979) Flavor composition of wines: a review, *CRC Critical Reviews* in *Food Science and Nutrition,* **12**(**1**), 59–111.

Schreier, P. and Drawert, F., Schwefelhaltige Aromastoffe, 1975 in F. Drawert (Ed), *Geruch- und Geschmackstoffe,* H. Carl Verlag, Nürnberg.

Seeber, R., Sferlazzo, G., Leardi, R., Dalla Serra, A. and Versini, G. (1991) Multivariate data analysis in classification of musts and wines of the same variety according to vintage year, *J. Agric. Food Chem.*, **39**(**10**), 1764–1769.

Segurel, M.A., Razungles, A.J., Riou, C., Trigueiro M.G.L. and Baumes, R. (2005) Ability of possible DMS precursors to release DMS during wine aging and in the conditions of heat-alkaline treatment, *J. Agric. Food Chem.*, **53**(**7**), 2637–2645.

Semmelroch, P. and Grosch, W. (1996) Studies on character impact odorants of coffee brews, *J. Agric. Food Chem.*, **44** (**2**), 537–543.

Shimoda, M. and Shibamoto, T. (1990) Isolation and identification of headspace volatiles from brewed coffee with an on-column GC-MS method, *J. Agric. Food Chem.*, **38**(**3**), 802–804.

Silva Ferreira, A.C., Rodrigues, P., Hogg, T. and Guedes de Pinho, P. (2003) Influence of some technological parameters on the formation of dimethyl sulfide, 2-mercaptoethanol, methionol, and dimethyl sulfone in port wine, *J. Agric. Food Chem.*, **51**(**3**), 727–732.

Simpson, R.F. (1979) Aroma composition of bottle aged white wine, *Vitis*, **18**, 148–154.

Singleton, V.L. (1987) Oxygen with phenols and related reactions in musts, wines, and model systems: observations and practical implications, *Am. J. Enol. Vitic.*, **38(1)**, 69–77.

Skouroumounis, G.K. and Sefton, M.A. (2000) Acid-catalyzed hydrolysis of alcohols and their beta-D-glucopyranosides, *J. Agric. Food Chem.*, **48(6)**, 2033–2039.

Strauss, C.R., Wilson, B., Gooley, P.R. and Williams, P.J. (1986) Role of monoterpenes in grape and wine flavour, in T.H. Parliment and R. Croteau (Eds), *Biogenertion of aromas,* ACS Symp. Series 317, Am. Chem. Soc., Washington DC.

Strauss, C.R., Wilson, B., Rapp, A., Güntert, M. and Williams, P.J. (1985) New monoterpene ethyl ethers in grape wines and brandies, *J. Agric. Food Chem.*, **33(4)**, 706–708.

Swiegers, J.H., Capone, D.L., Pardon, K.H., Elsey, G.M., Sefton, M.A., Francis, I.L. and Pretorius, I.S. (2007) Engineering volatile thiol release in Saccharomyces cerevisiae for improved wine aroma, *Yeast,* **24**(7), 561–574.

Swiegers, J.H. and Pretorius, I.S. (2007) Modulation of volatils sulfur compounds by wine yeast, *Appl. Microbiol. Biot.,* **74**(5), 954–960.

Tominaga, T., Furrer, A., Henry, R. and Dubourdieu, D. (1998a) Identification of a new volatile thiols in the aroma of *Vitis vinifera* L. var. Sauvignon Blanc wines, *Flavour Frag. J.*, **13**(3), 159–162.

Tominaga, T., Peyrot des Gachons, C. and Dubourdieu, D. (1998b) A new type of flavor precursors in *Vitis vinifera* L. Cv. Sauvignon Blanc: *S*-cysteine conjugates, *J. Agric. Food Chem.*, **46(12)**, 5215–5219.

Tominaga, T., Murat M.L. and Dubourdieu, D. (1998c) Development of a method for analyzing the volatile thiols involved in the characteristic aroma of wines made from *Vitis vinifera* L. Cv. Sauvignon Blanc, *J. Agric. Food Chem.*, **46(3)**, 1044–1048.

Ulrich, D., Hoberg, E., Rapp A. and Sandke, G. (1997) Flavour analysis in plant breeding – solid phase micro extraction of strawberry aroma compounds, in H.-P. Kruise and M. Rothe (Eds) *Proceed. 5th Wartburg Aroma Symp., Eisenach (Germany),* Eigenverlag Univ., Postdam.

Vas, G., Koteleky, K., Farkas, M., Dobó A. and Vekey, K. (1998) Fast screening method for wine headspace compounds using solid-phase microextraction (SPME) and capillary GC technique, *Am. J. Enol. Vitic.*, **49(1)**, 100–104.

Versini, G., Carlin, S., Dalla Serra, A., Nicolini, G. and Rapp, A. (2002) Formation of 1,1,6-trimethyl-1,2-dihydronaphthalene and other norisoprenoids in wine. Considerations on the kinetics, in *Carotenoid-derived Aroma Compounds,* ACS Symp. Series 802, Am. Chem. Soc., Washington DC.

Versini, G., Dalla Serra, A., Magni, P., Munari F. and Di Stefano, R. (1995) Optimization of a rapid method for wine aroma extraction and analysis in GC-FID and GC-NPD by large volume on column injection, in *Proceed. 2nd National Congress on Food Chemistry*, Giardini Naxos, Catania, La Grafica Editoriale, Messina pp. 187–198.

Versini, G., Moser, S. and Carlin, S. (2005) Methyl salicylate as a remarkable almost bound compound in some renowned Italian varietal wines, *Abstract Proceed. 'In Vino Analytica Scientia',* Montpellier.

Versini, G., Rapp, A., Dalla Serra A. and Nicolini, G. (1997) Use of bound forms profile to improve the variety discrimination capability of monoterpenes in wines from some floral and non-floral groups of cultivars: the case of Rhine Riesling crosses, in H.P. Kruise and M. Rothe (Eds) *Proceed. 5th Wartburg Aroma Symposium, Eisenach, Germany,* Eigenverlag Univ., Postdam.

Versini, G., Rapp, A., Reniero, F. and Mandery, H. (1991) Structural identification and presence of some p-menth-1-endiols in grape products, *Vitis*, **30**, 143–149.

Versini, G., Rapp, A., Scienza, A., Dalla Serra, A. and Dell'Eva, M. (1998) Evidence of some glycosidically bound new monoterpenes and norisoprenoids in grapes, in P. Schreier (Ed), *Proc. Bioflavour '87*. W. de Gruyter, Berlin.

Versini, G., Schneider, R., Carlin, S., Depentori, D., Nicolini, G. and Dalla Serra, A. (1999) Characterization of some northern Italian Passiti-wines through aroma and stable isotope analysis, in E. Lemperle (Ed), *12th International Enology Symposium*, Montreal, Canada, IVOBW, Breisach, Germany, pp 544–571.

Voirin, S.G., Baumes, R.L., Gunata, Y.Z., Bitteur, S.M. and Bayonove, C.L. (1992) Analytical methods for monoterpene glycosides in grape and wine. I. XAD-2 extraction and gas chromatographic-mass spectrometric determination of synthetic glycosides, *J. Chromatogr. A*, **590(2)**, 313–328.

Williams, P.J. (1982) Use of C_{18} reversed-phase liquid chromatography for the isolation of monoterpene glycosides and nor-isoprenoid precursors from grape juice and wines, *J. Chromatogr. A*, **235(2)**, 471–480.

Williams, P.J. and Francis, I.L. (1996) Sensory analysis and quantitative determination of grape glycosides—the contribution of these data to winemaking and viticulture, *in* G.R. Takeoka, R. Teranishi, P.J. Williams, and A. Kobayashi, (Eds) *Biotechnology for improved foods and flavors,*, ACS Symposium Series 637, Am. Chem. Soc., Washington DC.

Williams, P.J., Sefton, M. and Francis, I.L. (1992) Glycosidic precursors of varietal grape and wine flavour, in R. Teranishi, G.R. Takeoka, and M. Güntert (Eds), *Flavor precursors – Thermal and enzymatic conversions,* ACS Symp. Series 490, Am. Chem. Soc., Washington DC.

Williams, P.J., Sefton, M.A. and Wilson, B. (1989) Nonvolatile conjugates of secondary metabolites as precursors of varietal grape flavor components, in R. Teranishi, R.G. Buttery, and F. Shahidi, (Eds), *Flavor chemistry trends and developments,* ACS Symp. Series 388, Am. Chem. Soc., Washington DC.

Williams, P.J., Strauss, C.R., Wilson, B. and Massy-Westropp, R.A. (1983) Glycosides of 2-phenylethanol and benzyl alcohol in *Vitis vinifera* grapes, *Phytochemistry*, **22(9)**, 2039–2041.

Winter, M., Furrer, A., Willhalm, B. and Thommen, W. (1976) Identification and synthesis of two new organic sulfur compounds from the yellow passionfruit (*Passiflora edulis f. flavicarpa*), *Helv. Chim. Acta*, **59(5)**, 1613–1620.

Winterhalter, P. (1992) Oxygenated C_{13}-norisoprenoids – Important flavor precursors, in R. Teranishi, G.R. Takeoka, and M. Güntert, (Eds), *Flavor precursors – Thermal and enzymatic conversions,* ACS Symp. Series 490, Washington DC.

Winterhalter, P., Baderschneider, B. and Bonnländer, B. (1998) Analysis, structure, and reactivity of labile terpenoid aroma precursors in Riesling wines, in A.L. Waterhouse, and Ebeler, S.E. (Eds), *Chemistry of wine flavor,* ACS Symp. Series 714, Am. Chem. Soc., Washington DC.

Winterhalter, P., Sefton, M.A. and Williams, P.J. (1990) Two-dimensional GC-DCCC analysis of the glycoconjugates of monoterpenes, norisoprenoids, and shikimate-derived metabolites from Riesling wine, *J. Agric. Food Chem.*, **38(4)**, 1041–1048.

Yan, X., Wang, Z. and Barlow, P. (1993) Quantitative determination and profiling of total sulphur compounds in garlic health products using a simple GC procedure, *Food Chem.*, **47(3)**, 289–294.

6

Analysis of Pesticide Residues in Grape and Wine

Paolo Cabras and Pierluigi Caboni

6.1. Introduction

At the present time the world area designated for grape cultivation is about 8 million hectares with a global production of about 270 million hectoliters of wine.

The Mediterranean area leads the world in grape cultivation. Whilst France and Italy are the two main producers, in the last ten years Italy – with more than 58 million hectoliters of wine – has produced more than any other country. The area cultivated in Italy is about 800 thousand hectares and the percentage production of wine grape and table grape is around 92% and 8%, respectively.

The main grape parasites are powdery mildew (*Uncinula necator*), downy mildew (*Plasmopora viticola*), grey mold (*Botryitis cinerea*), and European grapevine moth (*Lobesia botrana*). A large number of pesticides are commonly used by farmers to control these diseases. Before going on the market, pesticides need to be registered. In Italy, the Ministry of Health controls the registration process. For each pesticide the following information is issued: the culture, the rate, the pre-harvest interval (PHI), and the maximum residue limit (MRL).

Hyphenated Techniques in Grape and Wine Chemistry Edited by Riccardo Flamini

Practically speaking, two of these parameters are verifiable: the culture and the MRL. It is worth noting that the legislator defines the term 'residue' as 'residues of active ingredients and relative impurity that are present in products destined to human and animal feeding, and resulting from the pesticide use inclusive of those substances of toxicological significance deriving from metabolism and degradation of the active ingredients'.

It is commonly considered that a pesticide is 'not present' if the residue is below 0.01 mg/kg which is accepted as legal zero. Bear in mind that the legal limit of the residue does not coincide with the toxicological limit and, for this reason, when this limit is exceeded it does not pose a risk to human health. The legal limit is fixed on toxicological data that do not determine any human health risk (NOEL: no observed effect level) and is commonly corrected by a safety factor of 100 and on residues determined in field experiments when the residue is below the toxicological limit (Cabras, 2003).

Field residues, if below the toxicological limit (acceptable daily intake), are used to set the legal limit. The residues of pesticides in the field experiment are affected by the environmental conditions such as temperature, wind, rain, solar irradiance etc. For this reason, residues limits (MRLs) can vary between countries because of the different climatic conditions, leading to EU trade difficulties.

Currently, the EU is working for the harmonization of the maximum residue limits of pesticides. Moreover, MRLs are set for raw food and not for processed food. In Italy the pesticides currently registered for use on grapes are listed in Table 6.1 with their MRL. In 1995, Italy started, for the first time, setting the maximum residue levels for wine (Table 6.2). In Italy, where there is no legal limit as to the levels of pesticides in processed food, the amount permitted in raw food per unit of processed food (e.g. 1.5 kg of grapes to obtain 1 L of wine), *and* the incidence of technological processes, should *both* be taken into account. In other countries such as the US, the maximum residue levels for processed food products, and therefore for the wine, are the same as those raw food, and therefore for wine grapes. In Italy, the main issue encountered in defining wine legal limits is the lack of a study protocol.

Since each active ingredient has its own peculiar behavior (Cabras and Angioni, 2000), residue changes during processing should be determined. In the absence of these data, the unique and safe reference is the MRL of the raw food.

Table 6.1 Pesticides registered on grapes in Italy. MRL: maximum residue limit.

PESTICIDE	MRL (mg kg^{-1})	PESTICIDE	MRL (mg kg^{-1})	PESTICIDE	MRL (mg kg^{-1})
abamectin	0.01	esfenvalerate	0.1	methoxyfenozide	1
acrinathrin	0.1	ethephon	0.1	metiram	2
alcalines solphites	10	etofenprox	0.05	myclobutanil	1
alphamethrin	0.3	etoxazole	1	oxadiazon	0.05
azadirachtin	0.5	famoxadone	0.02	oxyfluorfen	0.05
azinphos-methyl	1	fenamidone	2	paraquat	0.05
azociclotin	0.3	fenamiphos	0.5	penconazole	0.2
azoxystrobin	2	fenarimol	0.02	phosalone	1
benalaxyl	0.2	fenazaquin	0.3	phosetyl – al	2
benfuracarb	0.05	fenbuconazole	0.2	piperonyl butoxide	3
bifenthrin	0.2	fenbutatin oxide	0.2	pirimicarb	0.2
bifentrin	0.2	fenhexamid	2	pirimiphos-methyl	2
bromopropylate	2	fenitrothion	5	procymidone	5
bromuconazole	0.5	fenoxycarb	0.5	propargite	2
buprofezin	1	fenpropidin	0.2	propiconazole	0.5
calcium polysulfide	50	fenpropimorph	2	propineb	2
captan	10	fenpyroximate	0.05	propyzamide	0.02
carbaryl	3	flazasulfuron	0.3	pyraclostrobin	2
carbendazim	2	fluazifop-p-butyl	0.01	pyrethrins	1
chloropicrin	0.05	fluazinam	0.1	pyridaben	0.1
chlorothalonil	3	fludioxonil	2	pyrimethanil	3
chlorpropham	0.05	flufenoxuron	2	quinoxyfen	0.5
chlorpyrifos	0.5	flusilazole	0.1	rotenone	0.05
chlorpyrifos-methyl	0.2	fluvalinate	0.01	spinosad	0.2
clofentezine	1	folpet	0.5	spiroxamine	1
cyanamide	0.05	glufosinate ammonium	10	sulphur	50
cyazofamid	1	glyphosate	0.1	tebuconazole	1
cycloxidim	0.1	glyphosate trimesium	0.1	tebufenozide	0.5
cyfluthrin	0.3	hesaconazole	0.1	tebufenpyrad	0.3
cyhexatin	0.3	hexythiazox	0.5	teflubenzuron	1
cymoxanil	0.1	indoxacarb	0.5	tetraconazole	0.5
cypermethrin	0.5	iprodione	10	thiamethoxam	0.5
cyproconazole	0.2	iprovalicarb	2	thiodicarb	1
cyprodinil	5	kresoxim-methyl	1	thiram	3.8
deltamethrin	0.1	lambda cyalothrin	0.2	tiophanate-methyl	2
diazinon	0.02	lufenuron	0.5	tolyfluanid	5
dichlobenil	0.1	mancozeb	2	triadimenol	2
dichlorvos	0.1	maneb	2	trichlorfon	0.5
dicofol	2	mcpa	0.1	trifloxystrobin	3
diethofencarb	1	mecoprop	0.1	trifluralin	0.05
dimethomorph	0.5	mepanipyrim	3	vinclozolin	5
diquat	0.05	metalaxyl-m	1	white mineral oil	0
dithianon	0.6	metam-sodium	2	zeta cypermethrin	0.5
diuron	0.05	methidathion	0.5	ziram	2
dodine	0.2	methiocarb	0.05	zoxamide	5
endosulfan	0.5	methomyl	1		

Table 6.2 Maximum residue limits in grape and wine.

PESTICIDE	grape (mg kg^{-1})	wine (mg L^{-1})
azoxystrobin	2	0.5
bromuconazole	0.5	0.2
buprofezin	1	0.5
cyazofamid	1	0.05
cyproconazole	0.2	0.02
cyprodinil	5	0.5
dazomet	–	0.02
diethofencarb	1	0.3
etofenprox	1	0.1
etoxazole	0.02	0.01
fenamidone	0.5	0.5
fenazaquin	0.2	0.01
fenhexamid	5	1.5
fenpropidin	2	0.5
flazasulfuron	0.01	0.01
fluazinam	1	0.02
fludioxonil	2	0.5
hesaconazole	0.1	0.01
indoxacarb	0.5	0.02
iprodione	10	2
iprovalicarb	2	1
mepanipyrim	3	1
metalaxyl-m	1	0.2
metam-sodium	2	0.2
methoxyfenozide	1	0.05
myclobutanil	1	0.1
procymidone	5	0.5
pyrimethanil	3	2
quinoxyfen	0.5	0.01
spinosad	0.2	0.01
spiroxamine	1	0.5
tebuconazole	1	0.5
tebufenozide	0.5	0.1
tebufenpyrad	0.3	0.1
teflubenzuron	1	0.01
thiamethoxam	0.5	0.5
trifloxystrobin	3	0.3
ziram	2	0.2
zoxamide	5	0.5

Normally, the study protocol is conducted on the basis of grape residue levels at harvest time which is ordinarily well below the MRL. This means that if an active compound does not undergo a decrease during wine-making (e.g. pyrimethanil, Cabras *et al.*, 1997), starting from the grape with a maximum residue level of 3.0 mg/kg, the maximum residue level of the active compound in wine should be at 3.0 mg/L while it was fixed at 2.0 mg/L.

6.2. Analytical Methods

Residues of an active ingredient and the identification and quantification of metabolites and/or degradation products in grapes and wine are usually performed with a monoresidue analytical method. By the same approach, the pesticide fate after field treatment and during wine-making is monitored. For the legal control of MRLs, multiresidue methods are usually used for contemporaneously monitoring residue levels of different pesticides, in grapes or wine. The main characteristics of these methods are summarized in the Table 6.3.

Table 6.3 Characteristics of monoresidue and multiresidue methods for the analysis of pesticide residues.

Monoresidue methods

- Allow the determination of active ingredients and known metabolites as well as field experiment studies
- Detection limits are usually low
- Few interferences during the analysis
- High reliability in the compound identification step

Multiresidues methods

- Are indicated for the quantification of many active ingredients at the same time
- Detection limits are higher if compared with the monoresidue methods
- Several interferences during the analysis

6.2.1. Sampling

Sampling is crucial for the correct quantification of the pesticide's residue in food. For this reason, during official controls, samples are usually taken by authorized and specialist personnel. Residue levels are also affected not only during preparation but also by sampling. Depending on the sample size, an erroneous sampling can lead to standard deviations varying from 20 to 40% (Ambrus *et al.*, 1996; Ambrus, 1996). For official controls sampling, procedures used for pesticide quantification are reported in the EU Directive 2002/63/CE as well as in the Italian G.U. n.221 of 09/23/2003. The minimum amount of grape should be 2 kg (almost five bunches, each one weighing at least 250 g), per 0.5 L of wine.

The quantification of active ingredients involves different steps: sample preparation, extraction, partition, purification and determination.

6.2.2. Sample Preparation

When the vineyard is treated, the active ingredient is deposited on the grape surface. Because most pesticides are lipophilic they do not tend to reach the pulp which is a polar aqueous mixture of sugars and organic acids.

It should be taken into account that pesticides are not evenly distributed in the sample, and for this reason after grinding the sample it should be homogenized in representative aliquots. A homogenized sample of 50–100 g is commonly analyzed by multiresidue methods. In some cases, the sample amount can be lowered to 5–15 g (Hill *et al.*, 2000).

Another consideration to take into account is that, after grinding, the pesticide on the grape surface comes into contact with acidic must (pH 3.0–3.7). In this case some active ingredients, such as captan and folpet, are rapidly degraded (Cabras *et al.*, 1997; Angioni *et al.*, 2003), therefore the timeframe from grinding to the extraction step should be reduced. Moreover, due to this aspect, some authors reported low recoveries for these two active ingredients.

Extraction and purification

Extraction procedures involving selective non polar solvents are commonly used for monoresidue methods. In this case, active ingredients are quantitatively extracted from the matrix without interfering substances and extracts are directly injected into gas chromatography (GC) and liquid chromatography (LC) for the quantitative analysis.

Previously, pesticide extraction procedures from grapes and wine in the multiresidue analysis were carried out with polar solvents such as methanol and acetone (Holland and Malcolm, 1992). To increase the recovery of more polar analytes, the extracts were salted out with NaCl, Na_2SO_4 or $MgSO_4$, and then partitioned with non-polar solvents such as CH_2Cl_2, petroleum ether, etc (Holland and Malcolm, 1992).

Currently, the use of water insoluble solvents such as ethyl acetate that do not require partition is preferred. Unfortunately, using ethyl acetate, lower recoveries are observed for the most polar pesticides. If the raw extract is saturated with sodium sulphate, satisfying recoveries are observed for methamidophos, acephate and omethoate (Holland and Malcolm, 1992).

Lately, a method called ‘QuEChERS’ – using acetonitrile as the extraction solvent and $MgSO_4$ and NaCl for salting out – was proposed.

The grape extract or the wine is then purified by using a dispersive aminopropyl solid-phase extraction. These extraction techniques allow the removal of unwanted substances from grapes and wine such us sugars, organic acids, and other substances interfering in the pesticide quantitation (Anastassiades *et al.*, 2003).

For the analysis of pesticides in wine a valid alternative is represented by the use of a solid phase extraction (SPE) step. In this case, fewer polar pesticides are adsorbed on a C_{18} cartridge allowing the percolation of undesirable polar substances. For a higher pesticides recovery, the wine is commonly diluted 10-fold to reduce the eluting effect of ethanol on the analytes during the elution. Adsorbed active ingredients are then eluted with non polar solvents such as methylene chloride (Cabras *et al.*, 1992; Holland *et al.*, 1994). The great advantage of this technique is that it achieves separation of a wide number of pesticides.

These extraction procedures do not require a purification step when analysis is performed by a liquid or gas chromatographic system coupled with a tandem mass spectrometry technique (LC-MS/MS and GC-MS/MS).

Analysis

Quantitative analysis of pesticides is usually performed by high performance liquid chromatography (HPLC) and gas chromatographic (GC) techniques.

For both techniques, co-extracted substances can determine enhancement or reduction of the signal of the analyte known as a matrix effect. To avoid this effect, a sample should be prepared for analysis by using an extract obtained from non-treated grape or by the standard addition method. Since it is difficult to determine all the active ingredients in a single chromatographic run (150 are the active ingredients registered in Italy), they are commonly divided in chemical classes such as pyrethroids, chloroorganic, organophosporus, etc.

A wide number of analytical methods are available for the determination of pesticides in grapes and wine. Here we report some of the methods commonly used for the MRL control and for the study of pesticide behavior in field experiments and wine-making.

Figures 6.1 and 6.2 show the chromatograms relative to analyses of the most common active ingredients in standard solutions; the analytical conditions used are reported in Tables 6.4 and 6.5, respectively.

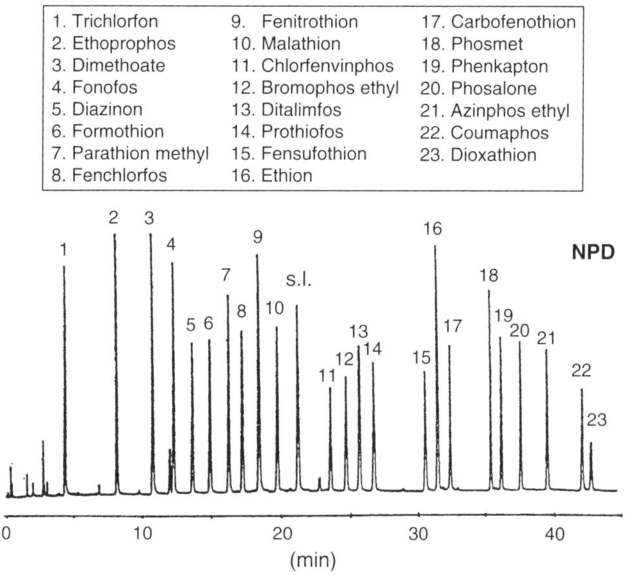

Figure 6.1 GC and nitrogen-phosphorous-detector (GC-NPD) analysis of a organophosporus insecticides standard solution

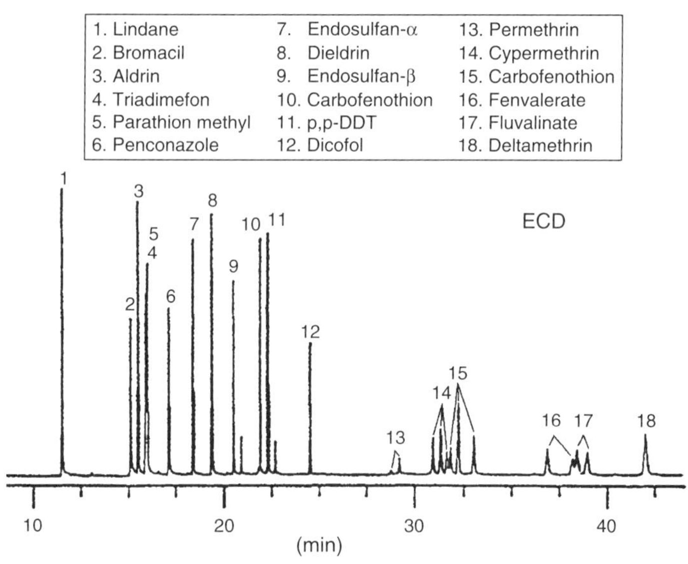

Figure 6.2 GC and electron-capture-detector (GC-ECD) analysis of a chloroorganic and pyrethroids pesticides standard solution. (From: Holland, P.T. and Malcolm, C.P. (1992) 'Multiresidue analysis in fruits and vegetables', in T. Cairns and J. Sherma (Eds) *Emerging Strategies for Pesticide Analysis*, Modern Methods for Pesticide Analysis Series, CRC Press, Boca Raton)

Table 6.4 Analytical conditions used in the GC-NPD analysis of organophosporus insecticides reported in Figure 6.1.

Column:	5% diphenyl – 95% dimethylpolysiloxane (25 m × 0.32 mm i.d.; film thickness 0.25 μm)
Injection mode:	splitless (60 sec)
Injected volume:	1 μL
Carrier gas:	He (flow rate 2 mL min^{-1})
Injector temperature:	240 °C
Oven temperature:	100 °C (isotherm for 1 min) → 150 °C (30 °C min^{-1}, isotherm for 2 min) → 205 °C (3 °C min^{-1}) → 240 °C (10 °C min^{-1}) → 260 °C (2 °C min^{-1}, isotherm for 10 min)
Detector:	NPD (make-up N_2)

Table 6.5 Analytical conditions used in the GC-ECD analysis of chloroorganic and pyrethroids pesticides reported in Figure 6.2 (Holland and Malcolm, 1992).

Column:	5% diphenyl – 95% dimethylpolysiloxane (25 m × 0.20 mm; film thickness 0.33 μm)
Injection:	splitless (60 sec)
Injected volume:	1 μL
Carrier gas:	He (flow rate 2 mL min^{-1})
Injector temperature:	220 °C
Oven temperature:	85 °C (isotherm 1 min) → 150 °C (40 °C min^{-1}, isotherm 2 min) → 250 °C (5 °C min^{-1})
Detector:	ECD

*Multiresidue method for the determination of pesticides residues in grape (QuEChERS)(*Anastassiades *et al.*, 2003*)*

Pesticides are extracted from grapes with acetonitrile using a single-phase extraction followed by liquid-liquid partitioning formed by addition of $MgSO_4$ and NaCl. Removal of residual water and clean up are performed simultaneously by using dispersive solid-phase extraction. The pesticide extract was analyzed by GC/MS.

The QuEChERS (**Qu**ick, **E**asy, **Ch**eap, **E**ffective, **R**ugged, **S**afe) method allows the extraction of a wide variety of pesticides from polar and basic (methamidophos, acephate, omethoate, imazalil and thiabendazole) to less polar, such as pyrethrins. Samples of homogenized grape (10 g) are extracted with 10 mL of acetonitrile. After shaking for 1 min, 4 g of anhydrous $MgSO_4$ and 1 g of NaCl are added and mixed. One milliliter of the acetonitrile aliquot is transferred to a 1.5 mL vial containing 150 mg anhydrous $MgSO_4$ and centrifuged to separate solids from solution.

For the sample cleanup, 1 mL of the upper acetonitrile layer is mixed with 25 mg of primary secondary amine (PSA) adsorbent and 150 mg of $MgSO_4$ and shaken for 30 seconds. The organic layer is centrifuged at 6000 rpm for 60 seconds and used for GC-MS analysis.

Recoveries of pesticides, determined operating MS analysis in selected ion monitoring (SIM) mode, are between 85% and 101% (mostly > 95%), repeatability is typically < 5%. Extracts are analyzed by GC-MS using the analytical conditions reported in Table 6.6. Pesticides with their retention time and ions used for quantification in SIM mode are reported in Table 6.7.

Table 6.6 GC-MS analytical conditions used for analysis of pesticides by the QuEChERS method.

Column:	35% diphenyl – 65% dimethylpolysiloxane (30 m × 0.25 mm i.d.; film thickness 0.25 μm)
Injection mode:	Splitless
Injected volume:	1.5 μL
Carrier gas:	He (flow rate 1.0 mL min^{-1})
Injector temperature:	250 °C
Oven temperature:	95 °C (isotherm for 15 min) → 190 °C (20 °C min^{-1}) → 230 °C (5 °C min^{-1}) → 290 °C (25 °C min^{-1}, isotherm for 20 min)
Detector:	MS transfer line 290 °C, full-scan *m/z* 50–450

Table 6.7 Pesticides with their retention time and ions used for quantification in SIM mode by performing GC-MS analysis with QuEChERS method.

Pesticide	Retention time (min)	Quantification ion (*m/z*)
Dichlorvos	5.3	185
Methamidophos	5.8	94
Mevinphos	6.6	127
Acephate	7.3	136
o-Phenylphenol	7.3	170
Omethoate	9.1	156
Diazinon	9.2	179
Lindane	9.8	181
Dimethoate	10.1	93
Vinclozolin	10.7	285
Chlorothalonil	11.5	266
Metalaxyl	11.5	206
Chlorpyrifos	12.1	197
Dichlofluanid	12.3	224

Carbaryl	12.4	144
Methiocarb	12.5	168
Fenthion	12.7	268
Cyprodinil	13.4	224
Captan	14.9	79
Folpet	15.1	260
Imazalil	15.7	201
Endosulfan sulphate	16.8	272
Dicofol	17.4	251
Phosalone	18.1	182
cis-Permethrin	18.5	183
trans-Permethrin	18.7	183
Azinphos-methyl	18.8	160
Coumaphos	19.6	362
Deltamethrin	23.1	181

Multiresidue method for the determination of pesticide residues in grapes with LC-MS/MS (Jansson *et al.*, 2004)

Pesticides are extracted from the grape with ethyl acetate and directly determined by LC-MS/MS analysis without further clean-up.

To perform sample preparation, 75 g of grape are extracted with 200 mL of ethyl acetate. After filtration an aliquot of 10 mL of the organic layer is evaporated to dryness, the residue is re-dissolved with 1.5 mL of methanol and the solution is injected for LC-MS/MS analysis without any further purification. Recoveries of pesticides range between 70–100%.

LC-MS analytical conditions are reported in Table 6.8; Table 6.9 reports the retention times and daughter ions relative to MS/MS experiments used for quantification of Italian registered pesticides.

Table 6.8 LC-MS conditions for multiresidue analysis of pesticides in grape.

Column:	C_{18} (100 × 3 mm; 4 μm)
Elution mode:	gradient between methanol and ammonium formate pH 4
Detector:	MS triple quadrupole equipped with ES ion source operating in both positive and negative mode
MS conditions:	nebulizing gas N_2 90 L h^{-1}, drying gas N_2 heated to 400 °C at 600 L h^{-1}; capillary voltage switched between +4.0 and −3.5 kV; source block temperature 120 °C; cone voltage between 10–70 eV; collision energy between 5–50 eV

Table 6.9 Retention times, molecular weight (MW) and daughter ions from MS/MS experiments used for quantification of Italian registered grape pesticides by multiresidue LC-MS/MS analysis.

Pesticide	Retention time (min)	MW	MS/MS (*m/z*)
Carbaryl	14.8	201.1	202 > 145
Carbendazim	10.9	191.1	192 > 160
Clofentezine	19.8	302.0	303 > 138
Fenoxycarb	18.8	301.1	302 > 116
Hexythiazox	21.4	352.1	353 > 228
Methiocarb	17.4	225.1	226 > 169
Methiocarb sulphone	11,2	257.1	258 > 122
Methiocarb sulphoxide	10.0	241.1	242 > 122
Methomyl	7.3	162.0	163 > 106
Thiodicarb	15.1	354.1	355 > 88

*Multiresidue pesticides analysis in wines by solid phase extraction and GC-MS (*Wong *et al.*, 2003*)*

The pesticides are extracted from wine using solid-phase extraction (SPE) on a polymeric cartridge and the interference compounds are removed with an aminopropyl-$MgSO_4$ cartridge. This multiresidue method can be applied for detection of organohalogen, organonitrogen, organophosphate, and organosulphur pesticides in wines by performing GC-MS analysis in SIM mode.

To perform sample preparation a volume of 20 mL wine is transferred to a 50-mL volumetric flask. HPLC-grade water (20 mL) is added to the wine, for a total volume of 40 mL, and vigorously mixed. For SPE, a HLB Oasis 200 mg C_{18} cartridge (Waters, Milford, MA) is rinsed with two column volumes of ethyl acetate/hexane 1:1 (v/v) solution, methanol, and HPLC-grade water. The 40-mL sample of diluted wine is loaded onto the SPE cartridge, then the sample flask is rinsed with approximately 10–15 mL of HPLC-grade water and loaded onto the cartridge; finally, the cartridge is dried under vacuum for 15 min. During the time the C_{18} cartridge is being dried, an aminopropyl (cleanup) cartridge (500 mg), attached to another vacuum manifold, is prepared by loading magnesium sulfate to fill approximately one-third of the cartridge volume. The magnesium sulfate-aminopropyl cartridge is conditioned with approximately 5 mL of ethyl acetate/hexane 1:1 (v/v) solution. The C_{18} cartridge is then attached by an adapter to the top of the cleanup cartridge. The tandem cartridges are eluted under gravity by successive passage of ethyl acetate/hexane

8:2, 5:5 and 2:8 solutions 5 mL each. The eluate is collected in a tube (approximately 15 mL) and evaporated to approximately 0.1 mL under nitrogen stream. One millilitre of 0.1% corn oil/ethyl acetate solution is added to the residue; the resulting solution is transferred to GC sample vials and finally added with an internal standard (see Table 6.12).

By this method, the pesticide recoveries in samples spiked with 0.10 mg/L of each compound were greater than 70% for 123 active ingredients (out of 153 pesticides added) in red wines, and for 128 active ingredients in white wines, respectively. Recoveries of 52 and 62 compounds higher than 90% were observed for extracted red and white wines, respectively. Thirty pesticides in red wines, and 25 pesticides in white wines, showed recoveries below 70%. These results are similar to those observed for the red and white wines spiked at low concentration (0.01 mg/L each): recoveries above 70% were found in 116 and 124 pesticide residues from extracted red and white wines, respectively.

Propargite generally has recoveries greater than 80% for high- and low-spiked concentrations in both red and white wines, with the exception of Acephate, Demeton-O and -S, Dichlorvos, Dicrotophos, Dimethoate, Mevinphos, Monocrotophos, Naled, and Omethoate. The organohalogenated pesticides such as the N-trihalomethylhalo compounds, Captafol and Folpet, the dicarboximide pesticides, Iprodione and Chlozolinate, and the organochlorine compound, Endrin aldehyde, showed recoveries of less than 70% for both wine types and at both concentrations. Recoveries in excess of 70% have been observed for most of these organonitrogen pesticides, such as the 1,3,5-triazines and amides (phenylamides, Napropamide, and Propyzamide), and most of the azole pesticides, such as Myclobutanil, Triadimefon, and its degradation product, Triadimenol. However, in addition to some of the polar organonitrogen pesticides previously mentioned, others, such as Chlorothalonil, Desmetryn, Fenpropimorph, Hexazinone, and Prochloraz, showed poor recoveries (<60%) in both red and white wines (data in Table 6.10).

GC-MS analysis is performed in SIM mode employing the analytical conditions reported in Table 6.11. Confirmation of the pesticide is established by the retention time of the target ion and the presence of three qualifier-to-target ion ratios (Table 6.12). Quantitative analysis is performed on the peak area ratio of the target ion divided by the peak area of the internal standard versus concentration of the calibration standards.

Table 6.10 Spike recoveries of pesticides registered in Italy and extracted from red and white spiked wines at 'High Spike' (0.10 mg/L) and 'Low Spike' (0.01 mg/L) level.

Pesticide	high spike ($0.10\ mg\ L^{-1}$) wine		low spike ($0.01\ mg\ L^{-1}$) wine	
	red	white	red	white
	Organohalogen			
benzilate				
Bromopropylate	89 ± 1	90 ± 1	90 ± 3	90 ± 2
dicarboximide				
Iprodione	47 ± 3	37 ± 2	73 ± 6	77 ± 5
Procymidone	88 ± 2	91 ± 2	79 ± 5	82 ± 6
Vinclozolin	83 ± 2	80 ± 3	83 ± 2	91 ± 4
N-trihalomethylhalo				
Captan	64 ± 3	80 ± 8	89 ± 4	109 ± 3
Folpet	n.d.	9 ± 4	n.d.	57 ± 2
Tolylfluanid	112 ± 5	91 ± 6	107 ± 4	92 ± 16
organochlorine				
Endosulfan-α	90 ± 1	91 ± 1	83 ± 2	86 ± 5
Endosulfan-β	95 ± 1	91 ± 1	86 ± 4	89 ± 5
	Pyrethroid			
Cypermethrin I	73 ± 2	82 ± 2	66 ± 4	80 ± 4
Cypermethrin II	74 ± 1	81 ± 2	71 ± 4	85 ± 4
Cypermethrin III	72 ± 2	82 ± 2	69 ± 4	83 ± 4
Cypermethrin IV	71 ± 1	81 ± 2	65 ± 4	78 ± 4
Deltamethrin	78 ± 4	84 ± 2	75 ± 4	89 ± 2
Fluvalinate tau-I	85 ± 4	82 ± 2	89 ± 11	86 ± 2
Fluvalinate tau-II	77 ± 3	83 ± 2	71 ± 4	84 ± 2
	Organonitrogen			
2,6-dinitroaniline				
Trifluralin	69 ± 6	70 ± 3	68 ± 4	68 ± 3
amide				
Propyzamide	93 ± 2	95 ± 1	93 ± 2	96 ± 2
anilinopyrimidine				
Cyprodinil	72 ± 3	62 ± 12	64 ± 7	62 ± 12
Pyrimethanil	87 ± 3	94 ± 4	79 ± 6	109 ± 2
Myclobutanil	86 ± 4	96 ± 4	108 ± 8	110 ± 9
Penconazole	91 ± 2	88 ± 5	91 ± 4	87 ± 3
Tebuconazole	83 ± 2	70 ± 14	80 ± 4	83 ± 7
Triadimenol	98 ± 1	89 ± 6	108 ± 10	98 ± 5
benzonitrile				
Chlorothalonil	53 ± 5	30 ± 4	81 ± 5	69 ± 3
Dichlobenil	67 ± 10	77 ± 6	62 ± 3	69 ± 4
carbamate/thiocarbamate				
Carbaryl	86 ± 5	71 ± 9	124 ± 12	100 ± 11
diphenyl ether				
Oxyfluoren	73 ± 3	74 ± 3	72 ± 6	79 ± 2
morpholine				
Fenpropimorph	6 ± 0.1	8 ± 6	21 ± 0.6	27 ± 5

Table 6.11 GC-MS conditions used in the multiresidue analysis of pesticides in wines.

Column:	5% diphenyl – 95% dimethylpolysiloxane (30 m × 0.25 mm i.d.; film thickness 0.25 μm)
Injection mode:	splitless
Injected volume:	2.0 μL
Carrier gas:	He constant pressure
Injector temperature:	250 °C
Oven temperature:	70 °C (isotherm for 2 min) → 150 °C (25 °C min^{-1}) → 200 °C (3 °C min^{-1}) → 280 °C (8 °C min^{-1}, isotherm for 10 min)
Detector:	MS/EI (70 eV); transfer line 280 °C; source 230 °C
MS conditions:	SIM mode

Table 6.12 Molecular weight (MW), retention time (t_R), target and qualifier ions used in the analysis of active ingredients registered in Italy. I.S., internal standard.

Pesticide	MW	t_R (min)	target ion	qualifier ion
acenaphthalene-d_{10} (I.S.)	164.3	8.39	164	162
Azinphos-methyl	317.3	29.68	160	132
Benalaxyl	325.4	26.75	148	91
Bromopropylate	428.1	28.71	341	183
Captan	300.6	21.35	79	80
Carbaryl	201.2	16.89	144	115
Chlorothalonil	265.9	14.90	266	264
Chlorpyrifos	350.6	19.25	197	199
Chlorpyrifos-methyl	322.5	16.61	286	288
chrysene-d_{12} (I.S.)	240.4	28.46	240	236
Cypermethrin I	416.3	32.79	181	163
Cypermethrin II	416.3	32.95	181	163
Cypermethrin III	416.3	33.07	163	181
Cypermethrin IV	416.3	33.16	163	181
Cyprodinil	225.3	20.66	224	225
Deltamethrin	505.2	36.07	181	253
Diazinon	304.3	14.47	179	137
Dichlobenil	172.0	6.77	171	173
Dichlorvos	221.0	5.83	109	185
Endosulfan-α	406.9	22.70	241	195
Endosulfan-β	406.9	25.24	195	237
Fenamiphos	303.4	23.65	303	154
Fenarimol	331.2	30.48	139	219
Fenitrothion	277.2	18.10	277	125
Fenpropimorph	305.5	19.28	128	129
Fludioxinil	248.2	24.25	248	127
Fluvalinate tau-I	502.9	34.88	250	252
Fluvalinate tau-II	502.9	35.02	250	252

Table 6.12 (Continued).

Pesticide	MW	t_R (min)	target ion	qualifier ion
Folpet	296.6	21.72	147	104
Hexaconazole	352.9	23.52	83	214
Iprodione	330.2	28.51	314	187
Metalaxyl	279.3	17.35	206	45
Methidathion	302.3	22.33	145	85
Myclobutanil	280.8	24.44	179	150
Oxadiazon	345.2	24.49	175	177
Oxyfluorfen	361.7	24.72	252	361
Penconazole	284.2	21.08	248	159
phenanthrene-d_{10} (I.S.)	188.3	13.80	188	189
Phosalone	367.8	29.71	182	367
Procymidone	284.1	22.89	96	283
Propargite	350.5	27.74	135	150
Propyzamide	256.1	14.05	173	175
Pyrimethanil	199.3	14.21	198	199
Tebuconazole	307.8	27.43	125	250
Triadimefon	293.8	19.39	57	208
Triadimenol	295.8	21.67	112	168
Trifluralin	335.3	11.35	306	264
Vinclozolin	286.1	16.73	212	198

Determination of captan and its metabolites in grape, must and wine by GC-MS and HPLC-DAD analysis (Angioni *et al.*, 2003)

Captan and its tetrahydrophthalimide (THPI) metabolites (Figure 6.3) are extracted from grapes, must, and wine with an acetone/petroleum ether 50:50 (v/v) mixture.

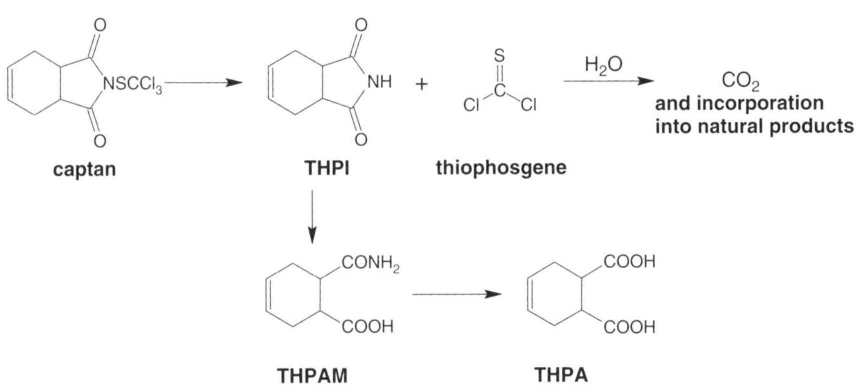

Figure 6.3 Captan and its main metabolites: tetrahydrophtalimide (THPI), tetrahydrophthalamic acid (THPAM), and tetrahydrophtalic acid (THPA)

To perform extraction, 5 g aliquot of homogenized sample (grapes, must, or wine) is weighed in a 30 mL screw-capped flask containing 4 g NaCl and 10 mL of acetone/petroleum ether 50:50 (v/v) solution, and the tube is agitated in a rotatory shaker for 15 min. The phases are allowed to separate, and an aliquot of the organic layer is used to perform GC or HPLC analysis.

The extract does not show interfering peaks and clean up is not necessary. Recoveries of analytes range between 70–100%.

GC analysis is performed by using an ion trap mass spectrometer (ITMS) equipped with a chemical liquid ionization system. Figure 6.4 shows the GC-MS chromatogram relative to THPI and captan analysis in a grape sample, analytical conditions are reported in Table 6.13. The

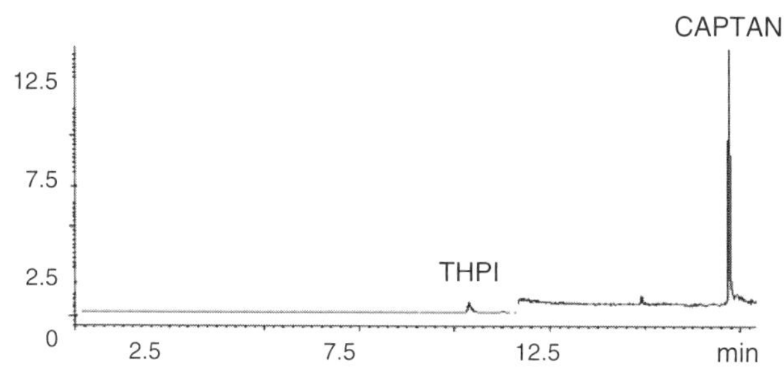

Figure 6.4 GC-MS chromatogram relative to analysis of THPI (0.13 mg/kg) and captan (0.76 mg/kg) in a grape sample. (Reproduced from *J. Agric. Food Chem.*, 2003, 51, 6761–6766, Angioni *et al.*, with permission of the American Chemical Society)

Table 6.13 Analytical conditions used in the THPI and captan analysis showed in the Figure 6.4.

Column:	5% diphenyl – 95% dimethylpolysiloxane (30 m × 0.25 mm i.d.; film thickness 0.25 μm)
Injection mode:	splitless
Injected volume:	4.0 μL
Carrier gas:	He (flow rate 1.0 mL min^{-1})
Injector temperature:	from 60 °C to 150 °C at 30 °C sec^{-1}
Oven temperature:	60 °C (isotherm for 1 min) → 240 °C (3 °C min^{-1})
Detector:	ITMS, transfer line 200 °C
IT conditions:	IT temperature 150 °C. Ionization mode: from 9 to 12 min liquid CI μSIS (methanol); from 12 to 18 EI SIS range mode (*m/z* 70–310). Quantitation ions: *m/z* 152 for THPI; *m/z* 79, 149 and 264 for captan.

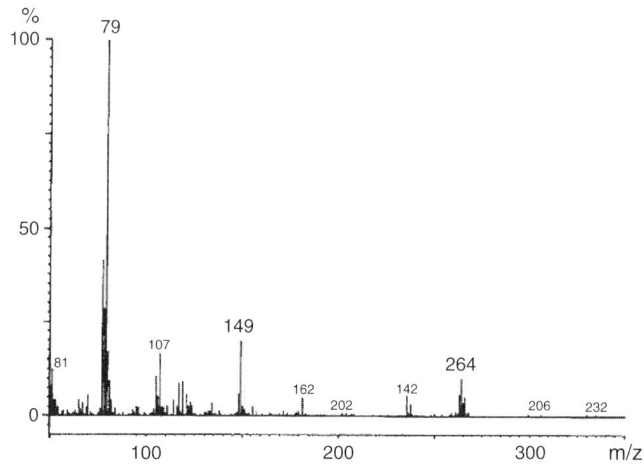

Figure 6.5 Mass spectrum of captan recorded in the GC-ITMS analysis operating in the EI mode. (Reproduced from *J. Agric. Food Chem.*, 2003, **51**, 6761–6766, Angioni *et al.*, with permission of the American Chemical Society)

EI mass spectrum of captan obtained by GC-MS/IT analysis is reported in Figure 6.5.

HPLC analysis is used for the determination of captan and its metabolites THPI, THPAM and THPA. Analysis is performed by using a C_8 (250×4.6 mm, 5 μm) column with a binary solvent mixture composed of A) acetonitrile and B) H_2SO_4 5×10^{-4} M and a gradient program of elution starting from A/B 45:55, linear gradient from 45% to 60% of A in 12 min, equilibration time to the initial conditions 6 min (flow rate 1.0 mL/min). Compounds are detected at wavelength 200 nm, injection volume 100 μL. Figure 6.6 shows the HPLC chromatogram of captan and metabolites.

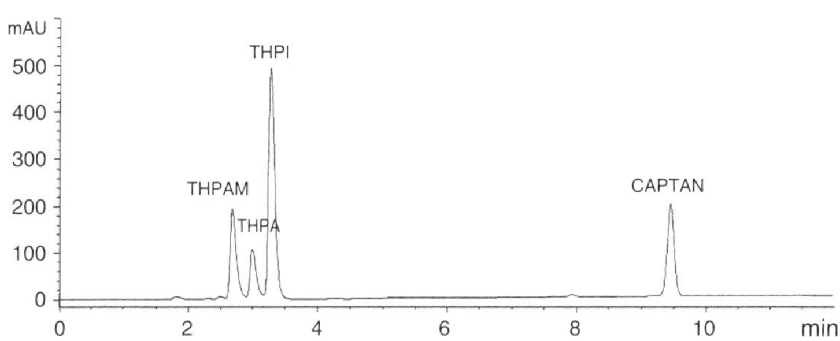

Figure 6.6 Chromatogram relative to the HPLC analysis of captan and its metabolites THPI, THPAM, and THPA. (Reproduced from *J. Agric. Food Chem.*, 2003, **51**, 6761–6766, Angioni *et al.*, with permission of the American Chemical Society)

It has been observed, that THPI residues in grapes are mainly due to captan degradation by the acid of the must, only a minor part is formed on the grape. The transformation can occur during either sampling or analysis. Model systems showed that the mechanism of disappearance of captan in grapes is mainly due to photodegradation and codistillation with water evaporation. During winemaking, captan is degraded quantitatively to THPI, and at the end of fermentation only THPI residues are present in wine. This degradation product is very stable and after one year its concentration is unchanged. As a consequence, the determination of the active ingredient in wine is unnecessary and only its metabolite should be considered.

Determination of zoxamide residues in grape and wine (Angioni *et al.*, 2005)

Zoxamide (Figure 6.7) is extracted from grape, must, wine and spirits with hexane and analyzed without clean-up. To perform extraction 5 g of homogenized sample is weighed in a 40 mL screw-capped flask containing 2 g of NaCl and 10 mL of hexane, and the tube is agitated in a rotatory shaker for 15 min. The phases are allowed to separate, and a aliquot of the organic layer is poured in a screw-capped tube containing 1 g of anhydrous sodium sulfate and used for GC analysis. Recoveries are in the range of 80–114%, with a relative standard deviation below 13% (Table 6.14). Analytical conditions are reported in Table 6.15.

Analysis is performed both in EI and CI mode using a GC-ITMS system. The experimental conditions are reported in Table 6.14.

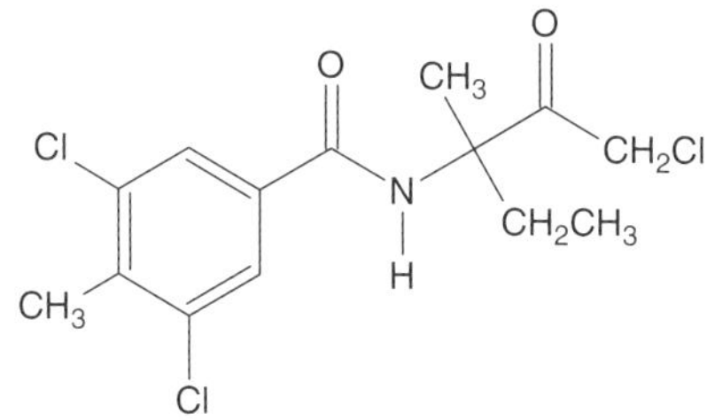

zoxamide (MW 336.64)

Figure 6.7 Chemical structure of zoxamide

Table 6.14 Recoveries (% ±SD) of zoxamide in grape must, wine, and spirits at different fortification levels (mg/kg).

Fortification level (mg kg^{-1})	Grape	Must	Wine	Spirits
0.05	112.1 ± 5.9	80.4 ± 3.2	89.0 ± 6.0	99.1 ± 2.5
0.10	92.6 ± 4.3	96.5 ± 4.7	91.0 ± 13.0	97.4 ± 6.3
0.50	87.0 ± 8.0	114.0 ± 3.0	86.0 ± 9.6	95.8 ± 4.6
2.00	106.2 ± 3.2	90.0 ± 3.0	98.6 ± 1.6	100.0 ± 13.0

Table 6.15 GC-ITMS analytical conditions used for analysis of zoxamide.

Column:	5% diphenyl – 95% dimethylpolysiloxane (30 m × 0.25 mm i.d.; film thickness 0.25 μm) 60 °C (hold 1 min), to 150 °C at 30 °C min^{-1} (hold 20 min),
Injector temperature:	splitless mode
Injected volume:	4.0 μL
Carrier gas:	He (flow rate 1.0 mL min^{-1} constant flow mode)
Injector temperature:	250 °C
Oven temperature:	65 °C (isotherm for 1 min) → 280 °C (10 °C min^{-1})
Detector:	ITMS operating in EI and CI (acetonitrile) mode trap, manifold and transfer line temperatures at 170, 100 and 200 °C, respectively multiplier voltage 1400 V emission current 30 μA (CI) and 80 μA (EI)
Detector conditions:	prescan ionization time 100 μs (CI) and 1500 μs (EI) EI selective ion storage range *m/z* 170–260 SIS *m/z* 300 for CI experiments MS/MS experiments in waveform resonant mode, parent ion *m/z* 299

6.3. Conclusive Remarks

LC/MS and GC/MS techniques allow a better understanding of the degradative behaviour of individual active ingredients and/or their metabolites during field experiments and wine-making. In the control of residues in grapes and wine there is no one method for the analysis of all pesticides. Usually, in every country the use of 40–50 active ingredients is sufficient for the 95% of control requirements of vines parasites. For this reason, multiresidue methods should be developed which take into account the most frequently used pesticides in the field, even with a survey at the manufacturing firm.

The use of LC/MS and GC/MS for multiresidue methods reduces the need for a purification step of grapes and wine extracts and circumvents any possible false positive. A description of MS applications in grape and wine pesticides analysis was recently reported by Flamini and Panighel (2006).

References

Ambrus, A. (1996) Estimation of uncertainty of sampling for the analysis of pesticide residues, *J. Environ. Sci. Health. B*, **31**(**3**), 435–442.

Ambrus, A., Solymosné, E.M. and Korsòs, I. (1996) Estimation of uncertainty of sample preparation for the analysis of pesticide residues, *J. Environ. Sci. Health. B*, **31**(**3**), 443–450.

Anastassiades, M., Lehotay, S.J., Štajnbaher, D. and Schenck, F.J. (2003) Fast and easy multiresidue method employing acetonitrile extraction/partitioning and "dispersive solid-phase extraction" for the determination of pesticide residues in produce, *J. AOAC Int.*, **86**(**2**), 412–431.

Angioni, A., Garau, V.L., Aguilera Del Real, A., Melis, M., Minelli, E.V., Tuberoso, C. and Cabras, P. (2003) GC-ITMS determination and degradation of captan during winemaking, *J. Agric. Food Chem.*, **51**(**23**), 6761–6766.

Angioni, A., Garau, A., Caboni, P., Russo, M., Farris, G.A., Zara, S. and Cabras, P. (2005) Gas chromatographic ion trap mass spectrometry determination of zoxamide residues in grape, grape processing, and in the fermentation process, *J. Chromatogr. A*, **1097**(**1–2**), 165–170.

Cabras, P. (2003) Pesticides: toxicology and residues in food and wine, in J.P.F. D'Mello (Ed), *Food Safety: Contaminants and Toxins,* CABI Pub.Oxon, UK.

Cabras, P. and Angioni, A. (2000) Pesticide residues in grapes, wine, and their processing products, *J. Agric. Food Chem.*, **48** (**4**), 967–973.

Cabras, P., Angioni, A., Garau, V.L., Melis, M., Pirisi, F.M., Farris, G., Sotgiu, C. and Minelli, E.V. (1997) Persistence and metabolism of folpet in grapes and wine, *J. Agric. Food Chem.*, **45**(**2**), 476–479.

Cabras, P., Angioni, A., Garau, V.L., Melis, M., Pirisi, F.M., Minelli, E.V., Cabitza, F. and Cubeddu, M. (1997) Fate of some new fungicides (cyprodinil, fludioxonil, pyrimethanil and tebuconazole) from vine to wine, *J. Agric Food Chem.*, **45**(7), 2708–2710.

Cabras, P., Tuberoso, C., Melis, M. and Martini, M.G. (1992) Multiresidue method for pesticide determination in wine by high-performance liquid chromatography, *J. Agric. Food Chem.*, **40**(5), 817–819.

Flamini, R. and Panighel, A. (2006) Mass spectrometry in grape and wine chemistry. Part II: The Consumer Protection, *Mass Spec. Rev.*, **25**(5), 741–774.

Hill, A.R.C., Harris, C.A. and Warburton, A.G. (2000) Effect of sample processing on pesticide residues in fruit and vegetables, in A. Fajgeli and A. Ambrus (Eds), *Principles and Practices of Method Validation,* Royal Society of Chemistry, Cambridge.

Holland, P.T. and Malcolm, C.P. (1992) Multiresidue analysis in fruits and vegetables, in T. Cairns and J. Sherma (Eds), *Emerging strategies for pesticide analysis,* Modern Methods for Pesticide Analysis Series, CRC Press, Boca Raton.

Holland, P.T., McNaughton, D.E. and Malcolm, C.P. (1994) Multiresidue analysis of pesticides in wines by solid-phase extraction, *J. AOAC Int.*, **77(1)**, 79–86.

Jansson, C., Pihlstrom, T., Osterdahl, B.G. and Markides, K.E. (2004) A new multiresidue method for analysis of pesticides residues in fruit and vegetables using liquid chromatography with tandem mass spectrometric detection, *J. Chromatogr. A.*, **1023(1)**, 93–104.

Wong, J.W., Webster, M.G., Halverson, C.A., Hengel, M.J., Ngim, K.K., and Ebeler, S.E. (2003) Multiresidue pesticide analysis in wines by solid-phase extraction and capillary gas chromatography-mass spectrometric detection with selective ion monitoring, *J. Agric. Food Chem.*, **51(5)**, 1148–1161.

7

Proteins and Peptides in Grape and Wine

Andrea Curioni, Simone Vincenzi and
Riccardo Flamini

7.1. Introduction

In recent years, protein studies have become a major topic in grape and wine science. As a matter of fact, important aspects of both grape biology and wine technology can be elucidated through the study of the nature and the properties of the grape and wine proteins. Most of the improvement in this type of research is certainly due to the tremendous progress of the analytical techniques for protein analysis which has occurred in the last few years (for a review see, for example, Glinski and Weckwerth, 2006). This chapter is devoted to the description of the most powerful analytical techniques that are applied to the study of proteins and peptides in grape and wine.

7.1.1. The Proteins of the Grape Berry

Although present in low amounts compared with other fruits, the grape berry contains a large variety of proteins. The most abundant proteins are synthesized after veraison and accumulate during

Hyphenated Techniques in Grape and Wine Chemistry Edited by Riccardo Flamini

ripening in conjunction with sugars (Tattersall *et al.*, 1997). Only recently, by exploiting a proteomic approach, several proteins have been identified in both the berry pulp (*cv.* Gamay, Sarry *et al.*, 2004) and skin (*cv.* Cabernet Sauvignon, Deytieux *et al.*, 2007). Of the approximately 300 spots detected on gels after Two Dimensional Electrophoresis (2DE) separation of the berry mesocarp proteins (Figure 7.1), 82 spots were analyzed by matrix-assisted laser desorption ionization-time of flight-mass spectrometry (MALDI-TOF-MS) and 66 different protein components were identified. Most of the identified proteins are involved in energy metabolism (34%), biotic or abiotic stress (19%) and primary metabolism (13%) (Sarry *et al.*, 2004) (Figure 7.2).

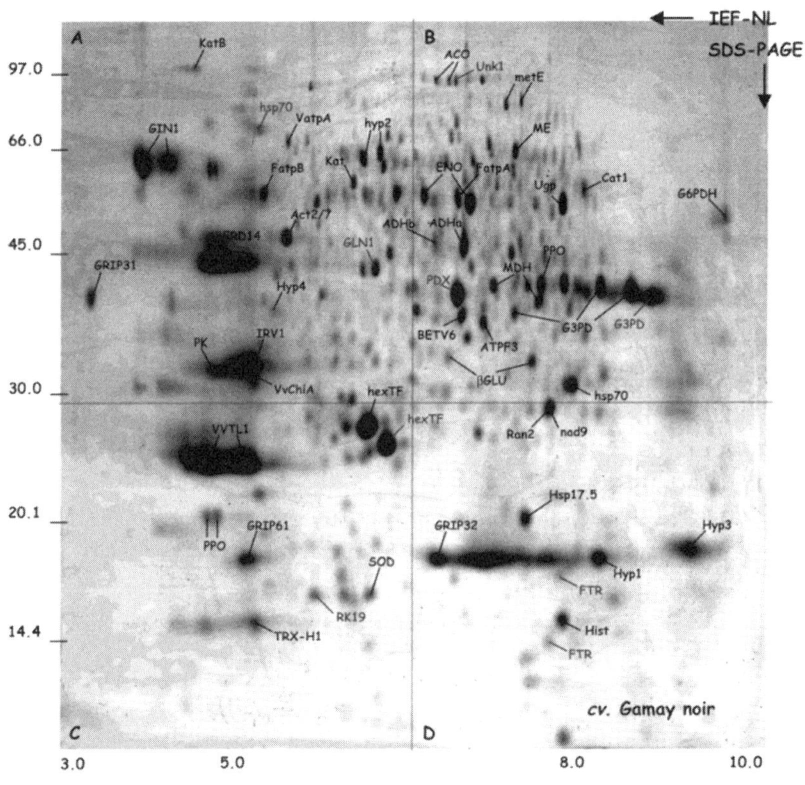

Figure 7.1 The complexity of the proteins of the grape berry mesocarp of *Vitis vinifera L. cv.* Gamay noir. 2-DE map of proteins identified by MALDI-TOF Peptide-Mass-Fingerprint (PMF) (Reproduced from *Proteomics*, 2004, 4, 201–215, Sarry *et al.*, with permission of Wiley-VCH)

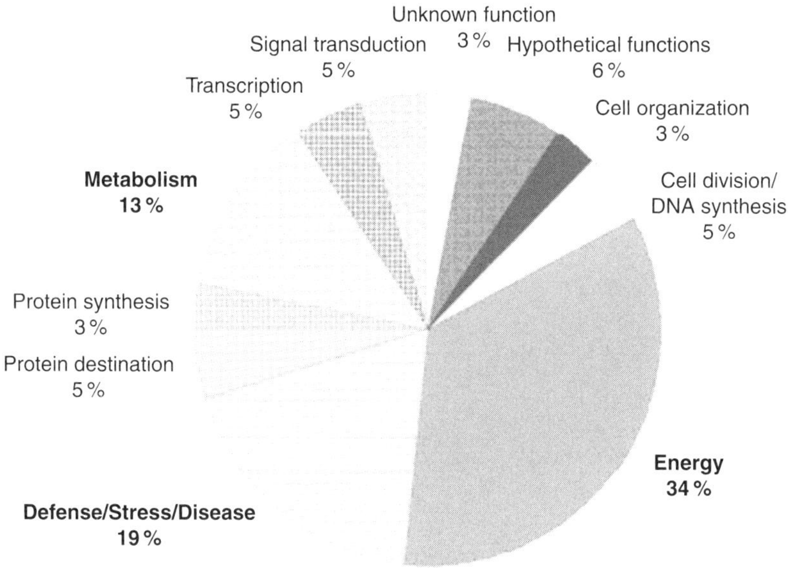

Figure 7.2 Functional distribution of the major proteins in the grapevine ripe berries, as separated by 2-D electrophoresis and identified by MALDI-TOF Peptide Mass Fingerprint (Reproduced from *Proteomics*, 2004, 4, 201–215, Sarry *et al.*, with permission of Wiley-VCH)

Also, the proteins of the grape skin have been recently analyzed by the proteomic approach at three different stages of development, showing that, at harvest, the dominant proteins are those involved in defense mechanisms, most of them being over-expressed at maturity (Deytieux *et al.*, 2007). The extraction procedure used for preparing the sample to be analyzed by proteomic approach has a strong influence on both the number and the type of the detectable proteins (Vincent *et al.*, 2006). Therefore, it is likely that the proteins identifiable in grapes using this approach are only a part of those present in the plant tissue.

In addition to the metabolic enzymes, other major proteins are present in the mature grape berry, most of them being Pathogenesis-Related (PR) proteins, including chitinases (PR3 family) and thaumatin-like proteins (TLP) (PR5 family), which are the dominant proteins of the ripe grape berry (Waters *et al.*, 1996), being present in different amounts in different *V. vinifera* cultivars (Table 7.1). Moreover, chitinases and TLP are abundant proteins also in *V. labruscana* (*cv.* Concord) berries (Salzman *et al.*, 1998). Several forms of chitinases are present in different tissues of *V. vinifera*, but the highest enzymatic activity is found in the berry (Derckel *et al.*, 1996). These proteins are involved in the defense

Table 7.1 Protein composition of juice from different grape cultivars. TLP, thaumatin like proteins. (Reproduced from *J. Agric. Food Chem.*, 2000, **48**(5), 1637–1643, Pocock *et al.*, with permission from the American Chemical Society).

Cultivar	°Brix	Protein concentration ($mg \cdot L^{-1}$)		
		Total protein	TLP	Chitinases
Muscat of Alexandria	20.06	251	119	118
Sauvignon Blanc	21.05	191	119	76
Sultana	21.00	86	23	44
Pinot noir	20.02	62	35	21
Shiraz	20.08	31	18	9

mechanisms of the plant against fungal pathogen attacks, being able to hydrolyze chitin, a structural component of the cell wall of the invading fungus.

Four chitinases have been purified from Muscat of Alexandria fruit and characterized by both sequence and mass spectral analysis, showing extensive sequence similarity (Waters *et al.*, 1998), although other cultivars contain a lower number of chitinase isoforms (Pocock *et al.*, 2000). These proteins show a Sodium Dodecyl Sulphate-Polyacrylamide Gel Electrophoresis (SDS-PAGE) mobility ranging from 32 to 34 kDa, although their real molecular weight (MW) is much lower (between 25 and 26 kDa), as determined by mass spectrometry (Pocock *et al.*, 2000). Despite the presence of several chitinase isoforms, which can derive from the expression of different genes (Busam *et al.*, 1997; Robinson *et al.*, 1997) or from limited protein degradation (Waters *et al.*, 1998), the most important chitinase isoform in grapes seems to be a class IV chitinase (containing a chitin-binding domain), which is highly expressed during berry ripening (Robinson *et al.*, 1997). Indeed, this is the sequence determined by the proteomic approach in mature Gamay berries (MW 27.2 kDa, isoelectric point p.I. 5.4, Sarry *et al.*, 2004) and in the whole grape cluster of *Cabernet sauvignon* (MW 27.53 kDa, p.I. 5.38, Vincent *et al.*, 2006).

The other major class of proteins of the grape berry are the Thaumatin-like proteins (TLP) (Tattersall *et al.*, 1997). In addition to the main TLP isoform called VVTL1, another minor TLP form can be detected (Pocock *et al.*, 2000). These proteins show, respectively, an apparent MW of 24 and 25 kDa by SDS-PAGE, but their MW, determined by electrospray ionization-mass spectrometry, (ESI-MS) is 21.272 and 21.260 kDa (Pocock *et al.*, 2000). Moreover, the MW of the TLP identified by the proteomic analysis of the whole *Cabernet sauvignon* grape cluster (p.I. 4.67) is 23.881 kDa (Vincent *et al.*, 2006) and that of the TLP identified in the *cv.* Gamay mesocarp (p.I. 5.1) is 24.0 kDa (Sarry *et al.*, 2004).

The PR-proteins of the grape berry are prone to be induced by pathogen attack (Busam *et al.*, 1997; Jacobs *et al.*, 1999), as well as by other environmental stimuli of chemical and physical nature, such as wounding (Derckel *et al.*, 1998), and the expression of similar proteins in other plants, such as tobacco (Singh *et al.*, 1987) and tomato (Yu *et al.*, 1998), are induced by salt and drought stress. Therefore, environmental and/or pathological factors prevailing during grape development and maturation can determine the level of both TLP and chitinases expression and accumulation in the berry, leading to different protein quantity and composition (Monteiro *et al.*, 2003). However, as regards the grape proteins, neither drought stress (Pocock *et al.*, 2000) nor the physical damage deriving from mechanical harvesting (Pocock *et al.*, 1998) seem to be involved in determining different PR-proteins expression levels in the berry.

In general, the grape seeds proteins profile shows signal clusters in the *m/z* 4000–5000, 5500–6500 and 12500–15000 ranges (Pesavento *et al.*, 2008).

7.1.2. The Wine Proteins and Peptides

Of the numerous proteins present in the grape berry, only a small part can survive the wine-making process. The low pH of the fermenting juice, the alcohol concentration and the presence of proteolytic activities cause extensive protein denaturation and degradation; as a consequence, only the proteins resistant to these factors are finally present in wine. For these reasons, wine contains only a limited number of protein types of grape and small quantity of proteins deriving from microorganisms, mainly the fermenting yeasts (Kwon, 2004). In any case, the majority of the proteins present in wines are from grapes (Ferreira *et al.*, 2000). Of these proteins, those having the characteristics needed to survive the fermentation process are the PR-proteins (Linthorst, 1991), in particular chitinases, and TLP, whose nature and function determine intrinsic resistance to proteolysis and denaturation (Waters *et al.*, 1992). Most of the knowledge on wine proteins derives from studies performed in white wines, where they have considerable technological importance (see below), whereas it is generally accepted that in red wines proteins are removed by the interactions with polyphenolic compounds. However, some authors found appreciable quantities of proteins (40–80 mg/L) also in red wines (Yokotsuka *et al.*, 1994).

The protein in wines can vary significantly, being affected by the protein content of the grape juice and also by the winemaking technologies used. Protein contents ranging from less than 1.0 mg/L to greater than 1.0 g/L have been reported in the literature, although the analytical methods used certainly influence the quantitative data obtained (Vincenzi *et al.*, 2005).

Proteins with MW ranging from 9 to 62 kDa and with different p.I.s (from pH 3 to pH 9) have been described in wines (Brissonet and Maujean, 1993; Hsu and Heaterbell, 1987). In addition, highly-mannosylated proteins of yeast origin (mannoproteins) with very high MW are also present in wines (Gonçalves *et al.*, 2002). However, the main protein components show MW around 20–30 kDa and p.I. in the range 4–6, being constituted by the chitinases and TLP deriving from the grape berry (Waters *et al.*, 1996). Moreover, grape invertase (MW 62-64 kDa) is also believed to be one of the most abundant proteins in wine (from 9 to 14% of the total protein content of a *Chardonnay* wine) (Dambrouck *et al.*, 2005). Despite the presence of the low number of bands detectable by SDS-PAGE, separation methods based on charge differences allow the detection, for each MW, of several polypeptides, indicating the existence of a micro-heterogeneity probably deriving from limited proteolysis or chemical modification of structurally-related proteins occurring during winemaking (Monteiro *et al.*, 2001). However, this idea has been recently revised by showing that the diversity of the PR proteins is reduced during berry development, which indicates that these proteins are synthesized in a wide variety of forms in the berry (Monteiro *et al.*, 2007).

Although present in low amounts in wine, proteins play a relevant technological role, especially in white wines, where they are responsible for the so-called 'protein casse'. This major defect is due to the presence of a visible haze, which forms after the wine is bottled and makes the wine unsuitable for sale, resulting in great economic losses. This type of haze is due to protein denaturing and aggregating during wine storage, giving rise to particles which remain suspended and/or precipitate. Although the presence of proteins is certainly necessary for formation of this type of haze, the mechanism involved in this phenomenon is not yet fully understood. Unfavourable storage conditions and the presence of compound(s) of non-protein nature (the so-called factor X), seem to be necessary for protein haze formation in white wines. Starting almost 50 years ago, a great amount of research has been done to identify and characterize the proteins responsible for wine turbidity and excellent reviews can be find on this topic (Ferreira *et al.*, 2002, Tattersall *et al.*, 2001). The latest findings indicate that, paradoxically, the proteins causing instability in wine are those that are stable in the conditions of winemaking, i. e. low pH, protease activities and the presence of ethanol. These proteins are typically the PR-proteins chitinases and TLP (see above) (Waters *et al.*, 1992 and 1996), whose intrinsic resistance is obviously related to their function in the defense of the plant from pathogen attacks.

In addition to their importance for white wine stability, proteins seems to be involved in other aspects of wine quality. For example, it is recognized that proteins can interact with aromatic compounds (Lubbers *et al.*, 1994), influence the perception of wine 'body' in the mouth, and, due to their surface properties, affect foam formation and stability in sparkling wines (Brissonet and Maujean, 1993).

The presence of peptides can also have an effect on wine quality. In general, these compounds are known to have several biological (antioxidant, antimicrobial, antihypertensive, etc.) and chemical-physical (surfactant) properties and also to affect the taste of some foods. However, peptides have received only a little attention, mainly because of the difficulties in their isolation from wines (Moreno-Arribas *et al.*, 2002).

Finally, it seems that part of the nitrogenous compounds, including proteins and peptides, can be transferred from the berry seeds and/or skin into the wine during maceration (Fukui *et al.*, 2002), but their role in wine is not known.

7.2. Extraction of Grape and Wine Proteins and Peptides

7.2.1. Sample Preparation

Grape proteins

The method of sample preparation for grape proteins analysis should be chosen according to the aim of the study. If grape proteins have to be studied in relation to wine characteristics and technology, it is convenient to start from the so-called 'free run juice', a material representative of the must, in which the proteins are supposed to be involved in the winemaking process and wine quality. The free run juice is obtained by recovering the liquid derived from squeezing the grape berries. In this case, only the proteins remaining soluble in the conditions arising from the preparation of the must will be present in the extract.

In contrast, if all proteins have to be extracted, e.g. for the study of the grape biology, it is necessary to adopt a multi-step method in order to minimize protein losses occurring with rupture of the cells and proteins interaction with phenolic compounds.

Proteins of the pulp

As a general rule, in order to detect the maximum quantity of the protein components belonging to the pulp it is necessary to introduce protective agents into the extraction buffer, such as reducing substances, ascorbic acid, PVPP, etc., which minimize protein modifications and losses (Tattersall *et al.*, 1997).

Recently, a method allowing detection of a high quantity of proteins with the proteomic approach has been developed (Sarry *et al.*, 2004). In this method, a solution containing 12.5% of trichloroacetic acid (TCA) in cold acetone and a reducing agent is used (Table 7.2). Using this approach only proteins in the pulp are extracted, and enzymes potentially involved in protein modification are inhibited.

Table 7.2 Protocol for protein extraction from the grape pulp (Sarry *et al.*, 2004).

1. Washing of fresh berries in tap water
2. Manual crushing in 1:10 (w/v) cold 12.5% TCA-acetone solution containing 28 mM 2-mercaptoethanol (skins and seeds have to be kept intact)
3. Filtration on 40 μm Miracloth mesh
4. Incubation at −20 °C for 60 min
5. Centrifugation at 10,000 g for 15 min
6. Washing twice with 85% ethanol

When only proteins remaining soluble after crushing the berries have to be studied, the so-called 'free-run juice' can be prepared. To prepare the free run juice, the suspension obtained from squeezing the berries is filtered through cloth and the suspended particles are separated by centrifugation. The soluble proteins can be concentrated by ultrafiltration on membranes with a cut-off lower than 10 kDa (Waters *et al.*, 1998). Proteins can be collected from the (concentrated) solution by the classical methods for protein precipitation, such as ammonium sulphate at 80% saturation, or organic solvents added to a final concentration of 80%. It is possible to remove phenolic compounds by solid phase extraction (SPE) passing the solution though a C_{18} cartridge. This step, however, leads to a loss of some proteins but does not affect the protein composition (Waters *et al.*, 1992). The recovered proteins can then be freeze-dried and stored for long periods of time, or re-solubilized with the solvent used for the analysis.

Proteins of the skin

The berry skin contains up to 13–15% of proteins on a dry weight basis (Igartuburu *et al.*, 1991) but, due to the high content of phenolic compounds, performing protein extraction from this tissue is difficult. Different methods of reducing protein-polyphenol interactions during extraction of plant tissues were reported in the literature (e.g. extraction in the presence of PVPP), but none of them seems to be appropriate for proteins of grape skin. However, the procedure recently reported by Deytieux *et al.* (2007), based on the phenol-extraction protocol of Saravan and Rose (2004), can provide good results in the extraction of the skin from red berries in terms of number of spots detected after 2DE. This protocol is summarized in Table 7.3.

Table 7.3 Protocol for protein extraction from grape skins (Deytieux *et al.*, 2007).

1. Separation of skins from 20 frozen berries
2. Grounding the skins to a fine powder with liquid nitrogen
3. Extraction with 3 volumes (v/w) of extraction buffer (0.1 M Tris-HCl pH 7.5, 5 mM EDTA, 1 mM phenylmethanesulphonylfluoride (PMSF), 2% 2-mercaptoethanol, 0.1 M KCl, 0.7 M sucrose, 1% PVPP) by stirring at 4 °C for 1 h
4. Addition of an equal volume of phenol-Tris-HCl pH 7.5 (commercially available ready to use solution) and agitation at 4 °C for 1 h
5. Centrifugation 9000 g at 4 °C for 30 min and collection of the lower (phenol) phase
6. Re-extraction of the aqueous phase for 30 min with 2 mL of extraction buffer plus 2 mL of phenol solution
7. Washing the phenol phase three times with an equal volume of extraction buffer
8. Addition to the phenol phase of 5 volumes (v/v) of 0.1 M ammonium acetate in methanol and incubation at −20 °C overnight
9. Centrifugation 9000 g at 4 °C for 30 min
10. Washing (and centrifugation) of the pellet once with 0.1 M ammonium acetate in methanol, twice with cold methanol and twice with ice-cold 80% acetone
11. Drying under nitrogen stream

Proteins of the seeds

In addition to the large quantity of lipids and polyphenols, grape seeds also contain significant amounts of proteins and peptides (Yokotsuka and Fukui, 2002) but few studies are reported in the literature. In this case, it is particularly necessary to use methods which allow mitigation of the effects of the high polyphenol content, such as performing extraction by a buffer with high pH (>10.0) and containing PEG, followed by precipitation with cold acetone (Famiani *et al.*, 2000) (Table 7.4).

Table 7.4 Protocol for protein extraction from grape seeds (Famiani *et al.*, 2000).

1. Grinding of seeds with liquid nitrogen
2. Extraction (0.05 g of powder) with 400 μL of ice-cold 0.5 M 2-amino-2-methyl-1-propanol (AMPS) (pH 10.8), 1% SDS, 1% PEG-6000 and 50 mM dithiothreitol (DTT)
3. Centrifugation at 12000 g for 5 min and collection of the supernatant
4. Proteins precipitation with 3 volumes of 80% acetone. Placing in liquid nitrogen for 10 minutes
5. Centrifugation at 12000 g for 5 minutes and collection of the protein pellet

Recently, a method for extraction of the grape seeds proteins finalized to perform MALDI-MS analysis, was proposed (Pesavento *et al.*, 2008). Grape seeds (e.g. 10 g), previously washed in water and dried, are powdered with liquid nitrogen. About 5 g of seed powder is defatted by suspension in 100 mL of n-hexane under stirring for 30 min. The organic solvent is removed and the residue left to dry at room temperature. The residue (500 mg) is suspended in 5 mL of water and dialyzed against double distilled water at 4 °C for 48 h. After centrifugation, the residue is extracted by suspending 10 mg of sample in 1 mL of water containing 0.1% trifluoroacetic acid (TFA), for 5 min under ultrasonic wave. The supernatant is collected and analyzed.

Wine proteins and peptides

Since the proteins in wine are very low (normally lower than 100 mg/L), concentration of the sample is performed prior to analysis. Moreover, it is also useful to reduce the content of interfering substances such as salts, acids and polyphenols. This can be done, either before or after the protein concentration, by dialysis with low cut-off membranes.

Several methods of wine sample concentration by precipitation of proteins, including use of ammonium sulphate, organic solvents (such as ethanol or acetone), trichloroacetic, sulfosalicylic or phosphotungstic acid, have been described (Moreno-Arribas *et al.*, 2002). In these cases, however, the presence of interfering compounds associated with the proteins cannot be excluded. Moreover, the drastic conditions of precipitation with acids can result in a strong protein denaturation, which impairs further analysis. In our experience, the most practical method allowing management of relatively large volumes of wine is to concentrate the wine to a minimal volume by ultrafiltration and to perform dialysis of the concentrate (in a ultrafiltration cell under nitrogen pressure or with centrifugal ultrafilter devices) on low cut-off membranes (i.e. 3.5 kDa). If a quantitative recovery of proteins is not necessary, it is possible to greatly reduce the polyphenols content by passing the sample through a C_{18} cartridge. The sample can then be freeze-dried and stored indefinitely.

An effective method for a quantitative protein recovery, which also allows the removal of phenolics, was described by Vincenzi *et al.* (2005) and is reported in Table 7.5.

Table 7.5 Protocol for protein extraction from wines (Vincenzi *et al.*, 2005).

1. Addition of 10 μL of 10% SDS to 1 mL of wine
2. Incubation at 100 °C for 5 min
3. Addition of 250 μL of 1 M KCl. Incubation at room temperature for 2 h
4. Centrifugation 14000 g at 4 °C for 15 min
5. Washing the pellet with 1 M KCl followed by centrifugation 14000 g at 4 °C for 15 min (three times)
6. Re-suspension of the protein pellet in a minimum volume of water or buffer

Protein recovered with this procedure can be quantified by the bicinchoninic acid assay (Smith *et al.*, 1985) or used for SDS-PAGE analysis (Vincenzi *et al.*, 2005).

Wine peptides are normally analyzed after a preparation step allowing their separation from the high MW components, mainly proteins and polysaccharides. This is achieved by ultrafiltration on membranes with appropriate cut-off and/or by gel filtration (e.g. on Sephadex LH-20 or G-10 gels) (Desportes *et al.*, 2000; Moreno-Arribas *et al.*, 1996 and 1998) of concentrated samples. This allows the isolation of one or more peptide fractions by removing the interfering compounds with different MW, such as salts, amino acids, phenols, organic acids and sugars.

7.3. Analysis of Grape and Wine Proteins and Peptides

7.3.1. Electrokinetic Methods

Electrokinetic methods certainly comprise some of the most powerful techniques for protein analysis. They are based on the property of charged molecules to migrate in an electric field according to their net charge (isoelectric focusing, IEF), net charge and size (native electrophoresis) or only size (SDS-PAGE). These techniques include both analytical and preparative systems.

Electrophoresis

At a given pH, proteins and peptides assume a net electrical charge which essentially depends on their amino acid composition. According to this charge they move at different rates when subjected to an electric field. If a molecular sieving effect is present in the migration medium, as in the case of gel electrophoresis, the migration rate of each polypeptide is determined by a combination of both the net charge and molecular size effects. However, migration based only on the molecular size can be obtained by denaturing the polypeptides in the presence of Sodium Dodecyl Sulphate (SDS), an anionic detergent which binds proteins at a constant rate and which confers to all proteins the same charge to mass ratio. In this way, by preventing the charge effects, it is possible to measure the (apparent) MW of a given protein by comparing its electrophoretic mobility with that of standard proteins of known MW.

These methods can be performed in slab gels (Polyacrylamide Gel Electrophoresis, PAGE) or in capillaries (Capillary Electrophoresis, CE).

Gel electrophoresis

In the case of gel electrophoresis, proteins migrate within a semi-solid support which is usually polyacrylamide. Several systems can be used, differing for both the electrophoretic conditions (pH, buffering system, presence of detergents and/or other additives) and the gel pore size regulated by varying the total concentration of polyacrylamide (often called T) and the percentage of the reticulating agent bis-acrylamide (often called C). The gel pore size is chosen in according to the MW range of the proteins which have to be analyzed; for analysis of grape and wine proteins 10–14% T is usually used.

Grape and wine proteins have been extensively studied by gel electrophoresis; in contrast, peptides have never been analyzed with this system, although by using the electrophoretic system of Schagger and von Jagow (Tris-Tricine SDS-PAGE) it would be possible to detect peptides with MW lower than 1 kDa (Schagger and von Jagow, 1987).

Sodium Dodecyl Sulphate-Polyacrylamide Gel Electrophoresis (SDS-PAGE)

SDS-PAGE is the classical electrophoretic procedure used for both grape and wine proteins separation based on the molecular size (Laemmli, 1970). Proteins in the sample must be denatured (commonly by heating at 100 °C) in the presence of a buffer (i.e. 0.25 M Tris-HCl, pH 6.8), containing the anionic detergent SDS (4%), a reducing agent such as 2-mercaptoethanol (about 2%) or dithiothreitol (DTT) (200 mM), glycerol (20%) and tracking dye Bromophenol blue (0.02%) (Laemmli 2X sample buffer). For wine and grape proteins it was observed that the omission of the reducing agent from the buffer results in a strongly modified SDS-PAGE pattern (Vincenzi *et al.*, 2005) (Figure 7.3). Suitable systems for the preparation of gels and buffers, as well as the conditions for the electrophoretic run and the gel staining, are available in the

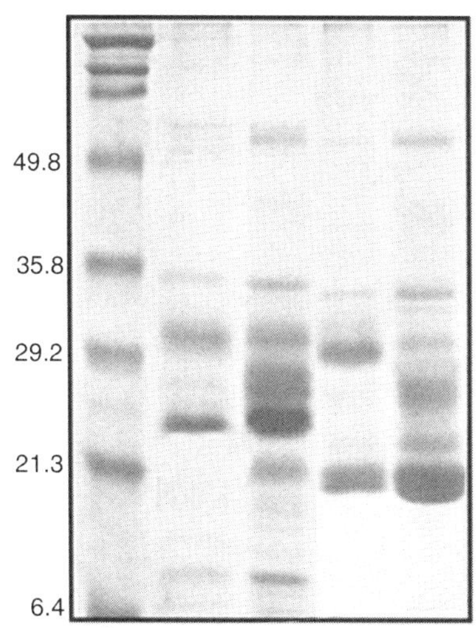

Figure 7.3 SDS-PAGE analysis of proteins from grape berries (1 and 3) and wine (2 and 4) under reducing (1 and 2) and non-reducing (3 and 4) conditions. In the lane St are loaded standard proteins of known Molecular Weight (MW expressed in kDa). Gel was stained with Coomassie Brilliant Blue (CBB) (Vincenzi and Curioni, unpublished)

literature (e.g. Marchal *et al.*, 1998 for grape proteins; Dorrestein *et al.*, 1995 for wine proteins). Usually, the electrode buffer is composed of 25 mM Tris-HCl, 192 mM glycine and 0.1% SDS at pH 8.8, the protein migration is performed under a constant electric field of 25 mA/gel (for mini-gels 1.5 mm thickness) until the tracking dye bromophenol blue reaches the bottom of the gel.

The gel can be stained by Coomassie Brilliant Blue (CBB) (1 g CBB, 400 mL ethanol, 50 mL acetic acid, brought to 1 L with water), which has a detection limit of about 1 μg/band. The gel is stained overnight and de-stained with 7% acetic acid in water. Furthermore, colloidal CBB staining is compatible with mass spectrometry analysis of protein spots separated by PAGE. This staining reveals a limited number of bands in both grape and wine extracts. A more sensitive staining with silver nitrate (detection limit of about 1 ng of protein/band) (Table 7.6) allows detection of a higher numbers of bands constituted of proteins present in relatively low concentrations (Dorrestein *et al.*, 1995).

Table 7.6 Procedure for silver staining after PAGE (Dorrestein *et al.*, 1995).

1. Soaking the gel in fixing solution (50% methanol, 12% acetic acid and 0.05% formaldehyde) for 12 h
2. Washing with 50% ethanol
3. Soaking with 200 mg · L^{-1} sodium thiosulfate solution for 1 min
4. Washing with water
5. Staining with 2 g · L^{-1} $AgNO_3$, 0.75 mL · L^{-1} formaldehyde and 4 mg · L^{-1} sodium thiosulphate
6. Washing with water and storage in 50% methanol-12% acetic acid

In addition to staining for total protein, electrophoretic gels can be used for detection of specific protein components. This can be done, for example, by treating the gel in conditions suitable to reveal enzymatic activities, such as chitinolytic (Vincenzi and Curioni, 2005), or staining for glycoproteins with the periodic acid-Schiff method (PAS) (e.g. Marchal *et al.*, 1996).

After PAGE separation, protein bands can also be transferred on membrane supports (Western blotting), which can then be probed with specific ligands such as lectins (lectinoblotting) (e.g. Hsu and Heaterbell, 1987) or antibodies (immunoblotting) (e.g. Dambrouck *et al.*, 2003).

A preparative SDS-PAGE system, exploiting the electroendosmotic flow, generated during electrophoresis, for the elution of the proteins from the gel, can be used for purification of nearly all the protein components of wine (Vincenzi *et al.*, 2003).

Native Polyacrylamide Gel Electrophoresis (N-PAGE)

'Native' electrophoretic systems can be exploited for separation of grape and wine proteins according to both their charge and mass properties. In this case, the relative migration rate of the different electrophoretic bands can be varied by varying the gel pore size (i.e polyacrylamide concentration) or pH of the electrophoretic buffers. To perform N-PAGE of acidic proteins (as grape and wine proteins) the Laemmli gel system (Laemmli, 1970) without SDS (neither in sample buffer nor in electrophoretic buffer), can be used (Hillier, 1976). This technique has been applied to must proteins for varietal differentiation (Moreno-Arribas *et al.*, 1999 and papers cited herein) (Figure 7.4).

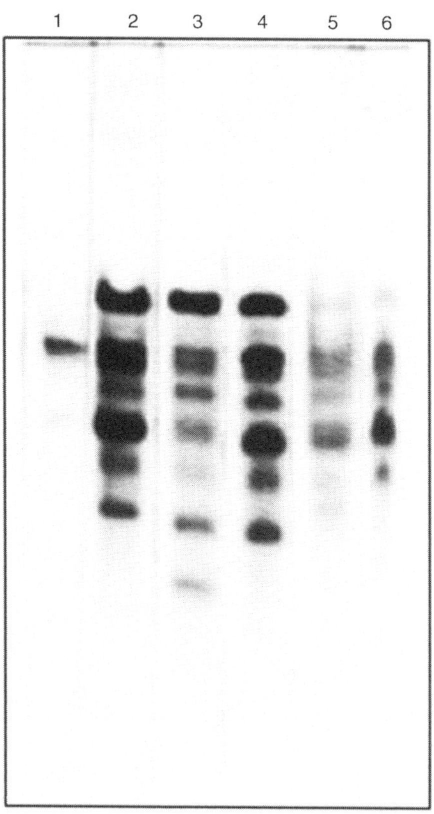

Figure 7.4 Native electrophoresis of proteins from different grape varieties (Reproduced from *J. Agric. Food Chem.*, 1999, **47**(1), 114–120, Moreno-Arribas *et al.*, with permission from the American Chemical Society)

Isoelectric focusing

In isoelectric focusing (IEF), the electric field forces proteins to migrate in a medium containing a pH gradient. Because the protein net charge depends on the pH value, each protein migrates to the point in which the pH is equal to its p.I. (where the net charge of the proteins is 0). This system allows resolution of proteins having minimal p.I. differences and results in the separation of a number of bands that is generally higher than other electrokinetic separations. The classical IEF system is based on pH gradients formed during the application of an electric field by particular zwitterionic compounds called carrier ampholytes (non-immobilized pH gradient). This procedure, used in different analytical systems such as agarose and polyacrylamide gels in both tube or slab, has often been applied to the must and wine proteins to perform their characterization on the basis of charge properties, and as tool for varietal differentiation (Luguera *et al.*, 1998; Pueyo *et al.*, 1993). Also, preparative free-flow IEF with the Rotofor apparatus (Bio Rad) was used for the fractionation of wine proteins (Marchal *et al.*, 1996). However, the IEF technique based on the use of ampholytes is difficult to perform and has serious problems of pH gradient instability and reproducibility. These drawbacks have been overcome recently by the much more powerful IEF on immobilized pH gradient (IPG-IEF). Immobilized pH gradient is produced by the use of a limited number of well-defined chemicals (the 'Immobilines') which are co-polymerized with the acrylamide matrix. A major advantage of IPG-IEF is that standardized pre-cast gels containing different pH ranges (broad, narrow or ultra-narrow) to be used in dedicated apparatuses, are commercially available. Before use, IPG strips must first be re-hydrated, generally under silicone oil. In order to load large quantities of protein, it can be beneficial to re-hydrate the strips directly with the protein sample solution. A suitable buffer for re-suspension of the protein sample before IEF contains 2 M thiourea, 7 M urea, 4% CHAPS (3-[(3-cholamidopropyl)-dimethylammonio]-1-propane sulfonate), and 0.1–1% ampholytes, although several modifications of the solution have been reported such as Triton-X100 (Sarry *et al.*, 2004) or tributylphosphine (Polesani, 2004) addition. Temperature (around 20 °C) should be controlled during IEF separation. Initially, IEF should be run at low voltage, then the voltage is increased when the current has settled down. Once the run is complete the strips can either be used immediately for staining or 2-D separation (see below) or frozen for later use. To date, this technique has been applied to the grape and wine proteins as the first dimension in 2-D separations (see below), although it can also be considered a promising tool for other applications.

Two dimensional electrophoresis

Two Dimensional-Polyacrylamide Gel Electrophoresis (2D-PAGE) is the most powerful tool to separate the different components of complex protein mixtures. Proteins are first separated according to their p.I. in IEF (first dimension), then a further orthogonal separation according to the MW is performed by SDS-PAGE (second dimension). In this way all the protein components of a biological sample can be virtually separated.

The strips deriving from the IEF must be equilibrated before SDS-PAGE to allow proteins to bind the detergent. The equilibration step is normally carried out in the presence of DTT to ensure complete protein unfolding.

2D-PAGE has been applied to analysis of grape pulp proteins (Sarry *et al.*, 2004), skin proteins (Deytieux *et al.*, 2007) and the entire grape cluster proteins (Vincent *et al.*, 2006). Several hundreds spots can be detected after gel staining (Figure 7.1). After 2D separation, the spots of interest can be excided from the gel, subjected to proteolytic digestion and the resulting peptides analyzed by mass spectrometry to achieve protein identification (see below).

2D-PAGE with IPG-IEF can also be applied to the analysis of wine polypeptides (Marshall and Williams, 1987; Polesani, 2004) (Figure 7.5). Detection of several spots with the same MW but differing in p.I. confirms that wine proteins are actually more heterogeneous than appears from a simple 1D-SDS-PAGE pattern (Monteiro *et al.*, 2001).

Capillary electrophoresis

Capillary electrophoresis (CE) also exploits the capacity of charged molecules, including proteins and peptides, to move in an electric field, but in this case separation occurs in a capillary tube. This technique has some advantages over the classical gel electrophoresis such as efficiency, short migration times, high resolution, automation and possibility of quantifying the separated proteins. Proteins and peptides eluting from the capillary are commonly revealed as peaks by UV detection, but direct connection of the outlet of the capillary to a mass spectrometer is also possible (Simò *et al.*, 2004). Fused silica capillaries are the most widely used. To prevent unspecific protein adsorption on the silanol groups, these capillaries can be coated with polymers of various type. Separation of proteins by CE can be achieved according to their mass/charge ratio

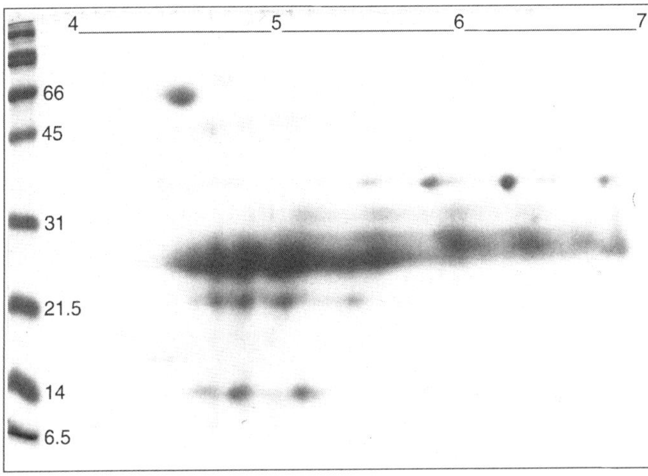

Figure 7.5 2D-PAGE of the wine proteins (*cv.* Incrocio Manzoni 6.0.13) (Polesani, 2004). Proteins from 800 μL of dialyzed/freeze-dried wine dissolved in 100 μL of 25 mM Tris-HCl buffer pH 9.0 containing urea 8 M (buffer A), then reduced with 5 mM tributylphosphine (90 min, room temperature), alkylated with iodoacetamide (20 mM, 90 min at room temperature) and diluted with 100 μL buffer A containing 3 M thiourea and 8% CHAPS. An IPG strip (7 cm, pH 4–7) was soaked in the sample (200 μL) for 12 h. IEF performed at 1000 V for 1 h and 4000 V for 6 h. Strip incubated in 10 mL of 0.1 M Tris-HCl buffer pH 6.8 containing 2% SDS, 5% 2-mercaptoethanol and 20% glycerol for 25 min and loaded on a SDS-PAGE gel

(Free Solution-CE, FSCE, also called Capillary Zone Electrophoresis, CZE), molecular size (SDS-CGE in capillaries filled with an acrylamide gel), or isoelectric point (CIEF).

CE on uncoated silica capillaries of grape proteins was used for varietal differentiation of Muscadine grapes, giving more than 20 well resolved peaks (Mazhar *et al.*, 2002).

Analysis of proteins of white and red wines can be performed by CGE in the presence of SDS, allowing determination of the molecular weight and the relative quantity of the separated proteins in about 1 h (Rodriguez-Delgado *et al.*, 2002). Another CE application includes the study of wine protein modifications during wine ageing, using uncoated capillary in basic (Luguera *et al.*, 1997) or acidic conditions (Moine-Ledoux *et al.*, 1992).

An extensive examination of the different conditions affecting CZE separation (sample preparation, type of loading buffer and pH of CE buffer) of wine proteins reveals as optimum conditions the sample preparation by either ultrafiltration or dialysis and dissolution and separation of the sample in conditions allowing stacking (i.e. dissolution of the protein sample in acidic buffer followed by running under basic conditions) as reported below (Figure 7.6).

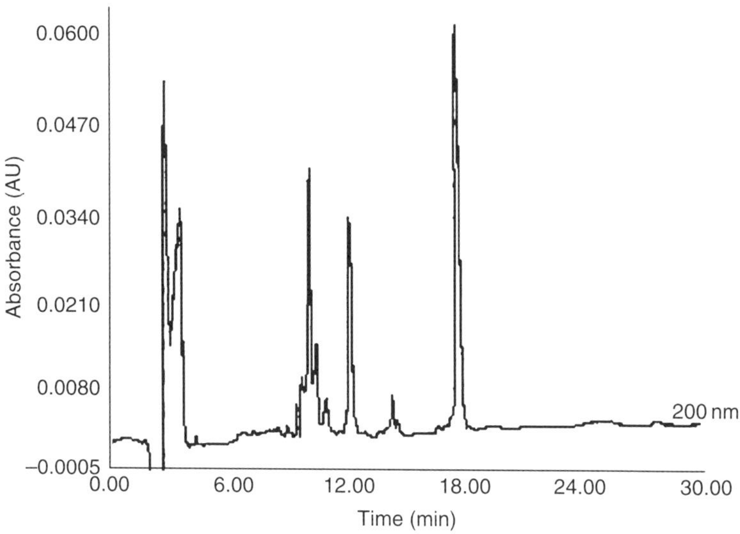

Figure 7.6 CZE of proteins from Sauvignon blanc wine. Wine dialyzed against 20 mM citrate solution pH 2.5, and loaded on non-polar (coated) DB-1 capillary (24 cm × 50 μm i.d). Injection: 4 or 2 sec at 5 psi; separation buffer: 0.3 M borate buffer pH 8.5; voltage: constant 10 or 12 kV; temperature: 20 °C; detection at wavelength 200 nm. (Reproduced from *AJEV*, 1999, **50**, 120–127, Dizy and Bisson, with permission of ASEV)

7.3.2. Methods by Liquid-Chromatography

Column separation of proteins and peptides is used for both preparative and analytical purposes. Generally, in the former case low- or medium-pressure systems are preferred, whereas in the latter, high-performance-liquid-chromatography (HPLC) is the method of choice. These systems are coupled with a spectrophotometric detector normally set at wavelengths 280 and (around) 220 nm for proteins and peptides, respectively. Coupling of HPLC with mass spectrometry allows structural identification of compounds after the separation.

The molecular characteristics exploited for the separation of proteins and peptides include molecular size (size exclusion chromatography, SEC), net charge (ion-exchange chromatography, IEC), polarity (reverse-phase chromatography, RPC), hydrophobicity (hydrophobic interaction chromatography, HIC) and capacity to bind specific chemical groups (affinity chromatography, AC).

Size exclusion chromatography

Size Exclusion Chromatography (SEC) is performed on gel columns and it is used for separation of grape and wine proteins according to their molecular size. Low-pressure SEC on Sephadex G-25 (Somers and Ziemelis, 1973) or Sephadex G-75 (Bayly and Berg, 1967) allows separation of the fraction of proteins from the compounds of higher (polysaccharides and nucleic acids) and lower (salts, organic acids and phenols) molecular size. Only high-resolution medium pressure (Superdex 75, Canals 1998; Pastorello *et al.*, 2002) or HPLC (BioSil TSK 125 GS2000SW, Modra and Williams, 1988) columns allow separation of the single proteins from both grape and wine. However, taking into account that most grape proteins share similar molecular weights (Monteiro *et al.*, 2001), separation by SEC is not sufficient to obtain pure components and, normally, it must be coupled to at least another separation step based on other protein parameters different from MW.

SEC of grape and wine proteins has to be performed on samples deprived of phenolics because these compounds tend to bind irreversibly the stationary phase. Phenolic compounds are removed by passing the protein extract through a C_{18} cartridge (see above). As an alternative to removing the interfering substances, it is also possible to pre-treat the sample with small amounts of the same stationary phase used for SEC.

Generally, the mobile phase used for SEC should have a pH similar to that of the wine, in order to avoid pH-dependent protein solubility changes. It should also contain some salt, typically NaCl (higher than 0.4 M, Somers and Ziemelis, 1973), in order to prevent possible protein aggregation or adsorption on the stationary phase due to electrical interactions.

In peptides analysis, SEC is generally used as a pre-fractionation step to remove high MW molecules such as proteins and polysaccharides. A Sephadex G10 column can be used to separate peptides in two fractions with MW lower and higher than 700 Da (Moreno-Arribas *et al.*, 1998).

Ion exchange chromatography

One of the most popular LC techniques used in grape and wine protein analysis is Ion Exchange Chromatography (IEC), which is based on protein separation according to their surface net charge (Monteiro *et al.*, 2001). The positive or negative net charge assumed at a given pH by the proteins allows their binding to charged groups present in the stationary phase by 'exchanging' the counter ion initially bound to the resin. By increasing the salt concentration of the elution solvent, this process can be reversed and the bound proteins are sequentially released according to the 'strength' of binding. Alternatively, proteins can be released by changing the pH of the mobile phase. Since proteins have positive net charge at the wine pH, the IEC most widely used is Anion Exchange with different types of resins, such as BioGel TSK DEAE-5PW (Sarmento *et al.*, 2000; Lagace and Bisson, 1990), Mono Q (Waters *et al.*, 1993) and Bioscale Q2 (Derckel *et al.*, 1996) with Fast Protein Liquid Chromatography (FPLC) system. However, good results using this system can also be obtained in the Cation Exchange mode using MonoS resins (Canals *et al.*, 1998; Mesquita *et al.*, 2001; Monteiro *et al.*, 1999).

An example of a chromatogram relative to the protein analysis of a free run juice by Anion Exchange Chromatography is shown in Figure 7.7 (Vincenzi *et al.*, 2005). Contrary to proteins, grape and wine peptides have not been studied by IEC.

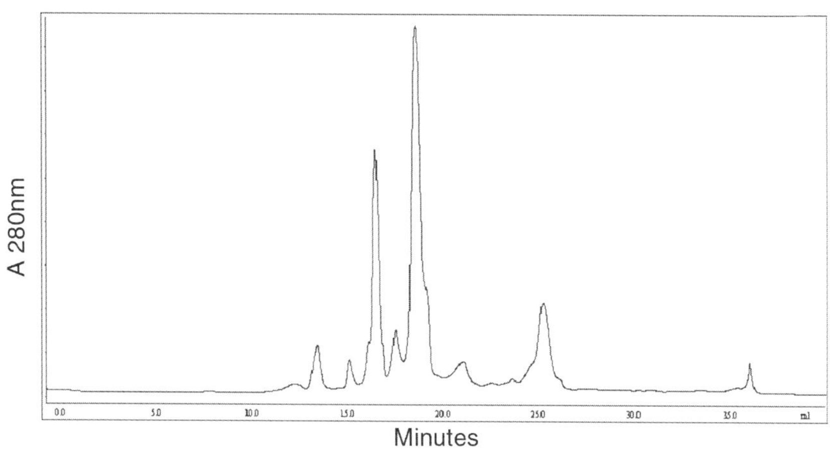

Figure 7.7 Anion Exchange Chromatography of the proteins from the free run juice of grape of *cv.* Incrocio Manzoni 6.0.13. Column Tricorn Mono Q 5/50 GLM; detection at wavelength 280 nm; buffer A: Tris-HCl 20 mM pH 8.5; buffer B: buffer A containing NaCl 1 M; gradient program: from 0 to 14% of B in 70 min, from 14 to 100% B in 20 min and from 100 to 0% B in 5 min; flow rate 0.6 mL/min (Vincenzi *et al.*, 2005)

Reverse phase chromatography

Reverse Phase Chromatography (RP) is a very common HPLC technique (RP-HPLC) to perform separation of both proteins and peptides. Stationary phases contain non-polar groups of different length, including butyl (C_4), octyl (C_8) or octadecyl (C_{18}) chains. Proteins and peptides interact with these groups according to their polarity and are eluted by increasing the concentration of non-polar organic solvents, such as acetonitrile, ethanol or methanol, in the aqueous solvent. To improve separation, an ion pairing agent (generally TFA) is added at low concentrations (typically 0.05–0.1%) to the mobile phase. During separation, proteins are exposed to organic solvents and can be denatured according to their sensitivity to these solvents.

Grape proteins have been separated on C_{18} (Santoro, 1995; Peng *et al.*, 1997) and C_8 (Tattersall *et al.*, 1997) stationary phases. A widely used column is the Vydac C_{18} (Peng *et al.*, 1997; Waters *et al.*, 1998), which allows separation of different forms of grape thaumatin-like proteins (VVTL) and chitinases (Chit) (Figure 7.8).

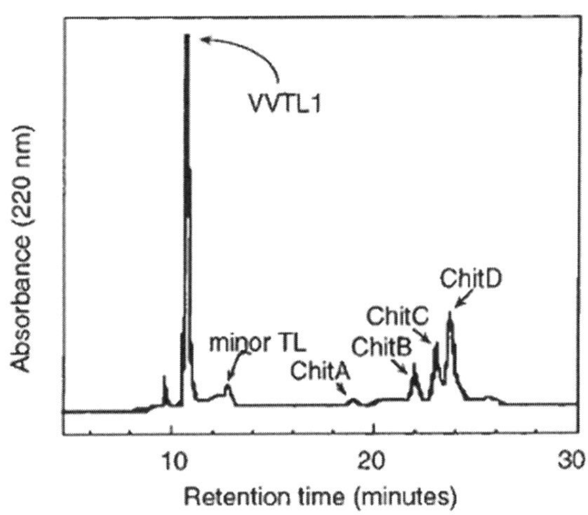

Figure 7.8 Reverse phase-HPLC chromatogram of grape proteins from analysis of 10-fold concentrated free run juice desalted on a Econo-Pac 10DG column; separation on semipreparative C_{18} (10 × 250 mm); detection at wavelength 280 nm; binary solvent composed of A) aqueous 0.05% TFA solution and B) acetonitrile+0.035% TFA; gradient program: from 29 to 61% of B in 7 min, from 61 to 70% B in 8 min, from 79 to 79% B in 1 min, from 79 to 94% B in 14 min, finally washing with 94% B for 5; injection volume 1 mL; flow rate 1.0 mL/min; temperature 40 °C. Peaks proteins identified by mass spectrometry as thaumatin-like proteins (VVTL) and chitinase (Chit). (Reproduced from *J. Agric. Food Chem.*, 2000, 48(5), 1637–1643, Pocock *et al.*, with permission from the American Chemical Society)

Despite the fact that RP-HPLC is the method most frequently used for peptide separation, only a few applications to grape and wine are reported in the literature (Desportes *et al.*, 2000; de Person *et al.*, 2004; Moreno-Arribas *et al.*, 1996 and 1998). After isolation from wine of the MW fraction corresponding to the peptides of interest (mainly performed by ultrafiltration or SEC), their separation is performed on a C_4 (recommended for large and hydrophobic peptides), C_8 or C_{18} (recommended for small and more hydrophilic peptides) column, using conditions similar to those reported for the proteins. Detection is normally performed in the wavelength range between 200 and 220 nm. Due to the presence of several interfering compounds absorbing in this range, post-column peptide derivatization with *o*-phataldehyde (OPA) and fluorescence detection can also be used (Acedo *et al.*, 1994).

RP-HPLC is also used for separation of peptides deriving from proteolytic digestion of the wine proteins separated on and excided from a electrophoretic gel slab. By this approach and performing nano-flow RP-HPLC coupled with MS/MS analysis, 20 different proteins were identified in a white wine (Kwon, 2004) (see below).

Other methods

Affinity chromatography on lectin-containing columns can be conveniently used to isolate glycosilated proteins from complex mixtures. The specificity of plant lectins for the different sugar residues allows selection of glycoproteins characterized by different terminal sugars. Affinity chromatography with Concanavalin A (a plant lectin specific for mannose and glucose residues) can be used to separate the highly mannosylated proteins (mannoproteins) deriving from yeast, which are involved in wine protein stabilisation (Waters *et al.*, 1993; Dupin *et al.*, 2000; Moine-Ledoux and Dubourdieu, 1999). To allow proteins to interact with Concanavalin A, it is necessary to use a buffer (e.g. Tris-HCl 20 mM) with pH in the range 5.0–7.4 and containing 1 mM $CaCl_2$, $MgCl_2$ and $MnCl_2$. NaCl (0.15-0.5 M) can be used to prevent unspecific interactions. Proteins elution is achieved with methyl-α-mannoside (0.1 or 0.5 M).

Another chromatographic technique useful to perform protein separation and for studying their physical-chemical properties, is Hydrophobic Interaction Chromatography (HIC). In this case, separation of proteins is essentially based on the different presence of hydrophobic groups in their surface. These differences are evidenced from the ionic strength of the solvent and no denaturation of the molecule occurs during analysis. Surprisingly, application of this technique to wine proteins is limited to just one paper (Brissonnet and Maujean, 1993), although hydrophobicity of proteins in wine can be important for their functionality.

Proteins of a Champagne based wine were characterized in relation to their effects on wine foaming by HIC using a Fractogel TSK Butyl 650 (S) column. Proteins were eluted by an optimized gradient system involving ammonium sulphate, water and aqueous 75% 2-propanol (Table 7.7). This system allowed recovery of 12 protein fractions differing in surface hydrophobicity (Figure 7.9).

Table 7.7 Optimized elution gradient for wine protein separation on Fractogel TSK Butyl 650 (S) column. (Reproduced from *AJEV*, 1993, **44**, 297–301, Brissonnet and Maujean, with permission of ASEV).

Time	Ammonium sulphate 1 M (%)	Water (%)	75 % (v/v) 2-propanol (%)
0	50	50	0
15	25	75	0
25	25	75	0
35	0	100	0
60	0	100	0
90	0	0	100
120	0	0	100

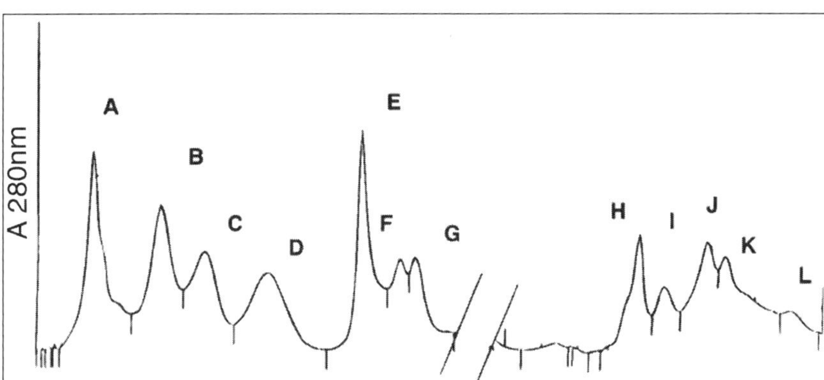

Figure 7.9 HIC of wine proteins dialyzed (6–8 kDa cut-off) and 100-fold concentrated with PEG; separation on Fractogel TSK Butyl 650 (S) (10 × 150 mm) column; flow rate 1.0 mL/min; detection at wavelength 280 nm; gradient program as reported in Table 7.7. (Reproduced from *AJEV*, 1993, **44**, 297–301, Brissonnet and Maujean, with permission of ASEV)

7.3.3. Grape and Wine Peptides and Proteins by Mass Spectrometry

Mass spectrometry (MS) techniques are based on measurement of the charge/mass (*m/z*) ratio of molecules ionized by electron impact, laser irradiation, or by interaction with electric fields. Several MS methods have been applied in the study of grape and wine proteins initially as off-line techniques, then coupled to LC and CE techniques.

By the use of electrospray (ESI), matrix-assisted laser desorption ionization (MALDI) and MS tandem (MS/MS) systems, structural information on the amino acid sequence of proteins is gleaned (Moreno-Arribas *et al.*, 2002). In ESI-MS, the sample is dissolved in a solvent and delivered into the ionization source of the mass spectrometer through a narrow metal capillary. A high voltage applied to the tip of the capillary generates charged droplets of the solution, as a consequence of the solvent evaporation $[M+H]^{+}$ or $[M+nH]^{n+}$ positive ions are released and analyzed according to their *m/z* ratios. ESI can be readily interfaced with LC and CE techniques. Nano-ESI utilizes a very low solvent flow rate that is brought about by the charge applied to the capillary rather than by a solvent delivery pump. The signal/noise ratio is enhanced compared with standard ESI, a small aliquot of sample (2 μL) sprayed for ca. 30 min provides a peptide mass map and several MS/MS sequence tag analyses on a single sample performed.

Capillary electrophoresis-MS (on-line coupling of CE to ESI-MS, CE-MS) provides high resolution separations of a wide range of proteins (Simó *et al.*, 2004). Two-dimension electrophoresis (2-DE) and MS analysis allow resolution and identification of several thousand proteins. The procedure, which can be automated, involves excision of the protein spots from the 2-DE gel, followed by enzymatic digestion with a protease (e.g. trypsin) and MS analysis (Ashcroft, 2003). 2-DE polyacrylamide gel electrophoresis (2D-PAGE) provides resolution based on both the size and mass differences.

MS data and genome database searching provides protein sequence information and identity. Further MS/MS analysis on each peptide after enzymatic digestion allows generation of a sequence tag of peptide, by a further database search with both the peptide molecular mass and sequence tag information, amino acid sequence of individual peptides and unambiguous identification of protein is achieved (Ashcroft, 2003). With these sequences an automated protein identification is possible by database searching using algorithms such as KNEXUS, SEQUEST, THEGPM, PHOENIX, or MASCOT. Each protein sequence from the database (e.g. SWISS-PROT, TrEMBL, NCBI) is virtually digested according to the specificity of the used protease, and the resulting peptides that match the measured mass of the peptide ion are identified. In the next step, the experimentally derived MS/MS spectrum of the peptide ion is compared to the theoretical spectra obtained by virtual fragmentation of candidate peptide sequences. Finally, a score is calculated for each peptide sequence by matching the predicted fragment ions to the ions observed in the experimental spectrum. Due to the fact that database searches can generate false positives and false negatives depending on the parameters used, a manual evaluation of the data is opportune to confirm the protein identification (Glinski and Weckwerth, 2006).

Use of MALDI coupled with time of flight mass analyzer (MALDI-TOF) is widely employed in proteins analysis. An acidic solution containing an energy-absorbing molecule (matrix) is mixed with the analyte and highly focused laser pulses are directed to the mixture. After desorption and ionization, molecules are accelerated by a high electrical field. The ions arrive at the detector in order of increasing m/z ratio. Due to robustness, tolerance to salt- and detergent-related impurities and ability to be automated, MALDI-TOF is regularly used to generate mass maps of proteins after enzymatic digestion (Ashcroft, 2003). Because large precipitates may not be effectively desorbed, dissolution of proteins can be achieved using detergents and urea, but they may interfere with analysis. Affinity MS (AMS) allows elimination of problems due to interfering components in the original mixture and increases the analysis sensitivity and selectivity (Hutchens and Yip, 1993). For analysis of peptides and small proteins α-cyano-4-hydroxycinnamic acid (CHCA) is commonly used as matrix; sinapinic acid (SA) is used instead for larger proteins analysis. MALDI-TOF provides good mass accuracy (0.01%) and sensitivity (proteins in femtomole range can be detect) and requires very little sample for analysis (Flamini and De Rosso, 2006, and references cited herein).

MS analysis of grape proteins

Grape proteins can be analyzed by LC/MS and MALDI-TOF, ESI-MS analysis is used as a complement to DNA methods in the variety characterization.

Grape juice usually contains proteins with range of masses 13–33 kDa. LC/ESI-MS is performed by loading the concentrated juice onto a C_8 reverse-phase LC column (250×1 mm, 5 mm) equilibrated in a mixture of 75% solvent A (0.05% TFA in water) and 25% solvent B (0.05% TFA in 90% v/v aqueous acetonitrile) and performing elution by linear gradient program: from 25 to 90% of solvent B in 60 min, then isocratic for 30 min at flow rate of 15 μL/min. The column is directly connected to ESI source or to a UV-Vis detector operating at wavelength 220 nm coupled on-line with a mass spectrometer. Analysis is performed in positive ion mode with ESI and orifice potentials set at 5.5 kV and 30 V, respectively. The curtain (N_2) and nebulizer (air) gases are set at 8 and 10 units, respectively. The mass spectrum, consisting of the multiple charge ions, is taken from the protein peak and is processed to determine the most probable MW of the proteins. The identity of individual proteins is determined on the basis of both LC elution order and MW. Distribution and intensities of the multiply charged molecular ions produced by ESI are directly related to the number of basic amino acids and the structural conformation of the protein (Hayasaka *et al.*, 2001).

Also protein trap-ESI-MS was used for proteins analysis in grape juice (Hayasaka *et al.*, 2001). The concentrated juice is loaded onto a protein trap (3 × 8 mm) cartridge directly connected to the mass spectrometer and proteins are eluted in one or two broad fractions. The cartridge is equilibrated in a mixture of 70% solvent A (formic acid/H_2O (2:98 v/v)) and 30% solvent B (2% (v/v) formic acid in 80% (v/v) aqueous acetonitrile) and proteins are eluted with the following gradient program: from 30 to 60% of B in 10 min, isocratic for 10 min, solvent B increased to 80% in 10 min, isocratic for 5 min. The cartridge is washed sequentially with 3 mL of 50% and 3 mL of 80% aqueous acetonitrile containing 2% formic acid before being reused. Protein trap-MS approach proved to enhance the detectability of *Vitis vinifera* thaumatin-like (VVTL) proteins 10-fold relative to LC/MS, even if with disadvantages in terms of complication of mass spectra due to the coelution of multiple proteins. These advantages, with respect to LC/MS, were not observed in the analysis of *Vitis vinifera* chitinase (VVCHI) proteins. The minimum amount of total proteins required for trap-MS is 150 ng.

Definition of different pathogenesis-related (PR) proteins in the ranges 21239–21272 Da and 25330–25631 Da proved potentially useful for varietal differentiation of grapes. The PR protein ESI-MS patterns of juice of the three different white grape varieties are reported in Figure 7.10, differences among varieties are evident (Hayasaka *et al.*, 2003).

MALDI-TOF analysis of grape seeds proteins can be used for differentiation of varieties (Pesavento *et al.*, 2008). For sample preparation 10 g

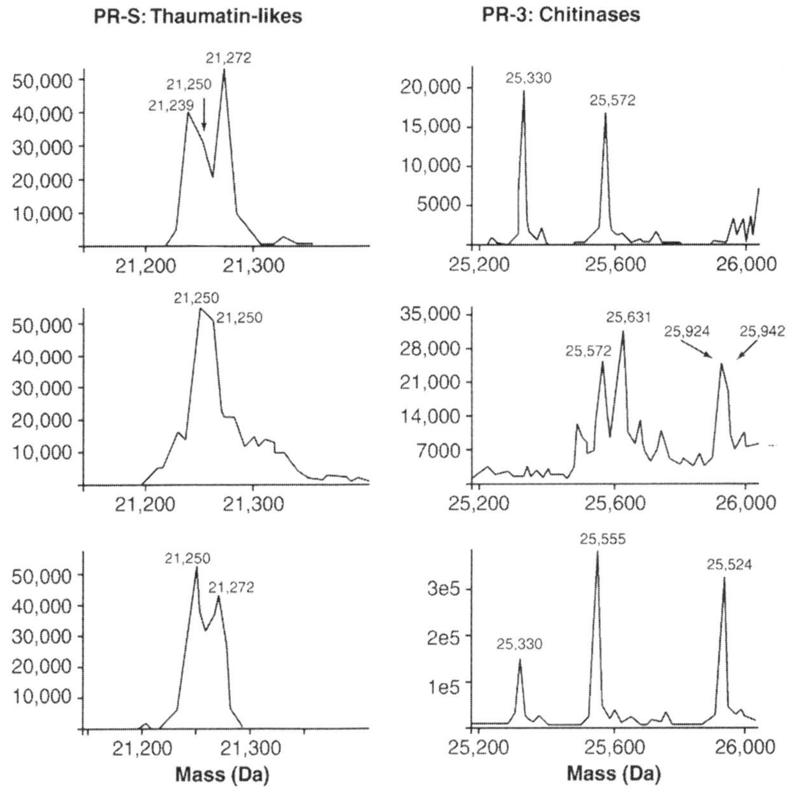

Figure 7.10 ESI-MS patterns of pathogenesis-related protein of juice from three different white grape varieties. (This figure was first published by the *Australian and New Zealand Wine Industry Journal*, May–June **18**(3), 2003, and is reproduced here with kind permission of the publisher, Winetitles www.winebiz.com.au)

of grape seed are powdered under liquid nitrogen. About 5 g of seed powder is defatted by suspension with 100 mL of n-hexane and stirred for 30 min; the organic layer is removed and the residue left to dry at room temperature. Then 500 mg of powder is suspended in 5 mL of water and dialyzed against double distilled water at 4 °C for 48 h, centrifuged for 15 min and the residue is used for analysis.

To compare different methods of extraction, an amount of 10 mg of the residue was suspended in 1 mL of three different solvents: H_2O+0,1% TFA, acetonitrile, methanol and acetonitrile (50:50 v/v) under ultrasonic wave for 5 min. The supernatants were collected and analyzed by MALDI using a 2,5-dihydroxybenzoic acid (DHB) solution prepared dissolving 10 mg of DHB in 1 mL of H_2O (0.1% TFA)/CH_3CN

(1:1 v/v). Analyses were performed for different red and white grape varieties, extraction by $H_2O + 0.1\%$ TFA solution gave the best results in terms of intensity of signals. Extraction with hexane and dialysis of samples is performed to reduce the chemical noise in the low m/z region of mass spectra. Figure 7.11 shows spectra of proteins in defatted and dialyzed grape seed powder samples of three different grape varieties.

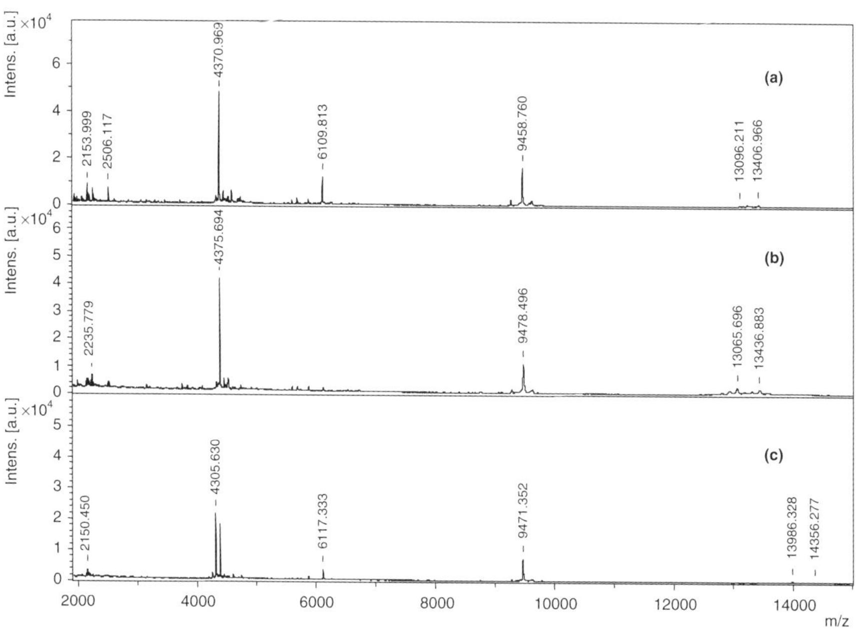

Figure 7.11 MALDI spectra of seed proteins extracted from Raboso Piave (a), Prosecco (b), and Malvasia Nera di Brindisi (c) (Measurements in the positive ion linear mode of ions formed by a pulsed nitrogen laser at $\lambda = 337$ nm with repetition rate 50 psec, ion source voltage 1: 25 kV, ion source voltage 2: 23,35 kV, ion source lens voltage: 10.5 kV. 5 μL of sample mixed with 5 μL of DBH matrix solutions, 1 μL of mixture deposited on the stainless-steel sample holder). (Pesavento *et al.*, JMS 2008 in press, DOI: 10.1002/jms.1295)

MS analysis of peptides and proteins in wine

One method of sample preparation proposed for MS/MS analysis of proteins is by performing enzymatic hydrolysis of gel pieces from SDS-PAGE (see paragraph 7.3.1) using trypsin, proteins are hydrolyzed at the carboxyl side of lysine and arginine residues. Protocol of enzymatic digestion and sample preparation is reported in Table 7.8.

Table 7.8 Protocol of enzymatic digestion and sample preparation for MS/MS analysis of proteins from gel pieces after SDS-PAGE (Kwon, 2004).

1. Gel pieces are dehydrated with acetonitrile and dried
2. Then rehydrated with modified porcine trypsin $10\,ng{\cdot}L^{-1}$ in 50 mM NH_4HCO_3 solution
3. In-gel digestion is carried on at 37 °C overnight
4. First extraction of tryptic peptides with acetonitrile/H_2O/TFA (50:45:5 v/v/v)
5. Second extraction of peptides with acetonitrile/H_2O/TFA (75:24.9:0.1 v/v/v)
6. Extracts are combined, dried and cleaned with C_{18} pipette tip

Analysis of low-MW peptides in wine can be performed by LC/ESI-MS. The wine sample is ultrafiltrated with a MW 1000 cut-off membrane and analyses are carried out operating in positive ion mode using the dipeptide Phe-Arg as internal standard. In champagne wine, Ile-Arg was found to be the most abundant peptide with concentration ranging from 2.2–7.0 mg/L, followed by Arg-Ile, Ile-Val, and Tyr-Lys (others dipeptides were found in concentration lower than 1 mg/L) (De Person *et al.*, 2004). For this analysis an alkyl-amide reversed phase (150×4.6 mm, 5 μm) LC column at 30 °C was used, with a binary solvent composed of (a) aqueous nonafluoropentanoic acid 2 mM pH 2.8 and (b) acetonitrile and gradient program from 0% to 10% of B in 4 min, 10% to 30% of B in 13 min, isocratic 30% of B for 13 min (flow-rate 1 mL/min). MS data of some peptides in wine are listed in Table 7.9 with the major collisionally activated dissociation (CAD) fragments.

The protonated molecule $[M+H]^+$ is the most abundant ion formed, only singly charged ions are observed. To perform MS/MS experiments of $[M+H]^+$ precursor collision energy from 10 to 50 eV, is used. The $[M+H]^+$ ion of dipeptides shows principally the ammonia loss, the y_1 ion and a_1 fragment. By increasing the collision energy the major CAD fragment ion abundance has a maximum ranging from 20 eV for most loss of the dipeptides, and 30 eV for the tripeptide Phe-Arg-Arg.

Nano-LC/ESI-MS has been successfully applied in analysis of peptides from wine proteins (Kwon, 2004). A volume of 2 μL of acetonitrile/H_2O/acetic acid (2:97.9:0.1 v/v/v) peptide solution (buffer A) is loaded into the capillary C_{18} column (50 mm × 75 μm i.d., 5 μm particle size, 300 Å pore diameter) and peptides are eluted with a gradient from 5% to 80% of buffer B [acetonitrile/H_2O/acetic acid (90:9.9:0.1 v/v/v)] in buffer A in 10 min at flow rate 0.3 μL/min. Fragmentation of the three strongest parent ions of full MS spectrum is performed, and *m/z* 700–1300 spectra recorded. The MS/MS spectra

Table 7.9 Mass-to-charge values of the protonated peptides investigated and their collisionally activated dissociation fragments. All fragments exceeding 2% of the total fragment ion abundance are reported. $[Im]^+$: $[H_2N=CH-R]^+$; R amino acid residue. In parentheses: typical sequence a, b, y fragment ions. Bold values correspond to loss of ammonia from $[A\ Im]^+$ or $[B\ Im]^+$. (*) Fragment obtained after the guanidino group loss by cyclisation of the Arg residue according to Dookeran *et al.* (1996). (**) Fragment obtained after the carboxylic group loss from the corresponding (*) fragment. MS conditions: selective reaction monitoring (SRM) mode with collision energy from 5 to 30 eV, curtain and collision gas N_2, ionspray temperature and voltage 400 °C and 5.2 kV, declustering potential 20 V, focusing potential 200 V, entrance potential 10 V, dwell time 250 ms (De Person *et al.*, 2004).

Peptide	*m/z*								
	$[M+H]^+$	$[M+H-NH_3]^+$	$[A+H]^+$	$[B+H]^+$ (y_1)	$[A+H-H_2O]^+$ (b_1)	$[B+H-H_2O]^+$	$[A\ Im]^+$ (a_1)	$[B\ Im]^+$	Other
Ile–Val	231						86	72	**69**
Val–Ile	231			132			72	86	**55**, 185
Arg–Ile	288	271	175		157			86	70**, **112**, 229
Ile–Arg	288			175			86		70**, 116*, 158
Lys–Phe	294	277	147		129		101		**84**, 259
Phe–Lys	294	277		147		129	120	101	**84**, 259
Lys–Tyr	310	293	147	182	129		101	136	247, 275
Tyr–Gln	310	293	182	147			136		107, 119, 130, 276
Tyr–Lys	310	293		147		129	136		91, 130
Phe–Arg	322	305		175			120		70**, 116*
	$[M+H]^+$	$[M+H-NH_3]^+$	$[M+H-H_2O]^+$	$[B+H]^+$	$[C+H]^+$ (y_1)	$[A+H-H_2O]^+$ (b_1)	$[A\ Im]^-$ (a_1)		Other
Phe–Arg–Arg	478		460	175	175		120		116*, 287, 322, 418
Lys–Met–Asn	392	375			133	129			357, 264, 260, 84

are acquired in a data-dependent mode that determines the masses of the parent ions, and fragments are used for protein identification. Figure 7.12 shows chromatogram and mass spectra relative to nano-LC/MS analysis of a peptide from a protein in wine.

In this analysis the criteria used for a manual confirmation of protein identification were: i) the major isotope-resolved peaks should match

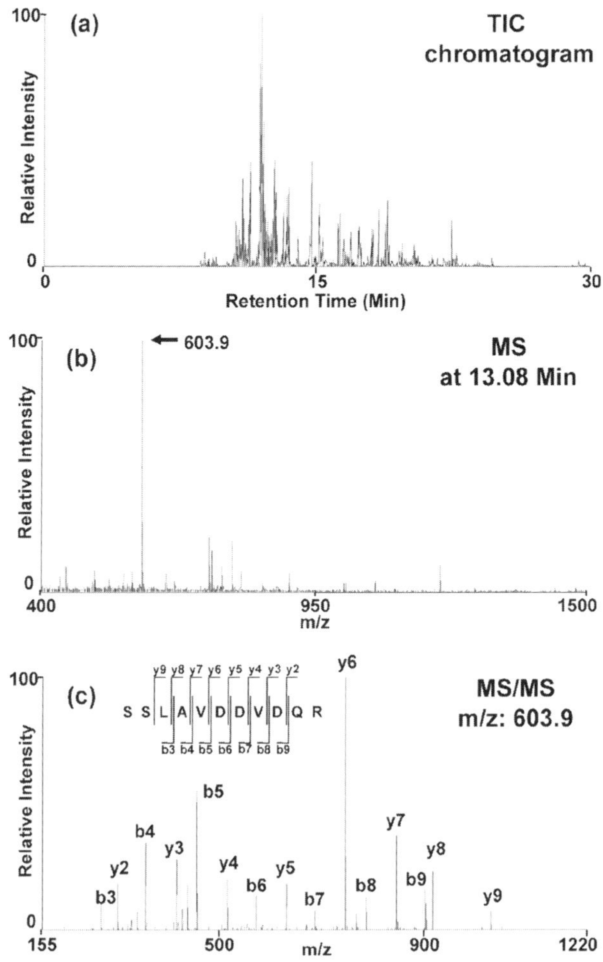

Figure 7.12 Nano-LC/MS analysis of a peptide from a protein in wine: (a) total ion current (TIC) chromatogram of the tryptic digest (MW range 60-75 kDa in SDS-PAGE). (b) Full MS spectrum at retention time 13.08 min (*m/z* 400–1500). (c) The MS/MS spectrum of *m/z* 603.9 identified the peptide SSLAVDDVDQR. (Reproduced from *J. Agric. Food Chem.*, 2004, **52(24)**, 7258–7263, Kwon, with permission from the American Chemical Society)

fragment masses of the identified peptide, ii) to consider y, b, and a ions and their water loss or amine loss peaks, iii) to emphasize the isotope-resolved peaks, iv) major of 7 isotope-resolved peaks are matched to theoretical masses of the peptide fragments, v) all redundant proteins are removed by confirming the unique peptides, vi) to confirm the unique peptides, all amino acid sequences of the identified proteins are listed and each peptide is examined (Kwon, 2004).

For MALDI-TOF characterization and analysis of wine proteins samples can be prepared by two different methods: i) wine and SA saturated acetonitrile/water/TFA solution are mixed and a volume of 2 μL of solution is applied to the sample holder and dried; ii) 50 mL of wine are lyophilized, the residue is dissolved in a H_2O/urea solution, proteins are precipitated with ethanol then again dissolved with urea. After a second precipitation, the residue is dissolved in an H_2O/TFA solution and mixed with SA. For the latter procedure, a much better resolution of the peak in the *m/z* 15000–18000 range is reported (Szilàgyi *et al.*, 1996). Analysis of higher MW proteins (10–100 kDa) is performed by sample dissolution in SA matrix, for analysis of 0–15 kDa mass range proteins CHCA is used (Weiss *et al.*, 1998). In general, wine proteins show peaks with masses between 7 and 86 kDa, with the 21.3 kDa the major proteins, and other significant peaks with masses 7.2, 9.1, 13.1, and 22.2 kDa, are observed. Formation of multiply charged ions and dimers can be influenced from the matrix and laser energy. Since desorption/ionization is dependent on the size and nature of individual proteins, it is not possible to make a direct comparison of relative intensities between different proteins. An accurate protein quantification is possible only by the use of internal standards very similar to each analyte.

Also, surface-enhanced laser desorption ionization-time of flight mass spectrometry (SELDI-TOF) was used for analysis of wine proteins and peptides (Weiss *et al.*, 1998). This affinity MS technique utilizes functional groups on inert platforms to capture molecules from the sample. The use of agarose beads containing iminodiacetate-chelated copper ion as functional groups (IDA-Cu), which interact with specific amino acid residues of wine proteins, induces formation of interactions with histidine, lysine, tryptophan, cysteine, aspartic acid, and glutamic acid. Wine proteins and peptides determined by SELDI-TOF show peaks quite similar to by MALDI-TOF. By coupling the two techniques, MALDI-TOF shows the greatest number of peaks, while SELDI-TOF provides an increased sensitivity as well as selectivity for some protein fractions.

References

Acedo, M.I., Pueyo, E. and Polo, M.C. (1994) Preliminary studies on peptides in wine by HPLC, *Am. J. Enol. Vitic.*, **45**, 167–172.

Ashcroft, A.E. (2003) Protein and peptide identification: the role of mass spectrometry in proteomics, *Nat. Prod. Rep.*, **20**, 202–215.

Bayly, F.C. and Berg, H.W. (1967) Grape and wine proteins of white wine varietals, *Am. J. Enol. Vitic.*, **18**, 18–32.

Brissonnet, F. and Maujean, A. (1993) Characterization of foaming proteins in a champagne base wine, *Am. J. Enol. Vitic.*, **44**, 297–301.

Busam, G., Kassemeyer, H.H. and Matern, U. (1997) Differential expression of chitinases in *Vitis vinifera* l. Responding to systemic acquired resistance activators or fungal challenge, *Plant Physiol.*, **115**, 1029–1038.

Canals, J.M., Arola, L. and Zamora, F. (1998) Protein fraction analysis of white wine by FPLC, *Am. J. Enol. Vitic.*, **49**, 383–388.

Dambrouck, T., Marchal, R., Cilindre, C., Parmentier, M. and Jeandet, P. (2005) Determination of the grape invertase content (using PTA-ELISA) following various fining treatments versus changes in the total protein content of wine. Relationships with wine foamability, *J. Agric. Food Chem.*, **53**, 8782–8789.

Dambrouck, T., Marchal, R., Marchal-Delahaut, L., Parmentier, M., Maujean, A. and Jeandet, P. (2003) Immunodetection of proteins from grapes and yeast in a white wine, *J. Agric. Food Chem.*, **51**, 2727–2732.

De Person, M., Sevestre, A., Chaimbault, P., Perrot, L., Duchiron, F., and Elfakir, C. (2004) Characterization of low-molecular weight peptides in champagne wine by liquid chromatography/tandem mass spectrometry, *Anal. Chim. Acta*, **520**, 149–158.

Derckel, J.P., Audran, J.C., Haye, B., Lambert, B. and Legendre, L. (1998) Characterization, induction by wounding and salicylic acid and activity against *Botrytis cinerea* of chitinases and beta-1,3-glucanases of ripening grape berries, *Physiol. Plantarum*, **104**, 56–64.

Derckel, J.P., Legendre, L., Audran, J.C., Haye, B. and Lambert, B. (1996) Chitinase of the grapevine (*Vitis vinifera*): five isoforms induced in leaves by salicylic acid are constitutively expressed in other tissues, *Plant Science*, **119**, 31–37.

Desportes, C., Charpentier, M., Duteurtre, B., Maujean, A. and Duchiron, F. (2000) Liquid chromatographic fractionation of small peptides from wine, *J. Chrom. A*, **893**, 281–291.

Deytieux, C., Geny, L., Lapaillerie, D., Claverol, S., Bonneau, M. and Donèche, B. (2007) Proteome analysis of grape skins during ripening, *J. Experim. Botany*, **58**, 1851–1862.

Dizy, M. and Bisson, L.F. (1999) White wine protein analysis by capillary zone electrophoresis, *Am. J. Enol. Vitic.*, **50**, 120–127.

Dookeran, N.N., Yalcin, T. and Harrison, A.G. (1996) Fragmentation reactions of protonated α-amino acids, *J. Mass Spectrom.* **31**, 500–508.

Dorrestein, E., Ferreira, R.B., Laureano, O. and Teixeira, A.R. (1995) Electrophoretic and FPLC analysis of soluble proteins in four Portuguese wines, *Am. J. Enol. Vitic.*, **46**, 235–242.

Dupin, I.V.S., McKinnon, B.M., Ryan, C., Boulay, M., Markides, A.J., Jones, G.P., Williams, P.J. and Waters, E.J. (2000) *Saccharomyces cerevisiae* mannoproteins that protect wine from protein haze: their release during fermentation and lees contact and a proposal for their mechanism of action, *J. Agric. Food Chem.*, **48**, 3098–3105.

Famiani, F., Walker, R.P., Tecsi, L., Chen, Z.H., Proietti, P. and Leegood, R.C. (2000) An immunoistochemical study of the compartmentation of metabolism during the development of grape (*Vitis vinifera* L.) berries, *J. Experim. Botany*, **51**, 675–683.

Ferreira, R.B., Monteiro, S., Piçarra-Pereira, M.A., Tanganho, M.C., Loureiro, V.B. and Teixeira, A.R. (2000) Characterization of the proteins from grapes and wines by immunological methods, *Am. J. Enol. Vitic.*, **51**, 22–28.

Ferreira, R.B., Piçarra-Pereira, M.A., Monteiro, S., Loureiro, V.B. and Teixeira, A.R. (2002) The wine proteins, *Trends in Food Sci. Technol.*, **12**, 230–239.

Flamini, R. and De Rosso, M. (2006) Mass Spectrometry in the Analysis of Grape and Wine Proteins, *Exp. Rev. Proteom*, **3**, 321–331.

Fukui, M., Okuda, T., Takayanagi, T. and Yokotsuka, K. (2002) Changes in amount of nitrogenous compounds from skins and seeds of four grape cultivars during extraction using juice or fermenting must-like model solutions, *J. Wine Research*, **13**, 203–215.

Glinski, M. and Weckwerth, W. (2006) The role of mass spectrometry in plant systems biology, *Mass Spectrom. Rev.*, **25**, 173–214.

Gonçalves, F., Heyraud, A. de Pinho, M.N. and Rinaudo, M. (2002) Characterization of white wine mannoproteins, *J. Agric. Food Chem.*, **50**, 6097–6101.

Hayasaka, Y., Adams, K.S., Pocock, K.F., Baldock, G.A., Waters, E.J. and Hoj, P.B. (2001) Use of electrospray mass spectrometry for mass determination of grape (Vitis vinifera) juice pathogenesis-related proteins: a potential tool for varietal differentiation, *J. Agric. Food Chem.*, **49**, 1830–1839.

Hayasaka, Y., Baldock, G.A., Pocock, K.F., Waters, E.J., Pretorius, I. and Hoj, P.B. (2003) Varietal differentiation of grape juices by protein fingerprinting, *The Australian and New Zealand Wine Industry J.*, **18**(3), 27–31.

Hillier, R.M. (1976) The quantitative measurement of whey proteins using polyacrilamide gel electrophoresis, *J. Dairy Res.*, **43**, 259–265.

Hsu, J.C. and Heatherbell, D.A. (1987) Isolation and characterization of soluble proteins in grapes, grape juice and wine, *Am. J. Enol. Vitic.*, **38**, 6–10.

Hutchens, T.W. and Yip, T.T. (1993) New desorption strategies for the mass spectrometric analysis of macromolecules, *Rapid Commun. Mass Spectr.*, 7, 576–580.

Igartuburu, J.M., Martín del Río, R., Montiel, J., Pando, E. and Luis, F.R. (1991) Study of agricultural by-products. Extractability and amino acid composition of grape (*Vitis vinifera*) skin proteins from cv. Palomino, *J. Sci. Food Agric.*, **57**, 437–440.

Jacobs, A.K., Dry, I.B. and Robinson, S.P. (1999) Induction of different pathogenesis-related cDNAs in grapevine infected with powdery mildew and treated with ethephon, *Plant Pathol.*, **48**, 325–336.

Kwon, S.W. (2004) Profiling of soluble proteins in wine by nano-high-performance liquid chromatography/tandem mass spectrometry, *J. Agric. Food Chem.*, **52**, 7258–7263.

Laemmli, U.K. (1970) Cleavage of structural protein during the assembly of head of bacteriophage T4, *Nature*, **227**, 680–685.

Lagace, L.S. and Bisson, L.F. (1990) Survey of yeast acid proteases for effectiveness of wine haze reduction, *Am. J. Enol. Vitic.*, **41**, 147–155.

Linthorst, H.J.M. (1991) Pathogenesis-Related proteins of plants, *Crit. Rev. Plant Sci.*, **10**, 123–150.

Lubbers, S., Voilley, A., Feuillat, M. and Charpentier, C. (1994) Influence of mannoproteins from yeast on the aroma intensity of a model wine, *Lebensmittel-Wissenschaft und Technologie*, **27**, 108–114.

Luguera, C., Moreno-Arribas, V., Pueyo, E., Bartolomé, B. and Polo, M.C. (1998) Fractionation and partial characterization of protein fractions present at different stages of the production of sparkling wines, *Food Chem.*, **63**, 465–471.

Luguera, C., Moreno-Arribas, V., Pueyo, E. and Polo, M.C. (1997) Capillary electrophoretic analysis of wine proteins. Modifications during the manufacture of sparkling wines, *J. Agric. Food Chem.*, **45**, 3766–3770.

Marchal, R., Berthier, L., Legendre, L., Marchal-Delahaut, L., Jeandet, P. and Maujean, A. (1998) Effects of *Botrytis cinerea* infection on the must protein electrophoretic characteristics, *J. Agric. Food Chem.*, **46**, 4945–4949.

Marchal, R., Bouquelet, S. and Maujean, A. (1996) Purification and partial biochemical characterization of glycoproteins in a Champenois Chardonnay wine, *J. Agric. Food Chem.*, **44**, 1716–1722.

Marshall, T. and Williams, K.M. (1987) High resolution two-dimensional electrophoresis of the proteins and macromolecular constituents of beer and wine, *Electrophoresis*, **8**, 493–495.

Mazhar, H., Basha, S.M. and Lu, J. (2002) Variation in berry protein composition of muscadine cultivars, *Am. J. Enol. Vitic.*, **53**, 87–91.

Mesquita, P.R., Piçarra-Pereira, M.A., Monteiro, S., Loureiro, V.B., Teixeira, A.R. and Ferreira, R.B. (2001) Effect of wine composition on protein stability, *Am. J. Enol. Vitic.*, **52**, 324–330.

Modra, E.J. and Williams, P.J. (1988) Are proteases active in wines and juices? *The Australian Grapegrower and Winemaker*, **292**, 42–46.

Moine-Ledoux, V. and Dubourdieu, D. (1999) An invertase fragment responsible for improving the protein stability of dry white wines, *J. Sci. Food Agric.*, **79**, 537–543.

Moine-Ledoux, V., Dulau, L. and Dubourdieu, D. (1992) Interprétation de l'amélioration de la stabilité protéique des vins au cours de l'élevage sur lies, *Journal International Scientifique Vigne Vin*, **26**, 239–251.

Monteiro, S., Piçarra-Pereira, M.A., Loureiro, V.B., Teixeira, A.R. and Ferreira, R.B. (2007) The diversity of pathogenesis-related proteins decreases during grape maturation, *Phytochemistry*, **68**, 416–425.

Monteiro, S., Piçarra-Pereira, M.A., Mesquita, P.R., Loureiro, B., Teixeira, A. and Ferreira, R.B. (2001) The wide diversity of structurally similar wine proteins, *J. Agric. Food Chem.*, **49**, 3999–4010.

Monteiro, S., Piçarra-Pereira, M.A., Tanganho, M.C., Rente, J.P., Loureiro, V.B., Teixeira, A.R. and Ferreira, R.B. (1999) Preparation of polyclonal antibodies specific for wine proteins, *J. Sci. Food Agric.*, **79**, 772–778.

Monteiro, S., Piçarra-Pereira, M.A., Teixeira, A.R., Loureiro, V.B. and Ferreira, R.B. (2003) Environmental conditions during vegetative growth determine the major proteins that accumulate in mature grapes, *J. Agric. Food Chem.*, **51**, 4046–4053.

Moreno-Arribas, M.V., Cabello, F., Polo, M.C., Martín-Alvarez, P.J. and Pueyo, E. (1999) Assessment of the native electrophoretic analysis of total grape must proteins for the characterization of *Vitis vinifera* L. cultivars, *J. Agric. Food Chem.*, **47**, 114–120.

Moreno-Arribas, M.V. and Polo, M.C. (2002) Analytical methods for the characterization of proteins and peptides in wines, *Anal. Chim. Acta*, **458**, 63–75.

Moreno-Arribas, V., Pueyo, E. and Polo, M.C. (1996) Peptides in must and wines. Changes during the manufacture of Cavas (Sparkling wines), *J. Agric. Food Chem.*, **44**, 3783–3788.

Moreno-Arribas, V., Pueyo, E., Polo, M.C. and Martin-Alvarez, P.J. (1998) Changes in the amino acid composition of the different nitrogenous fractions during the aging of wine with yeasts, *J. Agric. Food Chem.*, **46**, 4042–4051.

Pastorello, E.A., Farioli, L., Pravettoni, V., Ortolani, C., Fortumato, D., Giuffrida, M.G., Perono Garoffo, L., Calamari, A.M., Brenna, O. and Conti, A. (2002) Identification of grape and wine allergens as an endochitinase 4, a lipid-transfer protein, and a thaumatin, *J. Allergy Clin. Immunol.*, **111**, 350–359.

Peng, Z., Pocock, K.F., Waters, E.J., Francis, I.L. and Williams, P.J. (1997) Taste properties of grape (*Vitis vinifera*) pathogenesis-related proteins isolated from wine, *J. Agric. Food Chem.*, **45**, 4639–4643.

Pesavento, I.C., Bertazzo, A., Flamini, R., Dalla Vedova, A., De Rosso, M., Seraglia, R. and Traldi, P. (2008) Differentiation of Vitis vinifera varieties by MALDI-MS analysis of the grape seed proteins, *J. Mass Spectr.*, 43(2), 234–241.

Pocock, K.F., Hayasaka, Y., McCarthy, M.G. and Waters, E.J. (2000) Thaumatin-like proteins and chitinases, the haze-forming proteins of wine, accumulate during ripening of grape (*Vitis vinifera*) berries and drought stress does not affect the final levels per berry at maturity, *J. Agric. Food Chem.*, **48**, 1637–1643.

Pocock, K.F., Hayasaka, Y., Peng, Z., Williams, P.J. and Waters, E.J. (1998) The effect of mechanical harvesting and long-distance transport on the concentration of haze-forming proteins in grape juice, *Australian Journal of Grape and Wine Research*, **4**, 23–29.

Polesani, M. (2004) Caratterizzazione di chitina e chitosano per un loro possibile utilizzo nella chiarificazione dei vini bianchi, Master thesis, University of Verona.

Pueyo, E., Dizy, M. and Polo, C. (1993) Variety differentiation of must and wines by means of protein fraction, *Am. J. Enol. Vitic.*, **44**, 255–260.

Robinson, S.P., Jacobs, A.K. and Dry, I.B. (1997) A class IV chitinase is highly expressed in grape berries during ripening, *Plant Physiol.*, **114**, 771–778.

Rodriguez-Delgado, M.A., Malovanà, S., Montelongo, F.J. and Cifuentes, A. (2002) Fast analysis of proteins in wines by capillary gel electrophoresis, *Eur. Food Res. Technol.*, **241**, 536–540.

Salzman, R.A., Tikhonova, I., Bordelon, B.P., Hasegawa, P.M. and Bressan, R.A. (1998) Coordinate accumulation of antifungal proteins and hexoses constitutes a developmentally controlled defence response during fruit ripening in grape, *Plant Physiol.*, **117**, 465–472.

Santoro, M. (1995) Fractionation and characterization of must and wine proteins, *Am. J. Enol. Vitic.*, **46**, 250–254.

Saravan, R.S. and Rose, J.K.C. (2004) A critical evaluation of sample extraction techniques for enhanced proteomic analysis of recalcitrant plant tissues, *Proteomics*, **4**, 2522–2532.

Sarmento, M.R., Oliveira, J.C., Slatner, M. and Boulton, R.B. (2000) Influence of intrinsic factors on conventional wine protein stability tests, *Food Control*, **11**, 423–432.

Sarry, J.E., Sommerer, N., Sauvage, F.X., Bergoin, A., Rossignol, M., Albagnac, G. and Romieu, C. (2004) Grape berry biochemistry revisited upon proteomic analysis of the mesocarp, *Proteomics*, **4**, 201–215.

Schagger, H. and von Jagow, G. (1987) Tricine-sodium dodecyl sulfate-polyacrylamide gel electrophoresis for the separation of proteins in the range from 1 to 100 kDa, *Anal. Biochem.*, **166**, 368–379.

Simò, C., Elvira, C., Gonzalez, N., San Roman, J., Barbas, C. and Cifuentes, A. (2004) Capillary electrophoresis-mass spectrometry of basic proteins using a new physically adsorbed polymer coating. Some applications in food analysis, *Electrophoresis*, **25**, 2056–2064.

Singh, N.K., Bracker, C.A., Hasegawa, P.M., Handa, A.K., Buckel, S., Hermodson, M.A., Pfankoch, E., Regnier, F.E. and Bressan, R.A. (1987) Characterization of osmotin: a thaumatin-like protein associated with osmotic adaptation in plant cells, *Plant Physiol.*, **85**, 529–536.

Smith, P.K., Krohn, R.I., Hermanson, G.T. (1985) Measurement of protein using bicinchoninic acid, *Anal. Biochem.*, **150**, 76–85.

Somers, T.C. and Ziemelis, G. (1973) The use of gel column analysis in evaluation of bentonite fining procedures, *Am. J. Enol. Vitic.*, **24**, 51–54.

Szilágyi, Z., Vas, G., Mády, G. and Vékey, K. (1996) Investigation of macromolecules in wines by matrix-assisted laser desorption/ionization time-of-flight mass spectrometry, *Rapid Commun. Mass Spectr.*, **10**, 1141–1143.

Tattersall, D.B., van Heeswijck, R. and Høj, P.B. (1997) Identification and characterization of a fruit-specific, thaumatin-like protein that accumulates at very high levels in conjiunction with the onset of sugar accumulation and berry softening in grapes, *Plant Physiol.*, **114**, 759–769.

Tattersall, D.B., Pocock, K.F., Hayasaka, Y., Adams, K. van Heeswijck, R., Waters, E.J. and Høj, P.B. Pathogenesis related proteins – their accumulation in grapes during berry growth and their involvement in white wine heat instability. Current knowledge and future perspectives in relation to winemaking practices, in K.A. Roubelakis-Angelakis (Ed), *Molecular Biology & Biotechnology of the Grapevine*, Agritech Publications, New York, 2001, p. 183–201.

Vincent, D., Wheatley, M.D. and Cramer, G.R. (2006) Optimization of protein extraction and solubilization for mature grape berry cluster, *Electrophoresis*, **27**, 1853–1865.

Vincenzi, S. and Curioni, A. (2005) Anomalous electrophoretic behaviour of a chitinase isoform from grape berries and wine in glycol chitin-containing SDS-PAGE gels, *Electrophoresis*, **26**, 60–63.

Vincenzi, S., Mosconi, S., Zoccatelli, G., Dalla Pellegrina, C., Veneri, G., Chignola, R., Peruffo, A.D.B., Curioni, A. and Rizzi, C. (2005) Development of a new procedure for protein recovery and quantification in wine, *Am. J. Enol. Vitic.*, **56**, 182–187.

Vincenzi, S., Zoccatelli, G., Rizzi, C. and Curioni, A. One step purification of nearly all the proteins from wine by electroendosmotic preparative electrophoresis, in *VIIème Symposium International d'Oenologie*, Tec & Doc (Eds), Bordeaux, 2003, p. 538–541.

Waters, E.J., Hayasaka, Y., Tattersall, D.B., Adams, K.S. and Williams, P.J. (1998) Sequence analysis of grape (*Vitis vinifera*) berry chitinases that cause haze formation in wines, *J. Agric. Food Chem.*, **46**, 4950–4957.

Waters, E.J., Shirley, N.J. and Williams, P.J. (1996) Nuisance proteins of wine are grape pathogenesis related proteins, *J. Agric. Food Chem.*, **44**, 3–5.

Waters, E.J., Wallace, W., Tate, M.E. and Williams, P.J. (1993) Isolation and partial characterization of a natural haze protective factor from wine, *J. Agric. Food Chem.*, **41**, 724–730.

Waters, E.J., Wallace, W. and Williams, P.J. (1992) Identification of heat-unstable wine proteins and their resistance to peptidases, *J. Agric. Food Chem.*, **40**, 1514–1519.

Weiss, K.C., Yip, T.T., Hutchens, T.W. and Bisson, L.F. (1998) Rapid and sensitive fingerprinting of wine proteins by matrix-assisted laser desorption/ionisation time-of-flight (MALDI-TOF) mass spectrometry, *Am. J. Enol. Vitic.*, **49**(3), 231–239.

Yokotsuka, K. and Fukui, M. (2002) Changes in nitrogen compounds in berries of six grape cultivars during ripening over two years, *Am. J. Enol. Vitic.*, **53**, 69–77.

Yokotsuka, K., Nozaki, K. and Takayanagi, T. (1994) Characterization of soluble glycoproteins in red wine, *Am. J. Enol. Vitic.*, **45**, 410–416.

Yu, L.X., Djebrouni, M., Chamberland, H., Lafontaine, J.G. and Tabaeizadeh, Z. (1998) Chitinase: Differential induction of gene expression and enzyme activity by drought stress in the wild (*Lycopersicon chilense* Dun.) and cultivated (*L. esculentum* Mill.) tomatoes, *J. Plant Physiol.*, **153**, 745–753.

8

Elements and Inorganic Anions in Winemaking: Analysis and Applications

Roberto Larcher and Giorgio Nicolini

8.1. Introduction

8.1.1. Oenological Framework

The cationic and anionic mineral fraction making up wine ashes does not exceed 5000 mg/L (Eschnauer and Neeb, 1988) and is more frequently in the range of 2000–3000 mg/L. Of this, around 97% is essentially constituted of eight macroelements (K, Mg, Ca, Na, C, P, S, Cl), which are present with concentrations higher than 10 mg/L. The remaining 3% of the fraction, according to a classification proposed by Eschnauer *et al.* (2001), is represented by:

- micro- and trace elements with concentrations between 0.01 and 10 mg/L, including Si, Mn B, Rb, Zn, Sr, Fe, Cu, Al, F and I present in higher concentrations, and Sn, V, Ti, As, Ba, Pb, Br, Cr, Li, Ni, Co, Mo and Ag in lower concentrations;
- ultra-trace elements, with concentrations lower than 10 μg/L, which also includes rare earth and radioactive elements.

Hyphenated Techniques in Grape and Wine Chemistry Edited by Riccardo Flamini

Because of the role of the different elements in the chemical-physical stability of wine, mineral fraction, understood not just as an overall parameter expressed through the measurement of ashes and their alkalinity, but also as the quantification of individual elements, has been studied for decades. The main examples are precipitation in wine of tartaric acid salts with calcium and potassium, and precipitation caused by excessive content of iron (III) and phosphates ('iron casse'). This problem was remedied in Germany as early as 1923 by treatment of the wine with potassium ferrocyanide, and copper (I), for the elimination of which the use of sodium sulphide was proposed (Ribereau-Gayon, 1947). In addition to this, haze due to aluminium (Rankine, 1983), precipitation of calcium mucate in wines from botrytized grapes and of other elements (e.g. cadmium, manganese, mercury, lead, copper, tin and zinc) in the form of sulphides, may occur (e.g. Ribereau-Gayon, 1947; Garoglio, 1981; Ribéreau-Gayon *et al.*, 1980; Boulton *et al.*, 1995; Flanzy, 1998). Elements such as W, Zn, Cu, Co, Fe, Ni and Mn, in decreasing order, have the capacity to catalyze oxidative reactions. However, bearing in mind their concentrations, Fe and Zn primarily, and secondarily Cu and Mn, have a genuine catalytic role of technological significance. On the contrary, Al limits the oxygen consumption in wine (Vivas, 2002). It is likely that the capacity of many compounds to chelate Al^{3+} reduces their sensitivity to oxidation.

The elemental mineral composition of grapes, wines and derivates and by-products has been widely investigated, although the reasons for analyzing particular micro and trace elements have not always been completely clear, due to their limited technological significance (Ough and Amerine, 1988). A brief list of works integrating the publications cited in the commendable paper by Eschnauer and Neeb (1988), includes those by Margheri and Rigotti (1964), Ciusa and Giaccio (1972), Cerutti *et al.* (1981, 1984, 1990), Interesse *et al.* (1984, 1985) and Finoli *et al.* (1986). Eschnauer *et al.* (2001) estimated that at least 1000 papers were written between 1890 and 2000 on the content of trace and ultra-trace elements in wine, and around 350 regarding mineral elements. About 70 further studies focusing significantly on the elemental composition of grapes, wines and spirits and relative analytical methods, in addition to an equal number dealing with this topic in a marginal manner, have been published since 2000 according to the bibliographic research carried out using the Food Science and Technology Abstracts (FSTA) database.

Bringing together and summarizing the direct experience of the authors, deriving from everyday practice in the oenological laboratory, this chapter focuses on modern applications and the significance of the analysis of elements and inorganic anions in the oenological field. It updates data on grapes, wines and derivatives and the compositional effects of several winemaking treatments.

8.1.2. Elements

Soil is usually considered among the most important enrichment factors relating to the natural elemental content in grapes and wines, e.g. high contents of Na, and of B and V, in salty soils and in volcanic soils, respectively, are reported (Salaha-Moutsopoulou and Voudouri-Tsoukala, 1991; Savino *et al.*, 1994; Teissèdre *et al.*, 1996b; Conde *et al.*, 2002, Eschnauer, 1982).

Natural or artificial contamination often significantly increases the natural level of several elements in wine (Medina *et al.*, 1977; Seppi and Sperandio, 1978; Medina and Sudraud, 1980; Eschnauer, 1982; Pertoldi Marletta *et al.*, 1986; 1989; Prosperi and Piga, 1988; Lemperle and Lay, 1989a; Mc Kinnon *et al.*, 1992; Gulson *et al.*, 1992; Ough, 1993; Eschnauer and Scollary, 1995; Angelova *et al.*, 1999b; Médina *et al.*, 2000; Castiñeira Gómez *et al.*, 2004a) sometimes in relation to the addition of adulteration substances (e.g. preservatives, sweetening agents) as recalled by Eschnauer and Neeb (1988).

The potential toxicity of some elements – with particular reference to heavy metals and, among these, with particular attention being paid to lead – is one of the main reasons for investigating the content of such elements in grapes and wines (Castelli *et al.*, 1974; Basile and Tarallo, 1974; Crecelius, 1977; Stefani, 1978; Handson, 1984; Sherlock *et al.*, 1986; Gabrielli Favretto *et al.*, 1986; Elinder *et al.*, 1988; Smart *et al.*, 1990, Marini and Gallese, 1991; Teissèdre and Cabanis 1993; Teissèdre *et al.*, 1993; 1994; Ough, 1993; Šebecic *et al.*, 1998; Pinamonti *et al.*, 1999; Angelova *et al.*, 1999a). The contribution of wine with specific reference to the dietary uptake of some elements has been much more rarely investigated (e.g. Teissèdre *et al.*, 1996a; 1996b; Cabrera-Vique *et al.*, 2000; Terrés *et al.*, 2001), in particular in a context of positiveness. What is more, the real bioavailability of some elements, not only in wine but also along the whole soil-vine-grape-wine chain, has not been fully clarified. Tokalioglu *et al.* (2004) observed for 11 elements measured in 22 grape and corresponding polluted vineyard soils, a relationship between the content in soil and grapes for Cd, Pb, Zn, Mn and Fe. Soil-wine correlations in the content of several elements were also found by Almeida and Vasconcelos (2003). Karagiannidis and Nikolaou

(2000) observed that *arbuscular mycorrhizae* could reduce Pb and Cd uptake from soil, giving lower content of the two elements in grapes of Razaki cultivar grafted onto 110R rootstock. In wine, significant Pb, but also Sr and Ba, are linked to a dimer of rhamnogalacturonane II (Pellerin *et al.*, 1997; Green and Scollary, 2000), a pectic polysaccharide which can significantly condition their availability.

Depletion phenomena in the element content of wines are seen less frequently in everyday wine-making practice. In the case of heavy metals present in high concentrations, the only depletion interventions – without distinguishing between authorized and illegal procedures in the different countries – are chemical treatments with potassium ferrocyanide and Fessler's compound, and interventions with different chelate or ion exchange resins to mainly deplete Fe and Cu (Kern and Wucherpfennig, 1991; 1993; Fussnegger *et al.*, 1992; Mattivi *et al.*, 1994; Nicolini *et al.*, 2004c;), Pb (Mínguez *et al.* 1996; 1997), and various other elements in the preparation of rectified concentrated musts and Vermuts. A lowering in the content of some elements obtained biologically has been also observed with live and non-living biomass of *Saccharomyces cerevisiae* in laboratory conditions (Volesky and May-Phillips, 1995), and various micro-organisms have shown capacity for element accumulation (Norris and Kelly, 1977; Nakajima and Sakaguchi, 1986; 1993; White and Gadd, 1987; Volesky *et al.*, 1993; Blakwell *et al.*, 1995).

In relation to the health and technological aspects seen above, some cations – mainly Cu, Fe, Pb, Zn, K and Ca – are routinely monitored in quality control protocols and Hazard Analysis and Critical Control Points (HACCP) applied in wineries. Moreover, more and more restrictive 'technological' limits have been defined in wine transactions (e.g. Ni $<$ 8–10 μg/L toward Germany) between private companies by buyers, particularly big bottlers or importers. Sometimes such 'technological' limits overlap with the existing legal or maximum acceptable limits which to date can be summarized as follows: As 0.2 mg/L, Cu 1 mg/L, Zn 5 mg/L, Pb 0.15 mg/L, B 80 mg/L (as boric acid), Cd 0.01 mg/L, F 1 mg/L (3 mg/L in the countries where criolite can be used as insecticide) and Br 1 mg/L. The International Organisation of Vine and Wine (O.I.V.) defined the limit of 60 mg/L for Na exceeding the equivalent chloride concentration, but contents higher than 60 mg/L (expressed exclusively as total Na), as well as higher than 60 μg/L for Li, can sometimes lead control agencies to carry out deeper analysis for suspected use of wines denatured by addition of 5–10 g/L LiCl. For grapes, limits of 20 and 50 mg/kg have been established for Cu and S respectively.

8.1.3. Inorganic Anions

The inorganic anionic fraction of grapes and wine is essentially constituted of sulphates, chlorides, phosphates and nitrates and has been investigated to a much lesser extent in recent years as compared to the cationic fraction. These anions are absorbed by plants through the roots from the soil, which is the main source, depending on geo-pedological origin and any fertilization, correction and soil improvement measures (Currle *et al.*, 1983).

Sulphates

As regards sulphates, Leske *et al.*, (1997) observed overall average values of 260 mg/L in Australian must, with a maximum of 1197 mg/L and marked differences among different areas of origin of grape, varying between 157 mg/L and 393 mg/L. Ough and Amerine (1988), reviewing previous works regarding wines from 15 countries, reported K_2SO_4 concentrations in wines from 70 mg/L to 4.39 g/L. Values of between 200–1300 mg/L were reported by Flak and Schaber (1989) in Austrian wines, with the highest frequency between 500 and 800 mg/L, and contents up to 2200 mg/L (as SO_4^{2-}) were reported by Lemperle and Lay (1989a, quoting Zee *et al.*, 1983), with German wines between 160 and 400 mg/L. Artificial increases of sulphates can come from the use of fermentation adjuvants, acidifying treatments with gypsum (an old practice only allowed for *Jerez* wines), and oxidation of SO_2, more intense in wood barrels. Increases in sulphates, as well as other inorganic anions, can derive from the fraudulent addition of the corresponding acids. At all events, the final content of sulphates in wine cannot be higher than 2 g/L, expressed as K_2SO_4, with the exception of *Marsala*, liqueur-like wines and 'mistella' where the limit is 5 g/L (Reg. Eu. Cons. No. 1493/1999; Italian G.U. L. 82/2006). The role of sulphates in the formation of protein haze in white wine was recently proven by Pocock *et al.* (2007).

Phosphates

The content of phosphates in wines usually varies between 50 and 1000 mg/L (as PO_4^{3-}) with a lower concentration in white wines as compared to the red ones (Garoglio, 1981; Zee *et al.*, 1983; Ough and Amerine, 1988). Flak and Schaber (1989) did not find significant differences in Austrian wine, neither as regards the type of soil in the different areas of production nor as regards the rootstock used. The location of phosphates in the grape berry has an important role in conditioning the final content in the wines, seeds contributing significantly (Wucherpfennig and Chen Hsueh-err, 1983). Further increases may derive from the use of common fermentation adjuvants, mainly diammonium phosphate.

Chlorides

Being mainly geogenic, the natural content of chlorides in wine is closely related to the grape growing area. The natural NaCl content in wines usually lies in the range between 20 and 200 mg/L, but values up to 1800 mg/L can be found in arid areas of Europe, such as Spain (Ough and Amerine, 1988) and the Greek islands, also in connection with grape withering (Salaha-Moutsopoulou and Voudouri-Tsoukala, 1991), and from Australia (Leske *et al.*, 1997). According to Wurzinger *et al.* (1992), up to 50–60 mg/L NaCl can be expected in genuine Austrian wines, with higher values for wines designated as 'mit Praedikat', namely from grapes which are riper and of better quality than those for table wines, and from 'burgenländischen Seewinkels' near Vienna because of the salty soil and groundwater. For chlorides, the water quality used for irrigation is particularly important, as Savino *et al.* (1994) stressed the apparent contradiction of wines produced in areas relatively close to the sea with lower contents than wines produced in inland but irrigated with brackish waters. National limits are sometimes defined for chloride, but a European harmonization is in progress and has defined the limit of 1 g/L NaCl for wines and 2 g/L for *Marsala*, fortified wines, mute musts by alcohol addition and 'mistella' (Reg. Eu. Cons. No. 1493/1999; Italian G.U. L. 82/2006).

Nitrates

As for nitrates, summarizing the data from several authors, Rapp (1989) reports contents in juices ranging from 1–2 mg/L up to decidedly rarer cases with 26 mg/L. Such values are lower than those reported by Cerutti (1999) for several marketed vegetables. For wines, Lemperle and Lay (1989b) report mean values ranging from about 8 to 15 mg/L and a maximum of 53.7 mg/L NO_3^-. Although a specific legal limit has not been defined for nitrates in wines, content higher than 20 mg/L is sometimes considered to suggest the fraudulent addition of well water or nitric acid and can lead control authorities to carry out closer analysis (Garcia Moruno *et al.*, 1994).

Phosphites and phosphonates

In contrast with the other inorganic anions, knowledge of phosphites and phosphonates in grapes and wine is less extensive and widespread among winemakers. The inorganic and organic salts of phosphorous acid, the latter being available in the form of alkyl compounds or esters of phosphorous acid, are widely used for the control of fungal diseases in plants. Phosphites and phosphonates have both a direct and an indirect effect on fungi and *oomycetes*. They can inhibit the oxidative phosphorylation process in the metabolism of *oomycetes* and have an indirect effect, stimulating the plant's natural defense response against pathogen attack. As phosphites, they are readily absorbed by the leaves and are highly mobile within the plant. None of the plant enzymes described by Ouimette and Coffey (1989; 1990), and Smillie *et al.* (1989) are able to oxidize phosphites into phosphate. This means that phosphorous acid cannot be a substitute for phosphorus fertilization (McDonald *et al.*, 2001). Fosetyl-Al, which releases phosphonate as a product of breakdown, is a systemic fungicide often used against root pathogens for its mobility in the plant (Cohen and Coffey, 1986). Different legal limits for phosphites have been set or are under discussion, waiting for a standardized European decision. For grape for example, to date, a level of 50 mg/kg of phosphonic acid has been set in Switzerland (RS 817.021.23; 10/10/2006) and in Italy (D.M. 27/08/2004, G.U. 292 14/02/2004). A limit for wine has not yet been defined, but may be derived from the limit for grape, taking into account the yield in wine.

8.2. Traceability of *Vitis Vinifera* Related Products on the Basis of Geographical Origin

As has happened for several other products, such as milk, cheese, meat, vegetables and fruit juices discussed by Crews (1998) and Kelly *et al.* (2005), in addition to the various elements, isotope ratios were also measured in wines and spirits, partially changing the more traditional reasons for elemental analysis and focusing on the possibility of differentiating or ascertaining the geographical origin of the products. Interesting literature on this topic as regards wines is given in a recent paper by Castiñeira Gómez *et al.* (2004b). This can be integrated with the pioneering works of Siegmund and Baechmann (1978) and Scarponi *et al.* (1982), as well as in the 1990s with the papers of Herrero-Latorre and Medina (1990), Day *et al.* (1994) on grape musts, Greenough *et al.* (1997), and Martin *et al.* (1999) on Bordeaux wines, and with the recent papers of Suhaj and Korenovska (2006) on Slovak wines and Almeida and Vasconcelos (2001, 2003, 2004) on the discriminative capability of Sr isotopes, previously investigated by Horn *et al.* (1993) in the rock-soil-vine-wine system. Coetzee *et al.* (2005) investigated the isotopic ratios of elements in South African wines from Stellenbosch, Robertson and Swartland in the Western Cape region. European Community (EC) projects have also paid attention to this topic, for example with the SMT4-CT95-7500 and G6RD-CT-2001-00646-WINE DB contracts, the latter to establish a specific database for analytical parameters for wines from non-European Union countries. In the light of the interfering factors presented above, it is quite surprising that geographical discrimination among wines can be attained. In any case, all these studies and research projects have also had important repercussions of a commercial nature, having multiple objectives, including the safeguarding of the product, the producer and the consumer and the repression of frauds, or at least of acting as a significant form of prevention and a deterrent. With regard to this, deterrence is a particularly important aspect, since chemical analysis combined with a statistical approach, such as discriminant analysis, can significantly contribute to 'moralizing' the market through consortia self-control measures, even when the certainty of results ascertaining geographical origin is less than 100%.

8.3. Elemental Analysis

The methods officially used in the wine trade transactions are summarized in Table 8.1. Generally, the OIV methods are officially adopted in the European Union without significant technical changes. The methods reported are mainly colorimetric, titrimetric, or use Atomic Emission Spectroscopy (AES, e.g. Flame Spectrophotometry), Atomic Absorption Spectroscopy (AAS), Hydride Generation-AAS (HG-AAS), Electrothermal-AAS (ET-AAS) and Vapour Atomic Flourescence Spectrophotometry (VAF).

The main pros and cons of the techniques more extensively used for elemental analysis in the oenological field are shown in Table 8.2.

Table 8.1 Comparison of the most relevant characteristics of the analytical methods provided by OIV and AOAC.

OIV methods	AOAC methods
Arsenic (MA-F-AS323-01-ARSENI, MA-F-AS323-01-ASSAA)	
Reference method: after mineralization and reduction using the $NaBH_4$ of wine, it is measured as H_3As at 520 nm by hydride generation – AAS.	
Usual method: after mineralization of the sample and reduction to H_3As, it is colorimetrically quantified at 520 nm.	
Boron (MA-F-AS323-03-BORE) is colorimetrically measured at 420 nm after dealcoholization, colour removal, and reaction with azometine H reagent.	
Cadmium and lead (MA-F-AS322-10-CADMIU, MA-F-AS322-11-PLOMB): they are directly dosed by ETAAS at 228.8 and 283.3 nm respectively.	
Calcium (MA-F-AS322-04-CALCIU) is analyzed by AAS at 422.7 nm after opportune dilution of the sample and addition of $LaCl_3$.	
Copper (MA-F-AS322-06-CUIVRE) is quantified in the diluted wine by AAS at 324.8 nm.	**Copper** (970.18).
	Method I: the diluted samples are analyzed by AAS at 324.7 nm using alcoholized standards.
	Method II: the copper content is colorimetrically measured after reaction with Zn dibenzyldithiocarbamate using a spectrophotometer at 438 nm.
Iron (MA-F-AS322-05-FER).	**Iron** (970.19).
Reference method: it is quantified by AAS in the diluted and dealcoholized wine at 248.3 nm.	*Method I*: a diluted solution of wine is analyzed by AAS at 248.3 nm and the concentration is quantified in comparison with alcoholized standards.
Usual method: it is colorimetrically measured on the H_2O_2 oxidized sample after reaction with an ortophenantroline reagent.	*Method II*: the content of iron is measured after reaction with 2,4,6-tripyridyl-s-triazine with a spectrophotometer at 593 nm.

Table 8.1 (Continued).

OIV methods	AOAC methods
Magnesium (MA-F-AS322-07-MAGNES) is quantified in the diluted wine by AAS at 285 nm.	
Mercury (MA-F-AS323-06-MERCUR) is measured by vapour atomic fluorescence detection at 254 nm.	
Phosphorus (MA-F-AS321-04-PHOTOT) after oxidation and ashing of the sample and reaction with a molybdovanadate reagent, it is colorimetrically measured.	**Phosphorus** (962.11) is colorimetrically quantified at 400 °C after reaction of the mineralized sample with a molybdovanadate reagent.
Potassium and sodium (MA-F-AS322-02-POTASS, MA-F-AS322-03-SODIUM).	**Potassium and sodium** (963.13) sample is analyzed by flame spectrophotometry after dilution and addition of $LiNO_3$.
Reference method: The opportunely diluted and CsCl added sample is analyzed by AAS at 769.9 and 589 nm respectively.	
Usual method: the diluted sample is analyzed using a flame spectrophotometer.	
Silver (MA-F-AS322-09-ARGENT) is measured by AAS at 328.1 following ashing and oxidation.	
Total brome (MA-F-AS321-01-BROTOT). After ashing at 525°C, it is colorimetrically quantified as tetrabromosulfonphtalein at 590 nm.	
Total nitrogen (A MA-F-AS323-02-AZOTOT) is quantified after mineralization, alkalinization and ammonia distillation by acid-base titration (Kjeldhal method).	**Total nitrogen** (920.70) is quantified using the Kjeldhal method.
(MA-F-AS323-02-AZOTDU) is dosed by universal detector following total combustion of the organic substances, reduction of the gases produced and removal of CO_2 and H_2O (Dumas method).	
Zinc (MA-F-AS322-08-ZINC) is quantified by AAS in the diluted and dealcoholized wine at 213.9 nm.	

Table 8.2 Main advantages and disadvantages of the more extensively used techniques for elemental analysis.

	Flame Atomic Absorption Spectrometry (FAAS)	Electrothermal Atomic Absorption Spectrometry (ETAAS) e.g. Graphite Furnace Atomic Absorption Spectrometry (GFAAS)	Inductively Coupled Plasma Optical Emission Spectrometry (ICP-OES)	Inductively Coupled Plasma Mass Spectrometry (ICP-MS)	Anodic Stripping Voltammetry (ASV)
Advantages	Inexpensive; Rapid for few selected elements; Limited use with organic solvents	Relatively inexpensive; Small sample volume; Excellent sensitivity	General-purpose technique; Good dynamic range; Accommodates organic solvents; Multi-elemental technique	Small amount of sample; Excellent dynamic range; Multi-elemental technique; Isotope determination; Semi-quant scanning; Excellent sensitivity; Limited interferences	Low cost of consumables and equipment; High sensitivity
Disadvantages	Poor sensitivity; Single element determination at-the-time; Large amount of sample; Narrow linear range	Single element determination at-the-time; High operating costs as consumables; Very narrow linear range; Time-consuming technique; Not suited for organic solvents	Medium cost of the instrument; Limited sensitivity; Sample volume requirements; Spectral interferences	High cost of the instrument	Few quantifiable elements (Cu, Pb, Cd, Zn)

Today, Inductively Coupled Plasma Optical Emission Spectrometry (ICP-OES) is the best compromise between costs and technical performance for winery quality control laboratories, while Inductively Coupled Plasma Mass Spectrometry (ICP-MS), because of its extreme reliability and productivity, is more widely used in research laboratories.

8.4. Architecture and Components of an Inductively Coupled Plasma Mass Spectrometer

ICP-MS appeared with the first prototypes in 1974, but diffusely it arrived on the international markets at the beginning of eighties. Nowadays, after several thousand instruments have been installed, it can be considered to be a mature and highly powerful technique. ICP-MS has the capability to analyze a wide range of elements in a variety of sample matrices. It is widely employed in Geological and Environmental Sciences, Semi-Conductor Industries, Material Science, Medicine, Agriculture, and Food and Beverages Sciences.

Figure 8.1 shows the vertical section of an Inductively Coupled Plasma Mass Spectrometer.

8.4.1. Sample Introduction System

The physical status of the samples determines the choice of the analytical approach. In particular, liquid samples can be processed by nebulization

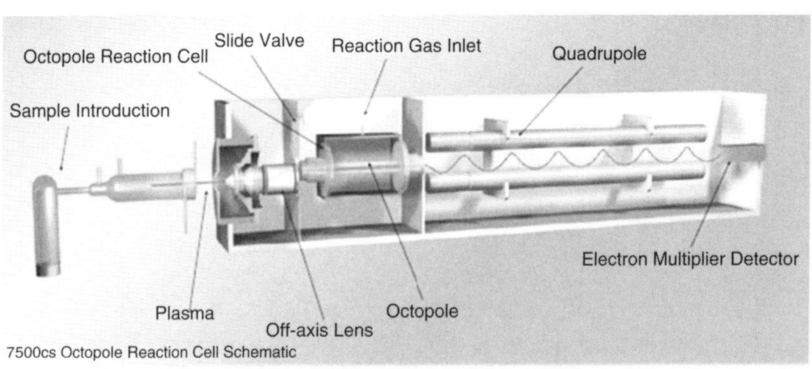

Figure 8.1 Vertical section of an Inductively Coupled Plasma Mass Spectrometer

or gaseous extraction. Solid samples can be analyzed after dissolution and/or mineralization, or by vaporization.

Liquid samples nebulization

The liquid sample introduction system most commonly used on an ICP-MS is very similar to that used on a flame Atomic Absorption Spectrometer or an ICP-OES. Liquid samples can be directly injected using a pneumatic nebulizer and a spray chamber.

Nebulizers

Conventional nebulizers
The ICP-OES nebulizers can aspirate up to 1–2% dissolved solids (a dry wine has 2–4% total extract). Some nebulizers, such as the Babington and cone-spray nebulizers, were studied to handle as much as 15–20% dissolved solids, but are not ideally conceived for use in ICP-MS. Pneumatic nebulizers are commonly used to generate an aerosol of samples with dissolved components below 0.2%. Pneumatic devices are made from glass or different kinds of polymers and use argon as nebulizer gas. The characteristics of some of the most popular pneumatic nebulizers are detailed below.

Concentric design, is indicated for clean solutions and can provide excellent sensitivity and stability, but the small orifices can be affected by occlusion problems, especially in the case of aspiration of large numbers of heavy matrix samples. In this device, the solution is aspirated through a capillary tube by the low-pressure condition produced by a high-speed gas flowing tangentially to the capillary axis. On the nebulizer tip the combination of low pressure and high gas flow breaks up the solution producing an aerosol.

Crossflow design. This nebulizer suffers from relatively reduced analytical sensitivity and precision if compared with the concentric one. It is devised for routine use and is probably the best choice for samples that contain a high concentration of dissolved solids or heterogeneous samples with small contents of undissolved matter. The aerosol is produced on the nebulizer tip where the drained sample collides with a perpendicular jet of argon gas.

Microflow design is based on the concentric principle. Typically it operates with higher gas pressure at less than 0.1 mL/min sample flow rates, while conventional nebulizers usually run at about 1 mL/min. This low flow rate is suitable in the case of applications with limited sample volume or with samples affected by memory effects, as is sometimes possible when analyzing wines with high extract.

Desolvating nebulizer
In this device the liquid sample is sprayed into a heated spray chamber, where the nebulizer gas transfers the aerosol through the membrane desolvator. An argon flow removes the solvent vapour from the exterior of the membrane. If compared to conventional pneumatic nebulizers, this system enhances analyte transport efficiency and limits solvent loading to the plasma. Oxide and hydride polyatomic ion interferences are significantly reduced, improving the detection limits by an order of magnitude.

Spray chambers

Because the plasma discharge is inefficient in dissociating droplets greater than ~10μm in diameter, the primary function of the spray chamber is to allow only the smallest droplets of about 5–10 μm to enter the plasma for dissociation, atomization, and finally ionization of the sample components. Its secondary aim is to reduce the effect of the peristaltic pump pulses on the nebulization process. Some ICP-MS spray chambers are also externally cooled (typically to 2–5 °C) to minimize the amount of solvent going into the plasma. Reduction of oxide species is the main benefit of this technical solution when volatile organic solvents are aspirated (e.g. ethanol in alcoholic beverages).

The *double pass spray chamber* (Scott-type) is the most common design, where the aerosol emerging from the nebulizer is directed into the central tube of the chamber. The larger droplets running the whole length of the tube emerge from it, and fall out by gravity through the drain exit. The fine droplets go back up passing between the chamber outer wall and the central tube and emerge from the spray chamber.

In the *cyclonic spray chamber* droplets are discriminated according to their size by means of a vortex produced by the tangential flow of the nebulizer. Smaller droplets are transferred with the sample aerosol into the ICP-MS, while the larger droplets collide with the walls and exit by gravity via a drain tube. If compared to the Scott design, the cyclonic spray chamber typically shows higher sensitivity.

Liquid samples gasouse extraction (Hydride/cold vapour generation system)

Hydride at ambient temperature (i.e. As, Bi, Ge, Pb, Sb, Se, Sn, Te) or cold vapour generation is a well established technique to enhance sensitivity by separating the analytes into gaseous form. This allows the increase of sample flow rates without loading plasma with water. Sensitivity can be enhanced by more than a factor of 100 and, additionally, some analytical interferences can be resolved. Hydride Generator systems operate with a gas-liquid separator which separates the gaseous phase from the liquid matrix. The modern designs use a porous membrane system in order to enhance gas/liquid separation and reduce noise/signal ratio. Often, the membrane systems can be plagued by clogging and ageing.

Solid sample introduction system

The direct analysis of homogenized solid material can be used if the sample of interest is hard to dissolve, or if the sample pre-treatment increases the risk of contamination or losses of the analyte. Solid samples (e.g. grape stones, skins) can be processed using lasers or heated cells to vaporize the sample. The vaporized sample can be directly introduced into the ICP-MS and measured. This technique is generally characterized by lower sample consumption and higher sensitivity. Solid sampling can be carried out for ICP-MS using electrothermal vaporization (ETV) or laser ablation (LA) sample introduction devices.

Electrothermal vaporization. With ETV/ICP-MS, a small amount of solid sample ($\sim$1 mg) is introduced into a graphite furnace and subjected to a multi-step temperature program for atomization. Additionally, a modifier can be used for thermal stabilization of the analytes or to improve the matrix mineralization. The vapours produced during the vaporization step are transferred into the plasma by argon carrier gas.

Laser ablation. LA/ICP-MS is highly versatile and can analyze any kind of solid material. Compared with conventional ICP-MS, the main advantages of this technique are the fast preparation of solid samples without time-consuming dissolution processes, the certainty of a quantitative analysis, and the reduction of potential contaminations during sample preparation (Longerich *et al.*, 1993; Jarvis and Williams, 1993).

Since only dry gasses are introduced into the plasma a decreased polyatomic interference can be observed.

8.4.2. ICP Torch

The plasma is generated by passing argon through a series of concentric quartz tubes (the ICP torch) that are wrapped at one end by a radio frequency (RF) coil. Energy is supplied to the coil by the RF generator couples with the argon to produce the plasma. RF voltage induces rapid oscillation and collision of argon ions enhancing the temperature of gas to ~10,000 K.

The liquid droplets containing the elements to be determined are dried and heated to gas. As the atoms pass through the plasma, they absorb energy from the plasma and release one electron to form a singly charged ion. The ions then pass through the interface toward the ion lens. The advantages of argon gas are its relatively limited cost and easy availability at very high purity. Moreover, it has a first ionization potential of 15.75 electron volts (eV), higher than the first ionization potential of most other elements (helium, fluorine and neon excluded), and lower than the second ionization potential of most other elements (calcium, strontium and barium excluded).

8.4.3. Mass Spectrometer

Interface

The interface consists of two devices called cones: the sampler and the skimmer cone. They have an opening at the apex of approximately 1 mm in diameter in the sampler, and 0.4 mm in the skimmer cone. The first cone is located next to the plasma and in direct contact with it. The skimmer cone is located several millimetres behind the sampler cone and is in contact with the detector zone. The cones are funnel-like devices and are typically made of nickel or platinum. They are screwed into a water-cooled housing to prevent damage from the plasma heat. The beam of ions and neutrals produced by the plasma at atmospheric pressure is extracted through the sampler cone orifice and further through the skimmer cone orifice into the mass spectrometer region. The orifices must also be small enough to preserve a consistent vacuum behind the interface and large enough to prevent occlusion when samples with high dissolved solids are analyzed. The region between the two cones is evacuated to a pressure of a few Torr in order to remove all the molecules using a combination of mechanical roughing and turbo-molecular pumps.

Ion lens

The ion lens, positioned immediately behind the interface, is responsible for focusing and getting as many of the ions as possible into the mass filter. Since the ions generated in the plasma are nearly all positively charged and have a tendency to repel each other, the lens charged with the same sign of the ions acts as a focusing device, keeping the ion beam from diverging. A particular device placed next to the lens and named 'Photon Stop' or 'Shadow Stop' prevents the photons and non-charged species emitted from the plasma from reaching the detector. The common approach is to place a metal disc in the beam path deflecting ions off-axis.

Collision/Reaction Cell (CRC)

Collision and reaction cell techniques have been used for many years in the study of organic and biological mass spectrometry, but only in the last few years in ICP-MS. The development of collision and reaction cells extended the capability of the technique by allowing the selective attenuation or removal of problematic spectral interferences. Today a variety of collision/reaction cells using various gasses (H_2, He, CH_4, NH_3 . . .) are available, virtually able to eliminate the problems associated with polyatomic interferences for most elements in food and beverage matrices. However, the simultaneous multi-element capability and maximum productivity of ICP-MS is partially reduced by the different CRC tuning conditions required to eliminate a specific interference in a specific matrix.

Quadrupole

This is the device generally used as mass filter in routine analytical instrumentation. It is a tunable filter constituted of four rods approximately 20 cm in length and 1 cm in diameter. It works by allowing only a single mass to charge ratio (m/z) to pass through the detector at a given time. The quadrupole is set up to a combination of voltages and radio frequencies to guide the ions with the selected m/z through the quadrupole into the detector. Ions characterized by a different *m/z* are ejected through the spaces among the rods and pumped out of the quadrupole. By changing the voltage and frequency settings, the quadrupole can sweep across the whole mass range of the Periodic Table.

Detector

The detectors used in ICP-MS are not significantly different from the photomultiplier tube used in AA and ICP-OES. The ions passing through the mass filter strike the active surface of the detector, known as dinode, and generate a measurable electronic signal.

8.4.4. System Controller

It is a dedicated computerized data system that manages all aspects of instrumental control and converts the measured signal intensities into concentrations and generates a final report.

In a quantitative analysis, the concentration of an element is determined by comparing the counts measured for a selected isotope of that element to an external calibration curve. ICP-MS instruments can measure the signal of a single isotope as well as isotope ratios, the latter used in a variety of applications including geological dating of rocks, determining the source of a contaminant, and food traceability studies. Isotope dilution methods can also be applied, enriching the sample with one of the isotopes of the element of interest, that behaves both as a calibration standard and an internal standard. By coupling a chromatographic technique (i.e. HPLC, GC, capillary electrophoresis) to the ICP-MS it is possible to discriminate the chemical form of an element and perform elemental speciation (i.e. Cr(III)/Cr(VI)).

8.5. Analysis of Inorganic Anions

The reference analytical methods for anions in wines and by-products are summarized in Table 8.3. They use titrimetric, gravimetric approaches or selective ion electrodes, but today the most performed and extensively used approach for quality control analysis in oenology is ion chromatography.

Table 8.3 Reference analytical methods for anions in wine and relevant by-products.

OIV methods	AOAC methods
Chlorides (MA-F-SA321-02-CHLORU). Chlorides are quantified by titration with standard $AgNO_3$, using the Ag-AgCl electrode as indicating electrode.	**Chlorides** (966.10) *Method I*: a portion of alkalized wine is evaporated to dryness and ignited at $\leq 500\,°C$. Ashes are extracted with hot water. Chlorides are quantified in two ways. With the first $AgNO_3$ is added to the acidified solution. The precipitate is dried, ignited at 700 °–800 °C and weighed. The second is a volumetric method with the use of $FeNH_4(SO_4)_2$ as an indicator and $AgNO_3$ and KSCN as standard solutions. *Method II*: they are quantified using titration with a $AgNO_3$ standard solution, using an Ag-AgCl indicating electrode.
Fluoride (MA-F-AS321-03-FLUORU) The concentration of fluoride is measured using a fluoride selective ion electrode. The addition of a buffer provides a high constant ionic strength, complexes iron and aluminium and adjusts the pH to reduce $HF{\cdot}HF_2$ formation. Matrix effects are minimized by using standard addition.	**Fluoride** (2003.03) – the concentration of fluoride is measured with the use of a fluoride selective ion electrode.
Sulphates (MA-F-AS321-05-SULFAT). A $BaCl_2$ solution is added to a portion of acidified and boiling wine. The precipitate is washed, dried, ignited at 700–800 °C and weighed.	**Sulphates** (955.26). $BaCl_2$ solution is added to a portion of acidified and boiling wine. The precipitate is washed, dried, ignited at 700–800 °C and weighed.

8.5.1. Ion Chromatography

Ion chromatography (IC) was developed in the 1970s by Small *et al.* (1975). It is a powerful technique for the measurement of ionic species, such as fluoride, chloride, nitrite, nitrate, and sulphate and cations like lithium, sodium, ammonium, and potassium. It is very useful for analysis of anions for which there are no other rapid analytical methods, and it is also commonly used for organic species such as amino acids, amines, and proteins. The most commonly used detection method is conductivity, which can be applied to the detection of inorganic cations and anions in solution. Overcoming the problems relating to the high concentration of the ionic eluent, the conductimetric suppression technique led to a remarkable improvement in the sensitivity of the method. Ion chromatography is a form of liquid chromatography that uses ion exchange resins as a stationary phase to separate atomic or molecular ions. The stationary phase surface displays ionic functional groups that interact with ions of opposite charge. Silica-based stationary phases are unstable in the presence of high salt concentrations and at extreme pH, while the polystyrene divinyl benzene cross-linked polymers are very stable.

Chromatographic separation requires the mobile phase, buffered appropriately by adjusting the pH or ionic concentration, to cause the dissociation of analytes to ions, so that they are free to interact with the stationary phase.

Aqueous solutions like wine, juice and spirits can be directly analyzed after removing interferences, filtration, and dilution. Solid samples are generally homogenized and extracted with water or buffered solutions. Organic liquids can also be extracted, obtaining aqueous solutions of ions for analysis.

8.6. Applications for Grape, Wine and Derivatives

Examples of analytical and technological experiences obtained in everyday practice in an oenology-specialized laboratory are presented. Some compositional aspects of grapes and wines and related analytical problems are updated and discussed, as well as the effects of some winemaking procedures.

8.6.1. Mineral Elements

Composition of grapes

Basic composition and exogenous contributions

The elemental composition of 36 grape samples harvested at technological ripeness from 10 varieties is reported in Figure 8.2. Concentration ranges agree with the other data reported in the literature (Cabanis, 1998; Sponholz, 1989).

The content of some elements is affected by the more or less significant exogenous contributions due to vineyard treatments or powder, soil and

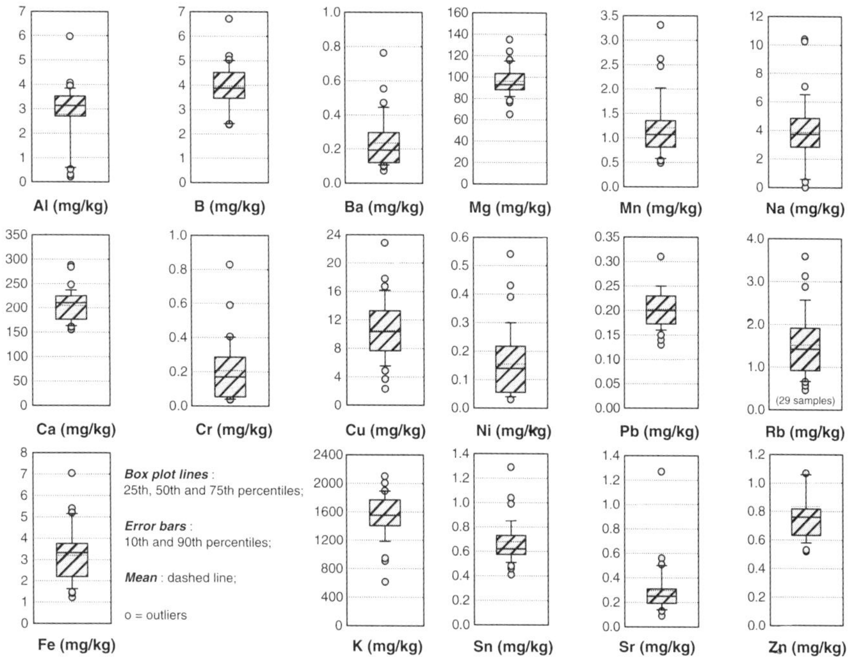

Figure 8.2 Box plots of the amounts of elements in 36 grape samples from 10 different varieties cultivated in northern Italy determined by ICP-MS

polluting deposits via air. This is evident from the data in Table 8.4 where unwashed control grapes were compared with the same grapes washed with a 2% ultra-pure nitric acid solution in high purity water before ICP-MS analysis. As confirmed by both parametric and non-parametric statistical tests, washing clusters significantly decreases the contents of six elements, Cu and Al to a more substantial extent. K, Mg, Pb and Sn also decreased, thus proving the exogenous contribution to their total content in grape.

Table 8.4 Effect of washing grape with 2% nitric acid high purity water solution. Median values are given (n.s. = not significant; $* = p < 0.05$; $*** = p < 0.001$).

element ($mg{\cdot}kg^{-1}$)	washed (n=24)	not washed (n=24)	sign.	element ($mg{\cdot}kg^{-1}$)	washed (n=24)	not washed (n=24)	sign.
Al	0.35	1.13	***	Ni	0.031	0.032	n.s.
Cu	1.44	11.19	***	Pb	0.028	0.054	*
Fe	2.14	2.37	n.s.	Rb	1.39	1.72	n.s.
K	2018	2356	*	Sn	0.004	0.083	*
Mg	74.7	90.4	*	Sr	0.199	0.246	n.s.
Mn	0.82	0.93	n.s.	Zn	0.526	0.558	n.s.
Na	1.36	1.74	n.s.				

Table 8.5 Cu content in wine grapes (mg/kg) and bottled wines (mg/L) from a single region in northern Italy.

	GRAPES						WINES (2002–2005)	
	1998	1999	2000	2004	2005	2006	Red	White
No. Samples	264	231	241	148	158	153	112	145
Min	1.6	1.5	1.9	1.6	1.2	1.1	0.01	0.01
25th percentile	5.8	4.1	5.7	5.9	4.7	3.4	0.07	0.08
Median	8.7	6.7	8.7	8.2	6.5	5.1	0.12	0.14
75th percentile	13.6	10.5	12.6	11.9	10.7	7.5	0.22	0.28
Max	41.8	29.9	32.6	28.7	25.1	19.7	1.34	1.65

Similar results were observed for washing treatments applied on an industrial scale to minimize pest residues, or carried out, for example by organic grape-growers, to reduce natural microbial contamination before crushing and deplete Cu, thus favouring the development of specific aroma compounds related to hyper-reductive winemaking techniques, e.g. 3-mercaptohexan-1-ol and 3-mercaptohexyl acetate and 4-mercapto-4-methylpentan-2-one. The trend toward Cu reducing treatments in the vineyard, as required to minimize soil pollution and sterilization, is clearly shown in particular by the maximum 75th percentile and median values measured in grapes (Table 8.5) where data for extensive sampling carried out in the same region by a public control body are presented. As regards the wines, the measured values are lower than in grapes, as expected, indicating good winemaking technology and no difference was observed in the distribution parameters of reds and whites. Only two samples exceeded the OIV legal limit for wine (1 mg/L).

The problem of residual sulphur

Quantification of the residual sulphur in grapes as a possible precursor of compounds formed during and after fermentation and liable to cause off-flavours is of remarkable interest, permitting better planning of anti-oidium treatments and the establishment of an adequate winemaking strategy. As for any industrial transformation, quality and process control in winemaking requires analytical approaches providing rapid measurement. Several methods have been proposed in the last few decades, applying different analytical techniques such as colorimetry, ICP-OES, AAS and HPLC (Cassidy, 1976; Wenzel *et al.*, 1980; Wenzel,

1980; Thomas *et al.*, 1993; Gump *et al.*, 1996). A new HPLC-DAD method was recently proposed (Larcher *et al.*, 2007), its performance and rapidity adequately meeting the needs of winery laboratories. Two grams of grape previously destemmed, are crushed with a mixer and homogenated with Ultraturrax (10000 rpm + 3 g diatomite) and extracted twice with 50 mL of *n*-hexane. The hexane fractions are combined and concentrated to 2 mL under vacuum at 40 °C, dried under nitrogen flow and recovered with 2 mL of acetonitrile. The analytical conditions are reported in Table 8.6 and a HPLC chromatogram is shown in Figure 8.3.

Table 8.6 Instrumental setting for HPLC-DAD analysis of residual sulphur on grapes.

parameter	description
Elution condition:	isocratic
Column:	C_{18}, 250 × 3 mm (5 μm)
Mobile phase:	100% acetonitrile
Flow rate:	0.7 mL·min^{-1}
Column temperature:	30 °C
Injection volume:	50 μL
Run time:	10 min

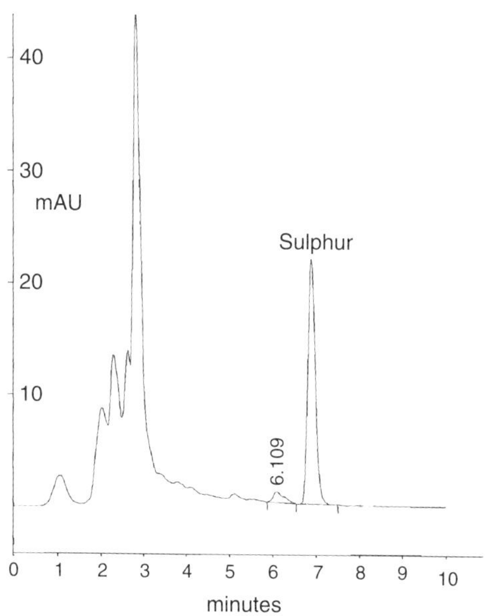

Figure 8.3 HPLC-DAD separation of sulphur in grape extract. The analytical conditions are reported in Table 8.6

The method is linear up to 10 mg/kg and the quantitation limit, assessed according Hubaux and Vos (1970), is 0.15 mg/kg.

Data from a survey of ripe grapes collected on delivery to the wineries for the 2005 vintage in Trentino are shown in Figure 8.4. Despite the seasonal climatic conditions having favoured dangerous *oidium* attacks and led to massive sulphide treatments, all the samples were far below the legal limit (50 mg/kg), showing the maximum value of 27 mg/kg. Only 16 samples (*ca.* 10%) resulted over 2 mg/kg, a level sometimes considered to be an indication of possible technological interference if not further depleted with juice settling.

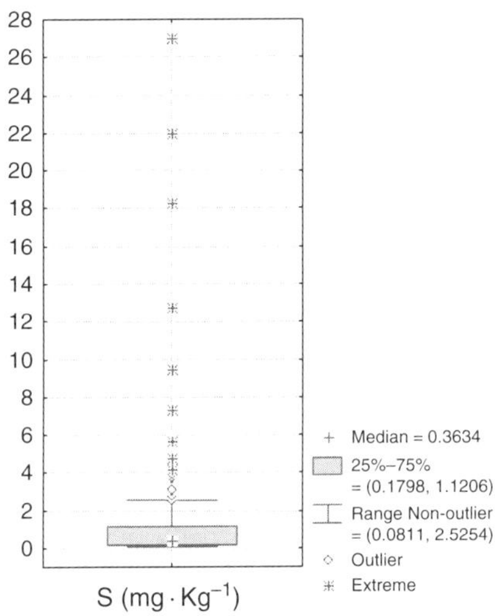

Figure 8.4 Distribution of the contents of residual sulphur in 159 grape samples of the 2005 vintage in Trentino (northern Italy)

Elemental composition of wine and technological interferences

A recent survey of commercially available wines (Larcher and Nicolini, 2001) substantially confirmed the typical ranges reported for each element by Eschnauer (1986), mostly showing values close to (Ba, Ca, Pb, Sn, Sr, Zn) or below (Ag, Cr, Li, Mn, Ni, V) the lower boundaries of the ranges reported by the German author. As compared to white wines, reds have significantly higher Ba, Mg and Ni (*ca.* +20–30%), B, Fe, K, Li, Pb and Sr (*ca.* +40–70%), and Rb (*ca.* +100%). Only Ca was lower (−13% *ca.*) in the reds, mainly as a result of the easier precipitation of salts at the higher pHs typical of such wines. Still whites have significantly lower content of Ba, B, Ca, Cr, K e Sr (*ca.* −10−30%) and Cu (*ca.* −60%) as compared to classic sparkling white wines (Nicolini *et al.*, 2003a), as a result of lower pomace pressing, accurate juice fractioning, higher stabilization and more intense precipitation and/or biosorption phenomena linked to two fermentations.

Over the last 50 years wine-making has seen a positive and significant decrease in the content of Ca, Mg, Fe, Pb and Zn in wines (Larcher and Nicolini, 2001), while there has been a trend towards an increase in K and related instability and pH problems, probably caused by changes in the levels of fertilization and irrigation, and in the management of green covering and canopy (Jackson, 1993; Fregoni, 1998). From here on we will focus on some winemaking practices in order to understand whether and how much they affect the elemental composition of wine and interfere with its 'geographical' characterization.

The effect of bentonites

Bentonite is one of the most important clarifying and fining agents in winemaking. Its use has been investigated frequently as regards the effects on the main elements, but few data are available for those in micro- and trace-quantities (Postel *et al.*, 1986; Enkelmann, 1988; McKinnon *et al.*, 1992; Leske *et al.*, 1995; Bauer *et al.*, 2001; Molina *et al.*, 2001). Bentonite fining (1 g/L) – carried out with nine different marketed products added to wine modified at 3 pH levels – resulted in statistically significant increases in most of the 44 elements measured and significantly lower contents of Cu, K, Rb and Zn. The increases were noteworthy (roughly 10-fold higher) for Ce, Gd, La, Nd, Pr and Y, and also very high for Be, Tl and U (4–6 fold higher); Sm, Eu, Dy, Ho, Er, Tm, Lu, Th, Yb and Hf contents reach measurable amounts only after bentonite treatment. In any case, significant differences among different bentonites were found for Al, Ba, Be, Ca, Co, Cr, Cs, Cu, Fe, Ga, Li, Mg, Mn, Pb, Rb, Sb, Sr, Tl, U, V and Zn, but no significant effect of pH between 3.0 and 4.0 was observed (Nicolini *et al.*, 2004a). An overall increase in rare earth elements (REE) caused by bentonite clarification is also reported by Rossano *et al.* (2007); on the other hand, these authors showed that filtration with cellulose powder only causes a small increase in the concentration of light REEs while filtration using silicate (SiO_2) releases in particular La, Ce, Pr, Nd, and Gd. As known and confirmed by the data in Table 8.7, redrawn by Nicolini *et al.* (2004a), the main oenological changes

Table 8.7 Changes observed in wine after addition of nine different bentonites (1g/L).

element	units	ICP	min	mean	max	element	units	ICP	min	mean	max
Be	$\mu g \cdot L^{-1}$	MS	1.3	5.9	11.8	Ca	$mg \cdot L^{-1}$	OES	2.6	18.3	26.3
Ce	$\mu g \cdot L^{-1}$	MS	2.1	5.8	9.2	Co	$\mu g \cdot L^{-1}$	OES	0.8	1.8	3.1
Cs	$\mu g \cdot L^{-1}$	MS	−0.4	0.3	1.4	Cr	$\mu g \cdot L^{-1}$	OES	−0.3	0.3	2.0
Ga	$\mu g \cdot L^{-1}$	MS	0.3	5.9	17.3	Cu	$mg \cdot L^{-1}$	OES	−0.17	−0.11	−0.08
Gd	$\mu g \cdot L^{-1}$	MS	0.3	1.0	1.8	Fe	$mg \cdot L^{-1}$	OES	0.17	0.45	0.67
La	$\mu g \cdot L^{-1}$	MS	0.9	3.4	6.0	K	$mg \cdot L^{-1}$	OES	−20.4	−8.6	2.6
Nd	$\mu g \cdot L^{-1}$	MS	0.8	3.3	5.5	Li	$\mu g \cdot L^{-1}$	OES	0.4	2.4	9.4
Pb	$\mu g \cdot L^{-1}$	MS	−9.4	3.8	23.0	Mg	$mg \cdot L^{-1}$	OES	−0.95	1.86	4.62
Pr	$\mu g \cdot L^{-1}$	MS	0.2	0.8	1.3	Mn	$mg \cdot L^{-1}$	OES	0.02	0.29	0.51
Sb	$\mu g \cdot L^{-1}$	MS	−0.1	0.2	0.9	Na	$mg \cdot L^{-1}$	OES	−12.0	12.7	29.4
TI	$\mu g \cdot L^{-1}$	MS	0.1	0.3	0.4	Ni	$\mu g \cdot L^{-1}$	OES	0.0	1.6	6.0
U	$\mu g \cdot L^{-1}$	MS	0.3	1.7	4.2	Rb	$mg \cdot L^{-1}$	OES	−0.24	−0.09	−0.03
Y	$\mu g \cdot L^{-1}$	MS	1.3	3.7	6.4	Sn	$\mu g \cdot L^{-1}$	OES	−5.1	1.6	5.9
Al	$mg \cdot L^{-1}$	OES	0.58	1.37	2.43	Sr	$mg \cdot L^{-1}$	OES	0.05	0.10	0.22
B	$mg \cdot L^{-1}$	OES	−0.06	−0.01	0.04	Va	$\mu g \cdot L^{-1}$	OES	−0.1	2.8	6.9
Ba	$mg \cdot L^{-1}$	OES	0.00	0.07	0.20	Zn	$mg \cdot L^{-1}$	OES	−0.03	−0.01	0.01

due to bentonite are in the content of Ca and Na. Speaking in terms of rate of oxygen consumption and potential catalytic activity in real wine (Vivas, 2002), the overall balance is not easily assessable, as Cu decrease and Al increase could counterbalance the increases of Mn and Fe, while the trend of Zn is not well defined.

The effect of yeast hulls

Biosorption is the sole mechanism for quick fixing of cations by non-living yeast cell walls (Blakwell *et al.*, 1995), a mechanism in which protein-polysaccharide complexes are involved. To understand the possible interference of such a mechanism with the mineral composition of wine, two amounts of yeast hulls (180 mg/L and 360 mg/L) were added to stabilized and sterile white and red wines (Nicolini *et al.*, 2004a). The yeast hulls used were provided by two different producers and triplicate experiments were carried out. Both untreated control wines and wines with yeast hulls were gently shaken for three hours, then centrifuged, and supernatant analyzed by ICP-MS. Al, B, Ba, Ca, Ce, Co, Cr, Cs, Cu, Fe, Ga, K, Y, La, Li, Mg, Mn, Na, Ni, Pb, Rb, Sb, Sn, Sr, U, V and Zn were over the relative DLs in all the samples, and the contents of Ce, Cu, Fe, La, Sb, U, V and Y were significantly depleted by the treatment, with a tendency for higher depletion with higher doses (Table 8.8).

Table 8.8 Average depletions (statistically significant, $p < 0.05$) in wine as a consequence of yeast hulls addition.

	Ce ($\mu g{\cdot}L^{-1}$)	Cu ($\mu g{\cdot}L^{-1}$)	Fe ($mg{\cdot}L^{-1}$)	La ($\mu g{\cdot}L^{-1}$)	Sb ($\mu g{\cdot}L^{-1}$)	U ($\mu g{\cdot}L^{-1}$)	V ($\mu g{\cdot}L^{-1}$)	Y ($\mu g{\cdot}L^{-1}$)
yeast hulls (180 $mg{\cdot}L^{-1}$; n = 12)	0.08	15.1	0.30	0.06	0.13	0.10	1.81	0.03
yeast hulls (360 $mg{\cdot}L^{-1}$; n = 12)	0.17	18.1	0.45	0.10	0.13	0.15	2.66	0.05

In percentage terms, the highest depletion was for U (24–36%), followed by Fe (17–25%), Ce (7–15%) and the other elements (<10%).

Tannin and gum arabic as winemaking additives

Oenological tannins are well known clarifying products originally authorized and used to precipitate proteins, thus minimizing the risk of protein haze, but in the last few years they have been added in wine in particular to take advantage of their usefulness as texturizers and colour stabilizers.

Gum arabic is an exudate made up of a mixture of polysaccharides and glycoproteins produced by the trees of different Acacia species growing in sub-Saharan Africa. It is an hydrocolloid traditionally widely used in oenology and the beverage industry as stabilizer, but more recently it has mainly been applied as texturizer. The elemental compositions of 21 tannins (Nicolini *et al.*, 2004b) and 11 arabic gums available on the Italian market – the former from 1993 to 2001 and the latter in 2006 – were studied to assess the hypothetical maximum elemental increases in wine, assuming that such products, used for the new technological purposes mentioned, may remain in wine. The contribution of these additives to the mineral composition of wine is shown in Table 8.9. As regards tannins, and in comparison with the mean values for wine in the literature, the data show that significant increases in some elements could be caused by the addition of certain products in particular for Sn (*ca.* 20-fold higher), U (*ca.* 10-fold higher), Pd (*ca.*, 6-fold higher), Pr and La (*ca.* two-fold higher). Among macroelements of technological or sensorial significance, 7 mg/L of Na could be given to wine by exogenous tannins. For all the other elements changes are insignificant, also in terms of healthiness (e.g. for Pb and Cd).

Gum arabic could significantly increase La, Pr and Th in wine (*ca.* 4–6 fold), also giving about a further 3 mg/L of Na.

Fermenting yeasts effect

While several factors increase the elemental content of wine, only a few among those not specifically directed to metal depletion, can contribute towards reducing the amounts of some elements. As regards this, a biosorption effect of heavy metals by *Saccharomyces cerevisiae* was observed by Volesky and May-Phillips (1995) on a laboratory scale. Changes caused by the yeast strain in the mineral composition of wines achieved under the same real winery conditions on semi-industrial scale in 11 white juices fermented with 4 different commercial strains, have been studied (Nicolini and Larcher, 2003). In particular, of the 20 elements measured, yeast strains were shown to affect the final content of Co, Cu, Mg, Na, Pb, Sr and Zn. It is reasonable to suppose this is related to the different genetic capabilities of the strains to produce H_2S as well as different metabolic needs, e.g. for Zn and Co, beside biosorption on cell walls. Technologically speaking, the most noteworthy average differences among strains were for Cu (0.08 mg/L) and

Table 8.9 Elemental content in arabic gums and tannins and maximum hypothetical increases in wine due to gum (500 mg/L dry weight) and tannin (200 mg/L dry weight) addition assuming absence of precipitations.

Element ($\mu g{\cdot}kg^{-1}$ dry weight)	Average contents in wine ($\mu g{\cdot}L^{-1}$) according to literature	bibl. ref.	GUM ARABIC (n = 11)					TANNIN (n = 21)		
			Powder form (n = 7)		Liquid form (n = 4)		Max.increase ($\mu g{\cdot}L^{-1}$ wine) caused by gum arabic (500 $mg{\cdot}L^{-1}$)	median	max	Max. increase ($\mu g{\cdot}L^{-1}$ wine) caused by tannin addition (200 $mg{\cdot}L^{-1}$)
			min	max	min	max				
Ag	13	f	0.6	4.6	0.3	0.8	0.002			
Al ($mg{\cdot}kg^{-1}$)	0.66 $mg{\cdot}L^{-1}$	a	2.48	32.74	1.04	4.56	0.016 $mg{\cdot}L^{-1}$	17.1	126	0.025 $mg{\cdot}L^{-1}$
As	80	c	0.9	49.9	0.9	19.3	0.025			
B ($mg{\cdot}kg^{-1}$)	4.70 $mg{\cdot}L^{-1}$	a	0.405	0.634	0.114	0.172	0.0003 $mg{\cdot}L^{-1}$	22	118	0.024 $mg{\cdot}L^{-1}$
Ba ($mg{\cdot}kg^{-1}$)	0.08 $mg{\cdot}L^{-1}$	a	1.418	85.90	0.378	3.79	0.043 $mg{\cdot}L^{-1}$	1.0	69	0.014 $mg{\cdot}L^{-1}$
Be	0.08	b	0.8	2.5	0.4	0.8	0.001	4	67	0.013
Bi	0.15	b	1.3	2.6	0.7	0.8	0.001			
Ca ($mg{\cdot}kg^{-1}$)	76.0 $mg{\cdot}L^{-1}$	a	4592	7714	1981	4200	3.857 $mg{\cdot}L^{-1}$	881	4000	0.800 $mg{\cdot}L^{-1}$
Cd	1	b	0.2	0.7	0.1	0.2	0.000	20	146	0.029
Ce	0.02	b	18.5	89.7	5.4	20.5	0.045	11	78	0.016
Co	2.50	a	27.2	72.9	8.3	11.0	0.036	124	442	0.088
Cr	22.1	a	6.9	405	8.4	130	0.203	173	1148	0.230
Cs	2.7	b	5.2	36.3	2.3	18.4	0.018	26.2	506	0.101
Cu ($mg{\cdot}kg^{-1}$)	0.26 $mg{\cdot}L^{-1}$	a	0.83	1.33	0.15	0.52	0.001 $mg{\cdot}L^{-1}$	0.8	3.0	0.001 $mg{\cdot}L^{-1}$
Dy	0.02	b	0.6	6.7	0.3	1.5	0.003	0	6	0.001
Er	0.01	f	0.3	3.6	0.2	0.8	0.002			
Eu	0.005	b	2.1	8.3	0.3	0.6	0.004	0	3	0.001
Fe ($mg{\cdot}kg^{-1}$)	1.93 $mg{\cdot}L^{-1}$	a	3.16	42.76	2.74	11.55	0.021 $mg{\cdot}L^{-1}$	12.4	163	0.033 $mg{\cdot}L^{-1}$
Ga	3	f						84	4754	0.951
Gd	0.02	b	1.3	8.5	0.5	1.8	0.004	1	9	0.002
Ge	0.1	f	1.2	18.8	3.1	7.7	0.009			
Hf	0.7	b	0.5	2.2	0.3	0.6	0.001	16	731	0.146
Hg	0.05	f	0.0	11.5	0.0	0.8	0.006			
Ho	0.004	f	0.1	1.3	0.0	0.3	0.001			
K ($g{\cdot}kg^{-1}$)	1.05 $g{\cdot}L^{-1}$	a	2.49	32.31	2.40	10.03	0.016 $g{\cdot}L^{-1}$	4.2	9.6	0.002 $g{\cdot}L^{-1}$
La	0.006	b	18.3	47.3	3.4	8.7	0.024	5.9	50	0.010
Li	6.74	a	14.5	43.0	6.3	10.5	0.021	198	3443	0.689
Mg ($mg{\cdot}kg^{-1}$)	78.9 $mg{\cdot}L^{-1}$	a	1423	2164	401	700	1.082 $mg{\cdot}L^{-1}$	369	1386	0.277 $mg{\cdot}L^{-1}$
Mn ($mg{\cdot}kg^{-1}$)	0.96 $mg{\cdot}L^{-1}$	a	1.50	7.71	0.62	1.69	0.004 $mg{\cdot}L^{-1}$	11.4	323	0.065 $mg{\cdot}L^{-1}$

Table 8.9 (continued).

Element ($\mu g \cdot kg^{-1}$ dry weight)	Average contents in wine ($\mu g \cdot L^{-1}$) according to literature	bibl. ref.	GUM ARABIC (n = 11)					TANNIN (n = 21)		
			Powder form (n = 7)		Liquid form (n = 4)		Max.increase ($\mu g \cdot L^{-1}$ wine) caused by gum arabic (500 $mg \cdot L^{-1}$)	median	max	Max. increase ($\mu g \cdot L^{-1}$ wine) caused by tannin addition (200 $mg \cdot L^{-1}$)
			min	max	min	max				
Mo	6.00	f	2.4	211	4.9	82.5	0.106			
Na ($mg \cdot kg^{-1}$)	13.4 $mg \cdot L^{-1}$	a	113	5620	20	1893	2.810 $mg \cdot L^{-1}$	573	36230	7.246 $mg \cdot L^{-1}$
Nd	0.04	b	8.4	45.5	2.9	10.5	0.023	5	40	0.008
Ni	24.4	a	56.8	337	15.4	76	0.168	931	3449	0.690
P ($mg \cdot kg^{-1}$)	180 $mg \cdot L^{-1}$	b	4.29	24.93	1.30	8.22	0.012 $mg \cdot L^{-1}$			
Pb	41.9	a	7.1	41.6	1.6	3.5	0.021	59	779	0.156
Pd	0.03	b	0.1	0.9	0.1	0.5	0.000	0	963	0.193
Pr	0.001	b	2.2	10.7	0.7	2.5	0.005	1	10	0.002
Rb ($mg \cdot kg^{-1}$)	2.41 $mg \cdot L^{-1}$	a	2.88	10.52	1.21	3.17	0.005 $mg \cdot L^{-1}$	3	102	0.020 $mg \cdot L^{-1}$
Sb	6	b	0.2	1.8	0.1	1.1	0.001	14	487	0.097
Se	4.6	e	6.0	31.1	6.7	16.3	0.016			
Sm	0.02	b	1.3	9.1	0.5	2.0	0.005	0	8	0.002
Sn	20.9	a	1.5	84.5	2.0	26.2	0.042	0	2436	0.487
Sr	277	a	42526	80049	13173	25038	40.0	1163	6501	1.300
Te	n.a.	f	0.1	0.3	0.0	0.1	0.000	0	29	0.006
Th	0.001	f	2.5	11.8	1.3	3.1	0.006			
Ti	70	d	102	1177	39	263	0.588			
Tl	0.1	b	0.6	2.4	0.2	0.9	0.001	9	108	0.022
U	0.003	f	0.3	6.9	0.9	2.5	0.003	3	145	0.029
V	46.8	a	5.3	1187	21.3	478	0.593	0	216	0.043
Y	0.06	b	4.8	33.4	2.1	7.3	0.017	4.2	25	0.005
Yb	0.04	f	0.1	3.0	0.1	0.6	0.001			
Zn ($mg \cdot kg^{-1}$)	0.51 $mg \cdot L^{-1}$	a	0.32	1.50	0.17	0.55	0.001 $mg \cdot L^{-1}$	1.89	12.4	0.003 $mg \cdot L^{-1}$

(a = Larcher & Nicolini 2001, *Ital. J. Food Sci.*, **13/2**, 237–245;
b = Eschnauer & Neeb 1988, in *Wine Analysis*, Linskens & Jackson (Eds.), Springer-Verlag, Berlin, 67–91;
c = Interesse *et al.*, 1985, Z. Lebens. Unters. Forsch. 181:470–474;
d = Interesse *et al.*, 1984, Z. Lebens. Unters. Forsch. 178: 272–278;
e = Bellanger *et al.*, 1992, *J. Int. Sci. Vigne Vin*, **26**(4) 287–293;
f = Eschnauer *et al.*, 2001, Bull. Erbsloeh;
n.a. = not available).

Zn (0.14 mg/L), as well as for Co (1.7 μg/L) if considered with respect to a very low 'normal' content in wine. In any case, it is interesting to note that even the winemaker's simple choice of the yeast strain to be used for fermentation may affect the final composition of wine elements.

The effects of a specific metal depletor

A vinylimidazole/vinylpyrrolidone co-polymer called Divergan HM rapidly and significantly depleted Cu, Fe, Zn, Al, Cr, B and Ag (Mattivi *et al.*, 2000; Eder *et al.*, 2001; Nicolini *et al.*, 2004c), being more effective when applied in juice than in wine. The highest depletion capabilities are for Cu and Fe (45–90%), but a significant reduction can also be obtained for Zn and Al (25–60%), in particular if the polymer is present in the wine during fermentation. Due to its characteristics, it can be considered an effective alternative to potassium ferrocyanide treatments, particularly against high levels of Cu in juice.

Grappa

This is the typical Italian grape-based pomace distillate. Produced and kept with a higher alcohol content, before sale it is diluted with water to a content of between 38% and 60% alcohol by volume (usually 42–43%) and filtered. As a consequence of water dilution, haze and turbidity of the product may occur. The elemental composition of grappa is related in particular to the stability of the product – mainly caused by Cu, Fe and related colour changes, and Ca and Mg precipitations with sulphates – and the presence of heavy metals eventually leached by distillation apparatus, tanks and pumps, wooden casks, filter pads and glass bottles. A survey of limpid and apparently stabile 58 grappa samples produced in northern Italy, some processed by very small distilleries, is shown in Table 8.10. The data show that the maximum values of several elements are clearly due to the cited leaching phenomena.

8.6.2. Inorganic Anions

Main inorganic anions

A recent investigation performed by ionic chromatography updated information both on the main inorganic anions in grape juice (obtained by crushing and pressing at 2.5 bar) and industrially produced wines

Table 8.10 Elemental contents in 58 grappa samples produced in northern Italy determined by ICP-MS. Elements in concentration above the detection limits in at least 10% of samples are reported.

	Al ($\mu g \cdot L^{-1}$)	As ($\mu g \cdot L^{-1}$)	B ($\mu g \cdot L^{-1}$)	Ba ($\mu g \cdot L^{-1}$)	Ca ($mg \cdot L^{-1}$)	Co ($\mu g \cdot L^{-1}$)	Cr ($\mu g \cdot L^{-1}$)	Cu ($mg \cdot L^{-1}$)	Fe ($\mu g \cdot L^{-1}$)	K ($\mu g \cdot L^{-1}$)	Li ($\mu g \cdot L^{-1}$)
Isotope	27	51	11	137	40	59	52	63	56	39	7
N_0 samples > DL	38	29	41	50	33	12	57	58	25	36	41
25th percentile	<9	<0.3	<10	<0.3	<0.06	<0.3	5.1	3.88	<10	<0.03	<0.3
Median	11.3	0.16	21.0	0.17	0.26	<0.3	6.7	6.24	<10	1.30	0.55
75th percentile	46.3	1.51	55.5	3.18	1.67	<0.3	9.0	9.53	33.9	10.7	0.95
Max	271	21.7	245	11.7	12.4	3.98	25.5	47.4	178	32.9	4.20

	Mg ($mg \cdot L^{-1}$)	Mn ($\mu g \cdot L^{-1}$)	Na ($mg \cdot L^{-1}$)	Ni ($\mu g \cdot L^{-1}$)	Pb ($\mu g \cdot L^{-1}$)	Rb ($\mu g \cdot L^{-1}$)	Se ($\mu g \cdot L^{-1}$)	Sn ($\mu g \cdot L^{-1}$)	Sr ($\mu g \cdot L^{-1}$)	V ($\mu g \cdot L^{-1}$)	Zn ($\mu g \cdot L^{-1}$)
Isotope	26	55	23	60	208	85	78	118	88	51	66
N_0 samples > DL	57	47	53	39	10	36	37	21	47	21	53
25th percentile	0.1	1.2	1.18	<0.3	<0.3	<0.3	<0.3	<0.3	0.5	<0.3	7.9
Median	0.39	9.4	3.18	1.0	<0.3	1.7	0.4	<0.3	2.4	<0.3	26.0
75th percentile	1.06	58.0	24.4	1.9	<0.3	10.4	0.6	0.48	12.0	0.78	45.7
Max	6.19	272	58.3	2052	199.5	37.4	2.4	1.38	45.4	15.6	1902

from the Trentino-Alto Adige region (Nicolini *et al.*, 2002). Results are reported in Table 8.11. The data show slightly higher values of nitrate in juice in comparison with the contents 1.83 ± 0.32 mg/L and 2.16 ± 0.32 mg/L observed in white and red grapes from Sicily, respectively. It was observed the vintage year also influences the anionic composition of juice. Compared with white wines, reds had roughly double the content of chloride and about 40–50% more sulphate and phosphate.

Table 8.11 Inorganic anion content in ripe grape juices and in white (W) and red (R) wines produced in Trentino-Alto Adige region (northern Italy). Analyses were performed by ionic chromatography.

Matrix	Anion ($mg \cdot L^{-1}$)	Wine colour	N°	Min	25° percentile	Median	75° percentile	Max
JUICE	Chloride (NaCl)		321	4.1	13.4	19.0	24.5	55.8
	Phosphate (PO_4^{3-})		309	42	190	260	326	598
	Sulphate (K_2SO_4)		321	92	198	236	278	547
	Nitrate (NO_3^-)		493	<0.2	2.36	3.61	5.23	19.5
WINE	Chloride (NaCl)	W	438	3.8	11.5	18.9	25.1	152
		R	183	8.1	27.3	39.1	52.0	146
		W+R	*621*		*14.7*	*22.4*	*35.4*	
	Phosphate (PO_4^{3-})	W	425	116	253	297	345	1110
		R	170	180	402	444	501	1021
		W+R	*595*		*270*	*327*	*426*	
	Sulphate (K_2SO_4)	W	438	176	345	420	502	855
		R	183	317	522	643	748	1003
		W+R	*621*		*377*	*462*	*590*	

Phosphites and phosphonates

A method of ion chromatography with conductivity detection of ions after eluent suppression for analysis of phosphites (phosphonate ion), was developed (Glenn *et al.*, 1990; Ouimette and Coffe, 1988). It is rapid and relatively simple and shows recoveries are higher than 95% and a detection limit in grapes of 0.5–1.0 ppm is reported.

A survey carried out between 2002 and 2006 on the phosphite contents in grapes and wines from Trentino (Italy) gave the results displayed in Figures 8.5 and 8.6, respectively. In both products, roughly 80% of samples had contents lower than 6 mg/L.

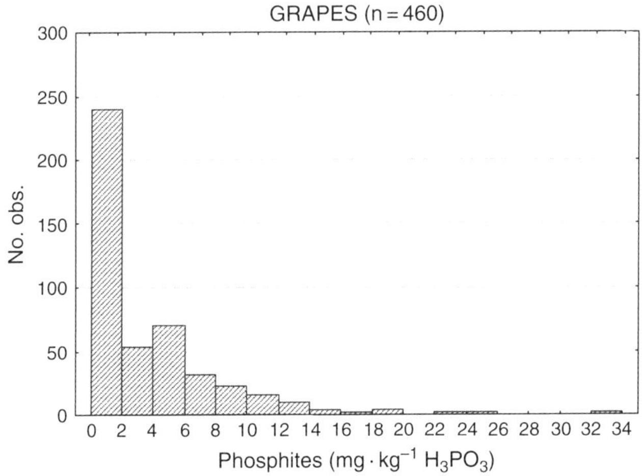

Figure 8.5 Distribution of the content of phosphites in grapes from Trentino (Italy)

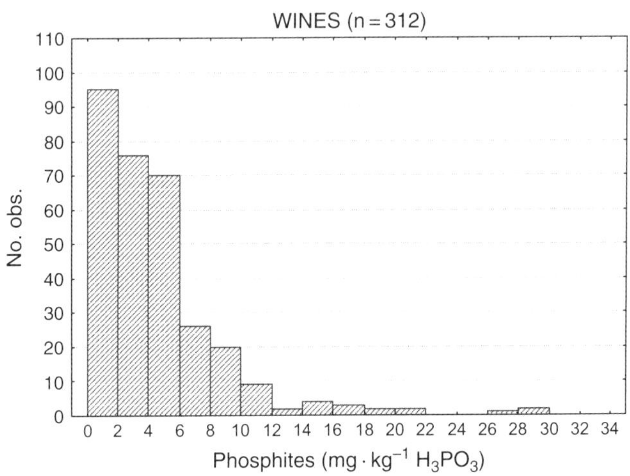

Figure 8.6 Distribution of the content of phosphites in wines from Trentino (Italy)

Data in Table 8.12 relative to phosphite in wine do not evidence differences in the distribution parameters of reds and whites.

Table 8.12 Phosphites in red and white wines from Trentino (Italy).

Phosphites ($mg \cdot L^{-1}$ as H_3PO_3)	No. of Samples	25th percentile	Median	75th percentile	90th percentile	Max
Red	129	<DL	3.60	5.50	9.13	20.0
White	183	<DL	3.88	5.81	9.70	29.8

8.6.3. Geographical Origin of Wines/Grapes

Element-based differentiation

Examples of geographical differentiation are also given for Italian sparkling wines processed according to the classical method in Trentino, Franciacorta and Oltrepò Pavese (Nicolini *et al.*, 2003b). Using parametric and non parametric statistical tests, the content of 15 elements in wines proved to be significantly related to the geographic provenance. In particular, Oltrepò sparking wines proved richer in almost all the measured elements (Figure 8.7), and Li, B, Sr and Rb mostly contributed to discriminate and re-attribute to the origin zones. Similarly, three grape-growing areas in Trentino were distinguished on the basis of the contents of Ba, Co, Li, Mn and Na in 67 commercially available and experimental *Müller-Thurgau* wines of various vintages (data not shown).

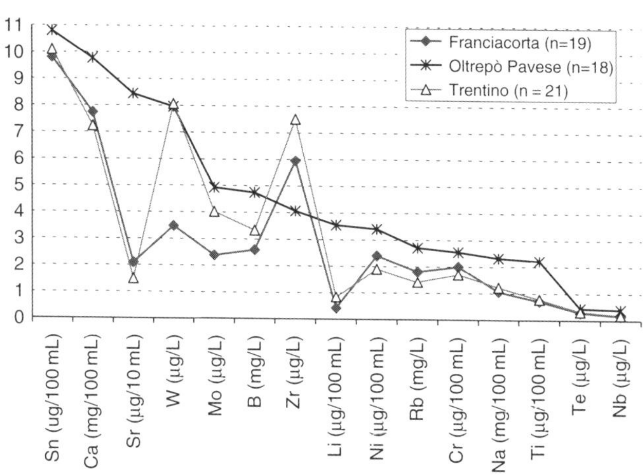

Figure 8.7 Elemental profile of sparkling Italian wines, limited to the elements whose contents are significantly different ($p < 0.05$), for the 3 production areas (Median values are shown. To stress the profile differences *per* geographic origin different measure units are used)

Lead isotope ratios (IRs) have also been used to distinguish wines from different geographical areas (Larcher *et al.*, 2003 and literature cited). However, in the case of genuine Italian wines, lead IRs proved not to be a very sensitive and effective tool for authenticating the origin of wines from different regions, although $^{206}Pb/^{207}Pb$ and $^{208}Pb/^{206}Pb$ could differentiate some macro-regions. In particular, high $^{208}Pb/^{206}Pb$ and low $^{206}Pb/^{207}Pb$ ratios characterized the wines from north-western Italy. The complexity of viticultural orography, population distribution, pollution sources and origin of Pb with different isotope compositions used by the petroleum companies can account for this partial differentiation.

Joint use of several parameters

Data on mineral elements, the stable isotope ratios of light bioelements and inorganic anions can be jointly used for distinguishing or ascertaining geographical origin, to improve discriminant performance. An example was given by Depentori *et al.* (1998) that using the isotope ratios (IRs) of hydrogen on methyl site of ethanol $(D/H)_I$ by Site-specific Natural Isotope Fractionation-Nuclear Magnetic Resonance (SNIF-NMR), oxygen ($^{18}O/^{16}O$), carbonium ($^{13}C/^{12}C$) and nitrogen ($^{15}N/^{14}N$) by Isotope Ratio Mass Spectrometry (IRMS), as well as the content of Na, Mg and chloride in about 1000 genuine Italian wines, differentiated their regional origin. Figure 8.8 shows the improvement in terms of correct reclassification percentage of samples to their geographical area achieved by this joint approach. The average correct reclassification increased from 43.5%, only using isotope measures, to 67.8%, when the mineral elements are also used. Similar results were observed by Day *et al.* (1994) for musts from different French regions, respectively. By determining Ca, Mg, Sr and Zn by AA they improved the overall classification of musts from 78% (with isotope data only) to 89% also considering the stable IRs measured by SNIF-NMR. In wines from the principal production zones in France, Day *et al.* (1995) achieved nearly 100% of correct classification using stable IRs measured by SNIF-NMR and Ba, Sr, Rb and Li by AA and ICP-OES.

Joint use of IRs (D/H; the isotopic ratio $\delta^{13}C‰ = 1000 \times (R_{sample} - R_{reference})/R_{reference}$, calculated on the reference international standard), some elements (Na, Sr, V, Rb) and anions (nitrate, chloride) gave positive results also for distinguishing mute and concentrated musts used for balsamic vinegar production from three Italian regions. Canonic

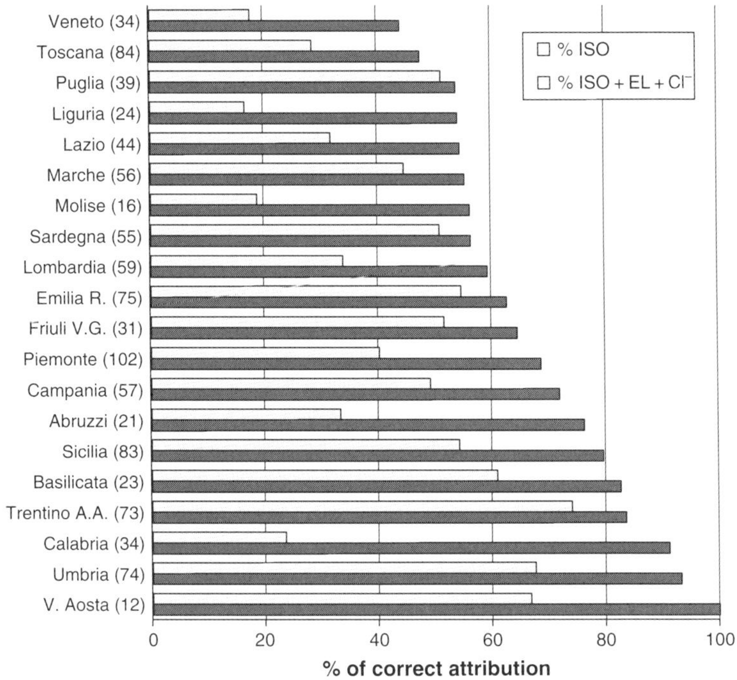

Figure 8.8 Percentages of correct re-attributions to the region of origin of Italian genuine wines using isotopic (ISO) and elemental (EL) parameters as well as chloride (Cl^-). In brackets, the number of samples analyzed is reported

discriminant analysis allowed a correct attribution of 85% of samples to their origin area and only one sample from the south was confused with the northern-B samples. 94% of southern samples were correctly attributed, while a lower performance was obtained for northern areas (Table 8.13). The plot of the canonic discriminant analysis is reported in Figure 8.9.

Table 8.13 Percentages of correct re-attribution to the geographic area of origin of musts used for balsamic vinegar.

	No. samples of certain origin	Percentage of correct re-attribution	North-A	South	North-B
North-A	28	79	22	2	4
South	17	94	0	16	1
North-B	23	87	1	2	20
Total	68	85	23	20	25

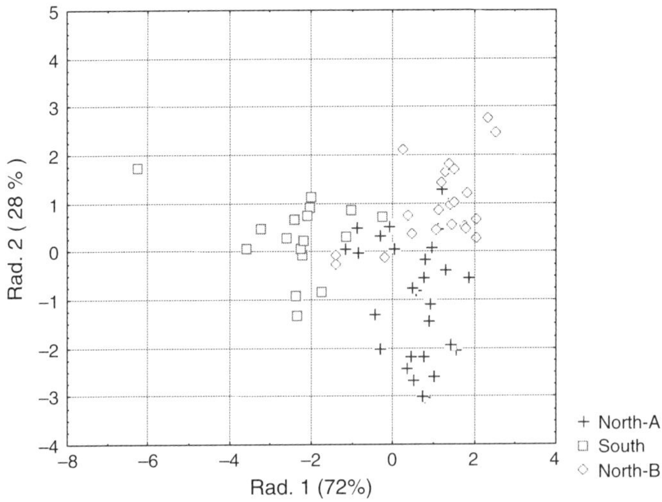

Figure 8.9 Plot of the canonic discriminant analysis carried out for mute and concentrated musts used for balsamic vinegar production from three Italian regions

References

Almeida, C.M.R. and Vasconcelos, M.T.S.D. (2001) ICP-MS determination of strontium isotope ratio in wine in order to be used as a fingerprint of its regional origin. *J. Anal. At. Spectrom.*, **16**, 607–611.

Almeida, C.M.R. and Vasconcelos, M.T.S.D. (2003) Multielement composition of wines and their precursors including provenance soil and their potentialities as fingerprints of wine origin. *J. Agric. Food Chem.*, **51**, 4788–4798.

Almeida, C.M.R. and Vasconcelos, M.T.S.D. (2004) Does the winemaking process influence the wine $^{87}Sr/^{86}Sr$? A case study. *Food Chemistry*, **85**, 7–12.

Angelova, V.R., Ivanov, A.S. and Braikov, D.M. (1999a) Heavy metals (Pb, Cu, Zn and Cd) in the system soil – grapevine – grape. *J. Sci. Food Agric.*, **79**, 713–721.

Angelova, V., Ivanov, A., Braikov, D. and Ivanov, K. (1999b) Heavy metal (Pb, Cu, Zn and Cd) content in wine produced from grape cultivar Mavrud, grown in an industrially polluted region. *J. Int. Sci. Vigne Vin*, **33**, 119–131.

Basile, G. and Tarallo, V. (1974) Influenza del traffico automobilistico sul contenuto in piombo dei vini. *Boll. Lab. Chim. Prov.*, **25(9)**, 185–188.

Bauer, K.H., Eschnauer, H.R. and Görtges, S. (2001) Indicator elements in wine analysis. The ultra-trace elements Beryllium and Zirconium. In *Abstract book 2nd Symp. "In Vino Analytica Scientia 2001"*, p. 30, Ecole Européenne de Chemie Analytique (Ed.).

Blakwell, K.J., Singleton, I. and Tobin, J.M. (1995) Metal cation uptake by yeast: a review. *Appl. Microbiol. Biotechnol.*, **43**, 579–584.

Boulton, R.B., Singleton, V., Bisson, L.F. and Kunkee, R. (Eds.), (1998) The Physical and Chemical Stability of Wine. In *Principles and Practices of Winemaking*, p. 320. Chapman & Hall, NY.

Cabanis, J.C. (1998) Substances minerals. In *Oenologie. Fondements scientifiques et technologiques*, p. 18. TEC&DOC, Paris.

Cabrera-Vique, C., Teissedre, P.L., Cabanis, M.T. and Cabanis, J.C. (2000) Manganese determination in grapes and wines from different regions of France. *Am. J. Enol. Vitic.,* **51**, 103–107.

Cassidy, M. (1976) A selective method for elementar Sulfur analysis by HPLC. *J. Cromatography,* **117**, 71–79.

Castelli, A., Cavallaro, A., Cerutti, G. and Fittipaldi, M. (1974) Pb, Cr, Cd, Zn, Mn, Cu, Fe di vini nazionali: analisi mediante spettrofotometria di assorbimento atomico. *Riv. Vitic. Enol.,* **27**, 247–257.

Castiñeira Gómez, M.M., Brandt, R., Jakubowski, N. and Andersson, J.T. (2004a) Changes of the metal composition in German white wines through the winemaking process. A study of 63 elements by Inductively Coupled Plasma – Mass Spectrometry. *J. Agric. Food Chem.,* **52**, 2953–2961.

Castiñeira Gómez, M.M., Feldmann, I., Jakubowski, N. and Andersson, J.T. (2004b) Classification of German white wines with certified brand of origin by multielement quantitation and pattern recognition techniques. *J. Agric. Food Chem.,* **52**, 2962–2974.

Cerutti, G. (1999) *Residui, additivi e contaminanti degli alimenti,* G. Cerutti (Ed.), Tecniche nuove S.p.A., Milano, pp. 66–75.

Cerutti, G., Ciappellano, S. and Testolin, G. (1990) Alluminio nei vini. Nota preliminare. *Vignevini,* **17**(5), 53–55.

Cerutti, G., Finoli, C., Vecchio, A. and Galkina Benelli, T. (1984) Microelementi in vini di varia età ed origine. *Vignevini,* **11**(**11**), 27–30.

Cerutti, G., Mannino, S. and Vecchio, A. (1981) Vini genuini italiani. Nota 2. Elementi minerali. *Rivista di Viticoltura e di Enologia,* **34**, 145–156.

Ciusa, W. and Giaccio, M. (1972) Il contenuto di oligo elementi nei vini a denominazione di origine controllata. *Boll. Lab. Chim. Prov.,* **23**, 446–468.

Coetzee, P.P., Steffens, F.E., Eiselen, R.J., Augustyn, O.P., Balcaen, L. and Vanhaecke, F. (2005) Multi-element analysis of South African wines by ICP-MS and their classification according to geographical origin. *J. Agric. Food Chem.,* **53**, 5060–5066.

Cohen, Y. and Coffey, M.D. (1986) Systemic fingicides and the control of Oomycetes. *Ann. Rev. Phytopathol.,* **24**, 311–338.

Conde, J.E., Estevez, D., Rodriguez-Bencomo, J.J., Garcia Montelongo, E.J. and Perez-Trujillo, J.P. (2002) Characterization of bottled wines from the Tenerife Island (Spain) by their metal ion concentration. *Ital. J. Food Sci.,* **14**, 375–387.

Crecelius, E.A. (1977) Arsenite and arsenate levels in wine. *Bull. Environ. Contam. Toxicol.,* **18**, 227–230.

Crews, H.M. (1998) Trace element analysis for food authenticity studies. In Ashurst P.R. & Dennis M.J. (Eds.) *Analytical methods of food authentication.* International Thomson Publ., pp 270–291.

Currle, O., Bauer, O., Hofäcker, W., Schumann, F. and Frisch, W. (1983) Mineralstoffhaushalt und Ernaehrungsstoerungen. In *Biologie der Rebe.* D. Meininger Verlag und Druckerei GmbH., pp. 248–277.

Day, M.P., Zhang, B.L. and Martin, G.J. (1994) The use of trace element data to complement stable isotope methods in the characterisation of grape musts. *Am. J. Enol. Vitic.,* **45**, 79–85.

Day, M.P., Zhang, B.L. and Martin, G.J. (1995) Determination of the geographical origin of wine using joint analysis of elemental and isotopic composition. II: Differentiation of the principal production zones in France for the 1990 vintage. *J. Sci. Food Agric.,* **67**, 113–123.

Depentori, D., Larcher, R., Monetti, A., Versini, G. (1998) The improvement of wine characterization through the integration of isotopic and inorganic ion analysis. *FIT-Symp. "The use of Isotopic Techniques for Food Analysis"* (Thematic network SMT4-CT95-7500 European Commission), Norwich.

Eder, R., Schreiner, A., Schlager, G. and Pischinger, K. (2001) Anwendung von metallselektiven Harzen zur Weinstabilisierung. *Comm. to Groupe d'Experts "Technologie du Vin"*, O.I.V., Paris, 27 Mars.

Elinder, C.G., Lind, B., Nilsson, B. and Oskarsson, A. (1988) Wine – an important source of lead exposure. *Food Addit. Contam.*, **5**, 641–644.

Enkelmann, R. (1988) Schwermetall-Abgabe von Weinbehandlungsmitteln. 1. Mitteilung: Bentonite. *Dtsch. Lebensm.-Rdsch.* **84(8)**, 243–247.

Eschnauer, H. (1982) Trace elements in must and wine: primary and secondary contents. *Am. J. Enol. Vitic.*, **33**, 226–230.

Eschnauer, H. (1986) Spurenelemente und Ultra- Spurenelemente in Wein. *Naturwissenschaften*, **73**, 281–290.

Eschnauer, H. and Scollary, G.R. (1995) Aluminium in Wein. *Vitic. Enol. Sci.*, **50(1)**, 24–30.

Eschnauer, H. and Neeb, R. (1988) Micro-Element Analysis in Wine and Grapes. In H.F. Linskens and J.F. Jackson (Eds.), *'Wine Analysis'*, Springer-Verlag, Berlin, pp. 67–91.

Eschnauer, H.R., Bauer, K.H. and Görtges, S. (2001) Den Spurenelementen des Weines auf der Spur. *Bulletin 2001, Erbslöh,* Geisenheim (seigmar.goertges@erbsloeh-geisenheim.de).

Finoli, C., Galkina–Benelli, T. and Vecchio, A. (1986) Microelementi del vino. *Imbottigliamento*, **9(11)**, 39–41.

Flanzy, C. (Ed.) (1998) *Oenologie. Fondements scientifiques et technologiques.* TEC & DOC, Paris.

Flak, W. and Schaber, R. (1989) Die quantitative Bestimmung von anorganischen Anionen in Wein mittels Einsäulenionenchromatographie, *Mitt. Klosterneuburg*, **39**, 175–179.

Fregoni, M. (1998) *Viticoltura di Qualità*, L'Informatore Agrario Edizioni, Verona, Italy.

Fussnegger, B., Maurer, R. and Detering, J. (1992) Unlösliche, komplexbildende Polymere als potentielle Substitutionsprodukte für Kaliumhexacyanoferrat (II) zur Schwermetallverminderung in Wein. *Vitic. Enol. Sci.*, **47(1)**, 8–23.

Gabrielli Favretto, L., Maraspin, M. and Vojnovic, D. (1986) Indagine sull'inquinamento da cadmio e piombo dei chicchi d'uva e del vino. *Riv. Soc. Ital. Sci. Aliment.*, **15(1–2)**, 63–68.

Garcia Moruno, E., Morando, A. and Buzzi, C. (1994) Contenuto in nitrati nei vini da tavola del Piemonte. *L'Enotecnico*, **30(1–2)**, 69–73.

Garoglio, R. (1981) Alterazioni dei vini: difetti e malattie. Ch. 18. In *Nuova Enologia*, AEB Edizioni, Brescia, Italy, p. 425.

Glenn, T.J., Biggins, M.R. and Magarey, P.A. (1990) Rapid, quantitative detection of phosphonate by Simple Ion-Exchange Chromatography using post-separation suppression. *Aus. Plant Path.*, **19(4)**, 139–140.

Green, A.M. and Scollary, G.R. (2000) Influence of metal ions on lead complexation in wine. *Aus. J. Grape Wine Res.*, **6**, 197–202.

Greenough, J.D., Longerich, H.P. and Jackson, S.E. (1997) Element fingerprinting of Okanagan Valley wines using ICP-MS: Relationships between wine composition, vineyard and wine colour. *Aus. J. Grape Wine Res.*, **3**, 75–83.

Gulson, B.L., Lee, T.H., Mizon, K.J., Korsch, M.J. and Eschnauer, H.R. (1992) The application of lead isotope ratios to determine the contribution of the tin-lead to the lead content of wine. *Am. J. Enol. Vitic.*, **43**, 180–190.

Gump, B., Wahlstrom, V. and Pham, R. (1996) Determination of Sulfur residues on grapes using Flame Atomic Absorption Spectroscopy. *Acta Horticulture*, **427**, 369–378.

Handson, P.D. (1984) Lead and Arsenic levels in wines produced from vineyards where lead arsenate sprays are used for caterpillar control. *J. Sci. Food Agric.*, **35**, 215–218.

Herrero-Latorre, C. and Medina, B. (1990) Utilisation de quelques éléments minéraux dans la différenciation des vins de Galice de ceux d'autres région d'Espagne. *J. Int. Sci. Vigne Vin*, **24**, 147–156.

Horn, P., Schaaf, P., Holbach, B., Hölzl, S. and Eschnauer, H. (1993) $^{87}Sr/^{86}Sr$ from rock and soil into vine and wine *Z. Lebensm. Unters. Forsch.*, **196**, 407–409.

Hubaux, A., Vos, G. (1970) Decision and detection limits for linear calibration curves. *Anal. Chem.*, **42**, 849–855.

Interesse, F.S., D'Avella, G., Alloggio, V. and Lamparelli, F. (1985) Mineral contents of some southern Italian wines. II. Determination of Li, Na, Mg, K, Ca, Co, As, Rb, Sr, Ag, Sb, Ba. *Z. Lebensm. Unters. Forsch.*, **181**, 470–474.

Interesse, F.S., Lamparelli, F. and Alloggio, V. (1984) Mineral contents of some southern Italian wines. I. Determination of B, Al, Si, Ti, V, Cr, Mn, Fe, Ni, Cu, Zn, Mo, Sn, Pb by Inductively Coupled Plasma Atomic Emission Spectrometry (ICP-AES). *Z. Lebensm. Unters. Forsch.*, **178**, 272–278.

Jackson, R.S. (1993) *Wine Science: Principles and Applications*, Academic Press, San Diego, CA, pp. 246–258.

Jarvis, K.E. and Williams, J.G. (1993) Laser ablation inductively coupled plasma mass spectrometry (LA-ICP-MS): A rapid technique for the direct, quantitative determination of major, trace, and rare-earth elements in geological samples. *Chem. Geol.*, **106**, 251–262.

Karagiannidis, N. and Nikolaou, N. (2000) Influence of arbuscular mycorrhizae on heavy metal (Pn and Cd) uptake, growth, and chemical composition of Vitis vinifera L. (cv. Razaki). *Am. J. Enol. Vitic.*, **51**, 269–275.

Kelly, S., Heaton, K. and Hoogewerff, J. (2005) Tracing the geographical origin of food: the application of multi-element and multi-isotope analysis. *Trends Food Sci. Technol.*, **16**, 555–567.

Kern, M. and Wucherpfennig, K. (1991) Entfernung von Eisen, Kupfer und Zink aus Weinen mit einene Chelatharz – ein Alternative zur Blauschönung. *Vitic. Enol. Sci.*, **46**(4), 69–77.

Kern, M.J. and Wucherpfennig, K. (1993) Anwendung selektiver Komplexbildner zur Schwermetallentfernung aus Wein. *Vitic. Enol. Sci.*, **48**(1), 20–26.

Larcher, R., Nicolini, G. and Pangrazzi, P. (2003) Isotope ratios of lead in Italian wines by Inductively Coupled Plasma Mass Spectrometry. *J. Agric. Food Chem.*, **51**, 5956–5961.

Larcher, R. and Nicolini, G. (2001) Survey of 22 mineral elements in wines from Trentino (Italy) using ICP-OES. *Ital. J. Food Sci.*, **13**, 237–245.

Larcher, R., Slaghenaufi, D., Ramponi, M., Grazzi, R. and Osler, R. (2007) Determinazione HPLC-DAD dello zolfo residuo in uve. Metodi rapidi per i controlli di processo in enologia. *Proc. "Innovazione ed eccellenza. Enoforum 2007"*, Piacenza, p. 69.

Lemperle, E. and Lay, H. (1989a) Mineralstoffe. In Würdig G. and Woller R. (Eds.), *Chemie des Weines*, Ulmer, Stuttgart, D., p. 555.

Lemperle, E. and Lay, H. (1989b) Nitrat. In Würdig G. and Woller R. (Eds.), *Chemie des Weines*. Ulmer, Stuttgart, D., p. 569.

Leske, P.A., Sas, A.N., Coulter, A.D., Stockley, C.S. and Lee, T.H. (1997) The composition of Australian grape juice: chloride, sodium and sulfate ions. *Aus. J. Grape Wine Res.*, **3**, 26–30.

Leske, P.A., Bruer, N.G.C., Capdeboscq, V. (1995) An evaluation of some characteristics of commercial bentonites. *Wine Industry Journal,* **10(1)**, 73–77.

Longerich, H.P., Jackson, S.E., Fryer, B.J. and Strong, D.F. (1993) The laser ablation microprobe-inductively coupled plasma-mass spectrometer. *Geosci. Canada*, **20**, 21–27.

Margheri, G. and Rigotti, M. (1964) Le sostanze minerali dei vini della provincia di Trento. *Riv. Vitic. Enol.*, **12**, 405–414.

Marini, S. and Gallese, G. (1991) Pb, Cu, Zn, Ni, Cr, Cd nei vini. Indagine sul mercato di Rimini. *Industrie delle Bevande,* **112**, 129.

Martin, G.J., Mazure, M., Jouitteau, C., Martin, Y.-L., Aguile, L. and Allain, P. (1999) Characterization of the geographic origin of Bordeaux wines by a combined use of isotopic and trace element measurements. *Am. J. Enol. Vitic.*, **50**, 409–417.

Mattivi, F., Nicolini, G. and Versini, G. (1994) Un nuovo polimero ad azione demetallizzante: risultati di alcune prove applicative sperimentali. *L'Enotecnico*, **30(10)**, 75–84.

Mattivi, F., Versini, G. and Nicolini, G. (2000) Interesting side-effects of the use of a PVI-PVP copolymer for removing heavy metals in white and red wines on polyphenols and colour characteristics. *Vitic. Enol. Sci.*, **55(2)**, 73–79.

McDonald, A.E., Grant, B.R. and Plaxton, W.C. (2001) Phosphite (phosphorous acid): Its relevance in the environment and agriculture and influence on plant phosphate starvation response. *J. Plant Nutr.*, **24**, 1505–1519.

McKinnon, A.J., Cattrall, R.W. and Scollary, G.R. (1992) Aluminum in wine. Its measurement and identification of major sources. *Am. J. Enol. Vitic.*, **43**, 166–170.

Médina, B., Augagneur, S., Barbaste, M., Grousset, F.E. and Buat-Ménard, P. (2000) Influence of atmospheric pollution on the lead content of wines. *Food Addit. Contam.*, **17**, 435–445.

Médina, B., Guimberteau, G. and Sudraud, P. (1977) Dosage du plomb dans les vins; une cause d'enrichissement: les capsules de surbouchage. *Conn. Vigne Vin*, **11**, 183–193.

Medina, B. and Sudraud, P. (1980) Teneur des vins en crome et en nickel. Causes d'enrichissement. *Conn. Vigne Vin*, **14(2)**, 79–96.

Mínguez, S., Hernandez, P., García, J., Médina, B., Bartra, E., Vilavella, M. and Espinás, E. (1997) Presence du plomb dans les vins et efficacité des techniques soustractives. *Comm. Groupe d'Experts 'Technologie du vin'*, OIV, Paris, 8–9 Sept.

Mínguez, S., Hernandez, P. and Gonzalo, M. (1996) Selective extraction of lead from wine with ionic exchange resins. In A. Lonvaud-Funel (Ed.) *Oenologie'95. Proc. 5e Symp. Int. d'Oenologie*, TEC&DOC, Paris, France, pp. 632–636.

Molina, R., Mingot, J., Giner, N., Revenga, E. (2001) Influencia de la clarificación con bentonita sobre ciertos metales pesados en el vino. *Tecnología del Vino*, **1(1)**, 39–47.

Nakajima, A. and Sakaguchi, T. (1986) Selective accumulation of heavy metals by microorganisms. *Appl. Microbiol. Biotechnol.*, **24**, 59–64.

Nakajima, A. and Sakaguchi, T. (1993) Accumulation of uranium by basidiomycetes. *Appl. Microbiol. Biotechnol.*, **38**, 574–578.

Nicolini, G., Larcher, R. and Fiorini, M. (2002) Indagine sulla composizione in anioni inorganici di mosti e vini. *Riv. Vitic. Enol.*, **55(2–3)**, 39–54.

Nicolini, G. and Larcher, R. (2003) Evidence of changes in the micro-element composition of wine due to the yeast strain. *Riv. Vitic. Enol.*, **56(4)**, 45–48.

Nicolini, G., Larcher, R. and Bontempo, L. (2003a) Caratterizzazione della composizione di base e del contenuto di elementi in micro-quantità ed in tracce in spumanti italiani elaborati con il metodo classico. *Riv. Vitic. Enol.*, **56(4)**, 29–44.

Nicolini, G., Larcher, R. and Bontempo, L. (2003b) Composizione di base e profilo minerale di spumanti classici italiani. *L'Enologo*, **39(12)**, 111–115.

Nicolini, G., Larcher, R., Pangrazzi, P. and Bontempo, L. (2004a) Changes in the contents of micro- and trace-elements in wine due to winemaking treatments. *Vitis*, **43**, 41–45.

Nicolini, G., Larcher, R. and Bontempo, L. (2004b) Micro- and trace-element composition of tannins used in winemaking. *J. Commodity Sci.*, **43(1)**, 21–29.

Nicolini, G., Larcher, R. and Mattivi, F. (2004c) Experiments concerning metal depletion in must and wine by Divergan HM™. *Mitt. Klosterneuburg*, **54**, 25–32.

Norris, P.R. and Kelly, D.P. (1977) Accumulation of cadmium and cobalt by *Saccharomyces cerevisiae*. *J. Gen. Microbiol.*, **99**, 317–324.

Ough, C.S. (1993) Lead in Wines – A Review of Recent Reports. *Am. J. Enol. Vitic.*, **44**, 464–467.

Ough, C.S. and Amerine, M.A. (1988) *Methods for Analysis of Musts and Wines. 2nd Edition*. John Wiley & Sons, Inc., NY, USA, pp. 288–296.

Ouimette, D.G. and Coffey, M.D. (1988) Quantitative analysis of organic phosphonates, phosphonate, and other inorganic anions in plants and soil by using high-performance ion chromatography. *Phytopathology*, **78**, 1150–1155.

Ouimette, D.G. and Coffey, M.D. (1989) Phosphonate levels in avocado seedlings and soil following treatment with fosetyl-A1 or potassium phosphonate. *Plant Dis.*, **73**, 212–218.

Ouimette, D.G. and Coffey, M.D. (1990) Symplastic entry and phloem translocation of phosphonate. *Pestic Biochem Physiol.*, **38**, 18–25.

Pellerin, P., O'Neill, M.A., Pierre, C., Cabanis, M.T., Darvill, A.G. Albersheim, P. and Moutounet, M. (1997) Complexation du plomb dans les vins par les dimères de rhamnogalacturonane II, un polysaccharide pectique du raisin. *J. Int. Sci. Vigne Vin*, **31(1)**, 33–41.

Pertoldi Marletta, G., Gabrielli Favretto, L. and Favretto, L. (1986) Cadmium in roadside grapes. *J. Sci. Food Agric.*, **37**, 1091–1096.

Pertoldi Marletta, G., Gabrielli Favretto, L. and Favretto, L. (1989) Chromium and nickel in roadside grapes. *Food Addit. Contam.*, **6**, 219–225.

Pinamonti, F., Nicolini, G., Dalpiaz, A., Stringari, G. and Zorzi, G. (1999) Compost use in viticulture: Effect on heavy metal levels in soil and plants. *Commun. Soil Sci. Plant Anal.*, **30**, 1531–1549.

Prosperi, S. and Piga, C. (1988) Contenuto in rame, zinco e piombo nei vini prodotti in Sardegna. *Riv. Vitic. Enol.*, **41**, 375–386.

Pocock, K.F., Alexander, G.M., Hayasaka, Y., Jones, P.R. and Waters, E.J. (2007) Sulfate – a candidate for the missing essential factor that is required for the formation of protein haze in white wine. *J. Agric. Food Chem.*, **55**, 1799–1807.

Postel, W., Meier, B. and Markert, R. (1986) Einfluss verschiedener Behandlungsstoffe auf den Gehalt des Weins an Mengen- und Spurenelementen. I. Bentonit. *Mitt. Klosterneuburg*, **36(1)**, 20–27.

Rankine, B. (1983) Aluminum haze in wine. *Austral. Grapegrower Winemaker*, **234**, 18.

Rapp, A. (1989) Ammonium und Nitrate. In G. Würdig and R. Woller (Eds.) *Chemie des Weines*, Eugen Ulmer Verlag, Stuttgart, pp. 76–77.

Ribéreau-Gayon, P. (Ed.) (1947) *Traité d'Oenologie*. Béranger, Paris.

Ribéreau-Gayon, P., Peynaud, E., Ribéreau-Gayon, P., Sudraud, P. and A. Amati A. (Eds.) (1980) Precipitazioni del vino. In *Trattato di Scienza e Tecnica Enologica. III Volume. La Vinificazione. Le Trasformazioni del Vino*, Edizioni AEB, Brescia, Italy, p. 372.

Rossano, E.C., Szilágyi, Z., Malorni, A. and Pocsfalvi, G. (2007) Influence of winemaking practices on the concentration of rare earth elements in white wines studied by Inductively Coupled Plasma Mass Spectrometry. *J. Agric. Food Chem.*, **55**, 311–317.

Salaha-Moutsopoulou, M.J. and Voudouri-Tsoukala, M. (1991) Chloride and sodium in wines. Problems in international exchanges. *Bull. O.I.V.*, **723–724**, 363–388.

Savino, M., Piracci, A. and Pastore, C. (1994) I cloruri ed il sodio nei vini bianchi pugliesi. *L'Enotecnico*, **30**(**4**), 49–55.

Scarponi, G., Moret, I., Capodaglio, G. and Cescon, P. (1982) Multiple discriminant analysis in the analytical differentiation of Venetian wines. 3. A reelaboration with addition of data from samples of 1979 vintage Prosecco wine. *J. Agric. Food Chem.*, **30**, 1135–1140.

Scherlock, J.C., Pickford, C.J. and White, G.F. (1986) Lead in alcoholic beverages. *Food Addit. Contam.*, **3**, 347–354.

Šebecić, B., Pavašić-Strache, D. and Vedrina-Dragojević, I. (1998) Trace elements in wine from Croatia. *Deutsche Lebensm. Rundschau*, **94**, 341–344.

Seppi, A. and Sperandio, A. (1978) I metalli dei vini. Teoria e verifica sperimentale della corrosione delle attrezzature metalliche di cantina. *Vini d'Italia*, **20** (**112**), 35–42.

Siegmund, H. and Baechmann, K. (1978) Application of numerical taxonomy to the classification of wines. *Z. Lebensm. Unters. Forsch.*, **166**, 298–303.

Small, H., Stevens, T.S. and Bauman, W.C. (1975) Novel ion exchange chromatographic method using conductimetric detection. *Anal. Chem.*, **47**, 1801–1809.

Smart, G.A., Pickford, C.J. and Sherlock, J.C. (1990) Lead in alcoholic beverages: a second survey. *Food Addit. Contam.*, **7**, 93–99.

Smillie, R., Grant, B. and Guest, D. (1989) The mode of action of phosphite: evidence for both direct and indirect modes of action on three Phytophthora spp. in plants. *Phytopathology*, **79**, 921–926.

Sponholz, W.R. (1989) Mineralstoffe. In G. Würdig and R. Woller (Eds.), *Chemie des Weines*, Ulmer, Stuttgart, D., p. 94.

Stefani, R. (1978) Il contenuto in piombo dei vini del Trentino-Alto Adige. *Vini d'Italia*, **36**, 343–345.

Suhaj, M. and Korenovska, M. (2006) Distribution of selected elements as wine origin markers in the wine-making products. *Czech J. Food Sci.*, **24**(**5**), 232–240.

Teissèdre, P.L. and Cabanis, J.C. (1993) Origine et réduction des teneurs en plomb dans les vins. *Rev. Fr. Oenologie, Cahiers Techniques*, **142**, 41–44.

Teissèdre, P.L., Cros, G., Krosniak, M., Portet, K., Serrano, J.J., and Cabanis, J.C. (1996b) Contribution of wine in vanadium dietary intake: geographical origin has a signicant impact on wine vanadium levels. In Ph. Collery, J. Corbella, J.L. Domingo, J.C. Etienne, and J.M. Llobet (Eds.), *'Metal Ions in Biology and Medicine', vol. 4*, John Libbey Eurotext, Paris, p. 183.

Teissèdre, P.L., Cabanis, M.T., Champagnol, F. and Cabanis, J.C. (1994) Lead distribution in grape berries. *Am. J. Enol. Vitic.*, **45**: 220–228.

Teissèdre, P.L., Cabrera, M. and Cabanis, J.C. (1996a) Wine contribution to chromium and nickel dietary intake and their biodisponibilities. In Ph. Collery, J. Corbella, J.L. Domingo, J.C. Etienne and J.M. Llobet (Eds.) *Metal Ions in Biology and Medicine vol. 4*, John Libbey Eurotext, Paris. p. 172.

Teissèdre, P.L., Cabanis, M.T., Daumas, F. and Cabanis, J.C. (1993) Evolution de la teneur en plomb au cours de l'elaboration des vins des Côtes du Rhône et de la Vallée du Rhône. *Rev. Fr. Oenologie, Cahiers Scientifiques*, **140**, 249–254.

Terrés, C., Navarro, M., Martín-Lagos, F., Giménez, R. López, H. and López, M.C. (2001) Zinc level in foods from southeastern Spain: relationship to daily dietary intake. *Food Addit. Contam.*, **18**, 687–695.

Thomas, C.S., Gubler, D.W., Silacci, M.W. and Miller, R. (1993) Changes in the elemental sulfur residues on Pinot noir and Cabernet Sauvignon grape berries during the growing season. *Am. J. Enol. Vitic.*, **44**, 205–210.

Tokalioglu, S., Kartal, S. and Gunes, A. (2004) Statistical evaluation of bioavailability of metals to grapes growing in contaminated vineyard soils using single extractants. *Int. J. Environ. Anal. Chem.*, **84**, 691–705.

Vivas, N. (2002) *Les oxidations et les réductions dans les mouts et les vins*. Editions Féret, Bordeaux, France, p. 120, p. 146.

Volesky, B., May, H. and Holan, Z.R. (1993) Cadmium biosorption by *Saccharomyces cerevisiae*. *Biotechnol. Bioeng.*, **41**, 826–829.

Volesky, B. and May-Phillips, H.A. (1995) Biosorption of heavy metals by *Saccharomyces cerevisiae*. *Appl. Microbiol. Biotechnol.*, **42**, 797–806.

White, C. and Gadd, G.M. (1987) The uptake and cellular distribution of zinc in *Saccharomyces cerevisiae*. *J. Gen. Microbiol.*, **133**, 727–737.

Wenzel, K. (1980) The HPLC determination of elemental sulfur in foodstuffs. *Lebensm. Unters. Forsch.*, **170**, 5–6.

Wenzel, K., Dittrich, H.H., Seyffardt, H.P. and Bohnert, J. (1980) Sulphur residues on grapes and in must and their influence on the H_2S formation. *Wein-Wiss*, **35**, 414–420.

Wucherpfennig, K. and Chen Hsueh-err, Bilanz der Inhaltsstoffe. (1983) Verhalten von Mineralstoffen beim Entsaften von Fruechten. *Fluessiges Obst*, **50**, 8–12, 15–22.

Wurzinger, A., Kohlhuber, S. and Bandion, F. (1992) Zur Beurteilung des Chloridgehaltes in Österreichischen Weinen. *Mitt. Klosterneuburg*, **42**, 10–12.

Zee, J.A., Szöghy, I.M., Simard, R.E. and Tremblay, J. (1983) Elemental analysis of Canadian, European and American wines by Photon Induced X-Ray Fluorescence. *Am. J. Enol. Vitic.*, **34**, 152–156.

Index

Hyphenated Techniques in Grape and Wine Chemistry Edited by Riccardo Flamini